my **revision** notes

CCEA GCSE

SCIENCE
DOUBLE AWARD

James Napier
Alyn G. McFarland
Roy White

Photo credit

Photo on page 346 is reproduced by permission of NASA/JPL-Caltech/Harvard-Smithsonian CfA

Hachette UK's policy is to use papers that are natural, renewable and recyclable products and made from wood grown in well-managed forests and other controlled sources. The logging and manufacturing processes are expected to conform to the environmental regulations of the country of origin.

Orders: please contact Hachette UK Distribution, Hely Hutchinson Centre, Milton Road, Didcot, Oxfordshire, OX11 7HH. Telephone: +44 (0)1235 827827. Email education@hachette.co.uk. Lines are open from 9 a.m. to 5 p.m., Monday to Friday. You can also order through our website: www.hoddereducation.co.uk

ISBN: 978 1 5104 0451 9

First published in 2017 by
Hodder Education,
An Hachette UK Company
Carmelite House
50 Victoria Embankment
London EC4Y 0DZ

www.hoddereducation.co.uk

Impression number 10

Year 2024

Cover photo © Sergiy Tryapitsyn/Alamy Stock Photo

Typeset in Bembo Std Regular 11/13 by Integra Software Services Pvt. Ltd, Pondicherry, India

Printed in India

A catalogue record for this title is available from the British Library.

Get the most from this book

Everyone has to decide his or her own revision strategy, but it is essential to review your work, learn it and test your understanding. These Revision Notes will help you to do that in a planned way, topic by topic. Use this book as the cornerstone of your revision and don't hesitate to write in it — personalise your notes and check your progress by ticking off each section as you revise.

Tick to track your progress 

Use the revision planner on pages iv to xii to plan your revision, topic by topic. Tick each box when you have:

- revised and understood a topic
- tested yourself
- checked your answers.

You can also keep track of your revision by ticking off each topic heading in the book. You may find it helpful to add your own notes as you work through each topic.

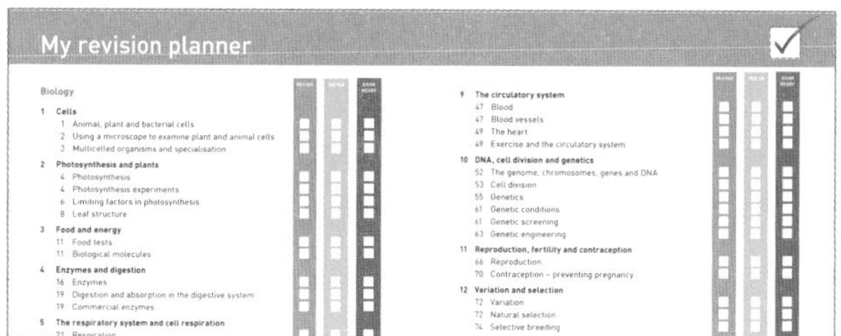

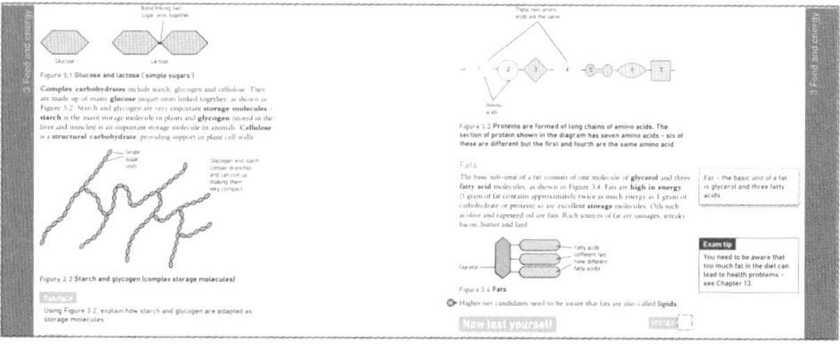

Features to help you succeed

Exam tips

Expert tips are given throughout the book to help you polish your exam technique in order to maximise your chances in the exam.

Now test yourself

These short, knowledge-based questions provide the first step in testing your learning. Answers Online (see right).

Definitions and key words

Clear, concise definitions of essential key terms are provided where they first appear.

Worked examples

Worked examples are given throughout the book.

Exam practice

Practice exam questions are provided for each topic.

Use them to consolidate your revision and practise your exam skills.

Online

Go online to check your answers **www.hoddereducation.co.uk/myrevisionnotes**

Level coding

If you are taking GCSE Double Award Foundation-tier you need to study only the material with no bars.

If you are taking GCSE Double Award Higher-tier you need to study the material with no bars, plus the material with the purple H bar.

My revision planner

Biology

1 Cells

1 Animal, plant and bacterial cells
2 Using a microscope to examine plant and animal cells
3 Multicelled organisms and specialisation

2 Photosynthesis and plants

4 Photosynthesis
4 Photosynthesis experiments
6 Limiting factors in photosynthesis
8 Leaf structure

3 Food and energy

11 Food tests
11 Biological molecules

4 Enzymes and digestion

16 Enzymes
19 Digestion and absorption in the digestive system
19 Commercial enzymes

5 The respiratory system and cell respiration

21 Respiration
23 Respiratory surfaces

6 Coordination and control

25 The nervous system
26 Hormones
28 The excretory system and osmoregulation
30 Plant hormones

7 Ecological relationships and energy flow

33 Fieldwork
35 Competition
35 Food chains and food webs
36 Energy flow
36 Nutrient cycles
38 Eutrophication

8 Osmosis and plant transport

41 Osmosis
43 Transpiration

REVISED | TESTED | EXAM READY

REVISED | TESTED | EXAM READY

9 The circulatory system

47 Blood

47 Blood vessels

49 The heart

49 Exercise and the circulatory system

10 DNA, cell division and genetics

52 The genome, chromosomes, genes and DNA

53 Cell division

55 Genetics

61 Genetic conditions

61 Genetic screening

63 Genetic engineering

11 Reproduction, fertility and contraception

66 Reproduction

70 Contraception – preventing pregnancy

12 Variation and selection

72 Variation

72 Natural selection

74 Selective breeding

13 Health, disease, defence mechanisms and treatments

76 Microorganisms and communicable diseases

77 The body's defence mechanisms

80 Antibiotics

81 Aseptic techniques

82 Non-communicable diseases

84 Cancer

Chemistry

REVISED | TESTED | EXAM READY

14 Atomic structure

85 The structure of atoms

85 Atomic number and mass number

85 Determining numbers of subatomic particles

86 Isotopes

87 Calculating relative atomic mass

88 Electronic configuration

91 Electronic configuration from the periodic table

REVISED | TESTED | EXAM READY

91 Compounds

92 Ions

15 Bonding, structures and nanoparticles

95 Types of bonding

95 Formation of ionic compounds

95 Formation of ions from atoms

99 Ionic bonding and structure

99 Properties of ionic compounds

100 Covalent bonding

102 Properties of molecular covalent substances

102 Metallic bonding

103 Giant covalent structures

104 Summary of bonding and structures

106 Uses and risks of nanoparticles

16 Symbols, formulae and equations

109 Valency and group number

109 The valencies of molecular ions

110 The valencies of transition elements

110 How to work out the formula of a compound

113 Atoms in formulae

113 Word equations and balanced symbol equations

117 Ionic equations

119 Half-equations

17 The periodic table

122 History of development of the periodic table

123 Groups and periods

123 Metals and non-metals

124 Names of groups

124 Solids, liquids and gases

124 Diatomicity

125 Group 1 (I)

128 Group 7 (VII)

131 Applying trends to unfamiliar elements

131 Group 0

132 Transition metals

REVISED TESTED EXAM READY

18 Quantitative chemistry I

134 Relative atomic mass

134 The mole

134 Relative formula mass (M_r)

135 Percentage of an element in a compound by mass

136 Interchanging M_r and moles

137 Measuring and calculating quantities

140 Limiting reactant

142 Percentage yield

19 Acids, bases and salts

144 Hazard symbols

144 Risk assessment

144 Indicators

145 Concentration of solutions

145 pH

146 Validity, reliability and accuracy

147 Formulae of acids and alkalis

147 Ions in acids and alkalis

148 Bases and alkalis

148 Reactions of acids

150 Observations during acid reactions

20 Chemical analysis

152 Melting and boiling point

152 Assessing purity

153 Determining state from melting and boiling points

154 Formulations

154 Separation techniques

160 Planning methods of separation

161 Basic information

161 Cation tests

21 Reactivity series of metals

164 Explaining the reactivity of metals

164 Reactions of metals with air

166 Reaction of metals with water

167 Reaction of metals with steam

168 Solutions

169 Solids

	REVISED	TESTED	EXAM READY

170 Placing an unfamiliar metal in the reactivity series
171 Extraction of metals from their ores

22 Redox, rusting and iron
173 Redox reactions
174 Redox and displacement reactions
176 The cause of rusting
176 Prevention of rusting
177 Extraction of iron in a blast furnace

23 Rates of reaction
180 Factors affecting rate of reaction
180 Reactions used in rate of reaction experiments
180 Measuring a change in mass
182 Measuring gas volume
183 Measuring the time for a reaction to be completed
183 Measuring production of a solid precipitate
183 Calculating rate from time
184 Planning an investigation
184 Explaining the effects of temperature and concentration on reaction rate
185 Investigating the effect of the presence of a catalyst
186 Metal with acid reactions
186 Catalysts

24 Equilibrium
188 Reversible reactions
188 Understanding dynamic equilibrium
188 Dynamic equilibrium reactions
189 Features of reversible reactions

25 Organic chemistry
190 General formulae for organic molecules
191 Drawing structural formulae for organic molecules
192 Alkanes
193 Fractional distillation of crude oil
195 Alkenes
196 Alcohols
197 Carboxylic acids
198 Functional groups
198 Complete and incomplete combustion

REVISED TESTED EXAM READY

199 Balancing equations for combustion of hydrocarbons
199 Combustion of alcohols
199 Testing the products of combustion
200 Chemistry of alkanes
200 Chemistry of alkenes
201 Addition polymerisation
203 Chemistry of alcohols
203 Chemistry of carboxylic acids
205 Greenhouse effect
205 Acid rain

26 Quantitative chemistry II

207 Water of crystallisation and degree of hydration
207 Empirical formula and molecular formula
208 Determining formulae of simple compounds
210 Determining degree of hydration by heating to constant mass
212 Heating to constant mass
213 Solution calculations
214 Determining moles of solute in a solution
215 Atom economies

27 Electrochemistry

217 What is electrolysis?
218 How electrolysis works
218 Electrolysis of molten ionic compounds
220 Extraction of aluminium from its ore

28 Energy changes in chemistry

222 Reaction profile diagrams
223 Understanding energy change values
224 Activation energy
225 Explaining energy changes in terms of bonds
226 Calculating energy changes from bond energies

29 Gas chemistry

232 Gases in the atmosphere
232 Preparation of the gases
232 Collection of gases
233 Nitrogen
233 Ammonia

REVISED | TESTED | EXAM READY

234 Hydrogen
236 Oxygen
238 Carbon dioxide

Physics

REVISED | TESTED | EXAM READY

30 Motion
241 Motion in a straight line
243 Vectors and scalars
244 Motion graphs

31 Forces
247 Balanced and unbalanced forces
251 Mass and weight
251 Free fall
252 Vertical motion under gravity
253 Hooke's law
254 Pressure
256 Centre of gravity
257 Moments and levers

32 Density and kinetic theory
261 Density
263 Kinetic theory

33 Energy
265 Energy forms
265 Energy resources
268 Energy flow diagrams
269 The Sun
270 Work
271 Work and energy
271 Power
272 Measuring power
273 Efficiency
274 Gravitational potential energy
274 Kinetic energy

34 Atomic and nuclear physics
277 The structure of atoms
278 Nuclear radiation

REVISED | TESTED | EXAM READY

279 Dangers of radiation
280 Nuclear disintegration equations
281 Radioactive decay
283 Uses of radiation
284 Background radiation
285 Nuclear fission
287 Nuclear fusion
287 What is ITER?

35 Waves

289 Longitudinal waves
289 Transverse waves
290 Describing waves
291 The wave equation
291 Graphs and waves
292 Echoes
295 Electromagnetic waves

36 Light

298 Reflection of light
300 Refraction of light
302 Dispersion of white light
304 Lenses
305 Image in a concave (diverging) lens
305 Image in a convex (converging) lens
306 Ray diagrams

37 Electricity

310 Why are metals good conductors?
311 Standard symbols
311 Cell polarity
312 The relationship between charge and current
313 Resistance
318 Voltage in series circuits
319 Voltage in parallel circuits
321 Hybrid circuits

38 Electricity in the home

326 Electrical energy
326 Electrical power
327 The three equations for power

REVISED TESTED EXAM READY

328 Paying for electricity

329 One-way switches

330 Two-way switches

330 How to wire a three-pin plug

331 The earth wire

331 Double insulation

332 The live wire

39 Magnetism and electromagnetism

333 Magnetic field pattern around a bar magnet

333 Magnetic field pattern due to a current-carrying coil

336 Alternating and direct currents

40 Space physics

340 The Solar System

340 Artificial satellites

341 The life cycle of stars

343 Planet formation

343 The galaxies

343 Formation and evolution of the Universe

344 Red-shift

345 Cosmic microwave background radiation

Answers at www.hoddereducation.co.uk/myrevisionnotesdownloads

Countdown to my exams

6–8 weeks to go

- Start by looking at the specification — make sure you know exactly what material you need to revise and the style of the examination. Use the revision planner on pages iv to xii to familiarise yourself with the topics.
- Organise your notes, making sure you have covered everything on the specification. The revision planner will help you to group your notes into topics.
- Work out a realistic revision plan that will allow you time for relaxation. Set aside days and times for all the subjects that you need to study, and stick to your timetable.
- Set yourself sensible targets. Break your revision down into focused sessions of around 40 minutes, divided by breaks. These Revision Notes organise the basic facts into short, memorable sections to make revising easier.

REVISED ☐

2–6 weeks to go

- Read through the relevant sections of this book and refer to the exam tips and key terms. Tick off the topics as you feel confident about them. Highlight those topics you find difficult and look at them again in detail.
- Test your understanding of each topic by working through the 'Now test yourself' questions in the book. Look up the answers online (see below).
- Make a note of any problem areas as you revise, and ask your teacher to go over these in class.
- Look at past papers. They are one of the best ways to revise and practise your exam skills. Write or prepare planned answers to the exam practice questions provided in this book. Check your answers online at **www.hoddereducation.co.uk/myrevisionnotesdownloads**
- Try out different revision methods. For example, you can make notes using mind maps, spider diagrams or flash cards.
- Track your progress using the revision planner and give yourself a reward when you have achieved your target.

REVISED ☐

One week to go

- Try to fit in at least one more timed practice of an entire past paper and seek feedback from your teacher, comparing your work closely with the mark scheme.
- Check the revision planner to make sure you haven't missed out any topics. Brush up on any areas of difficulty by talking them over with a friend or getting help from your teacher.
- Attend any revision classes put on by your teacher. Remember, he or she is an expert at preparing people for examinations.

REVISED ☐

The day before the examination

- Flick through these Revision Notes for useful reminders, for example the exam tips and key terms.
- Check the time and place of your examination.
- Make sure you have everything you need — extra pens and pencils, tissues, a watch, bottled water.
- Allow some time to relax and have an early night to ensure you are fresh and alert for the examinations.

REVISED ☐

1 Cells

The **cell** is the basic building block of animals and plants. Bacteria are formed of single cells.

Cell – the basic building block of all living organisms. Plants and animals are formed of millions of cells, but a bacterium is formed of only one cell.

Animal, plant and bacterial cells

REVISED

Table 1.1 summarises the main features of animal, plant and bacterial cells (see also Figure 1.1).

Table 1.1 Animal, plant and bacterial cells

Structure	Function	Present in: animal cells	plant cells	bacterial cells
Cell membrane	Forms a boundary to the cell and is selectively permeable, controlling what enters and leaves	Yes	Yes	Yes
Cytoplasm	Site of chemical reactions	Yes	Yes	Yes
Nucleus	Control centre of the cell containing genetic information in the form of chromosomes; surrounded by a nuclear membrane	Yes	Yes	No
Nuclear membrane	Boundary of nucleus; controls what enters and leaves the nucleus	Yes	Yes	No
H Mitochondria	Sites of cell respiration	Yes	Yes	No
Cell wall	Made of cellulose – a rigid structure that provides support	No	Yes (cellulose)	Yes (non-cellulose)
Vacuole (large permanent)	Contains cell sap and provides support	No	Yes	No
Chloroplasts	Contain chlorophyll; the place where photosynthesis takes place	No	Yes	No
Plasmids	Small circular rings of DNA	No	No	Yes

Exam tip

The only structures present in all three types of cell are **cell membranes** and **cytoplasm**.

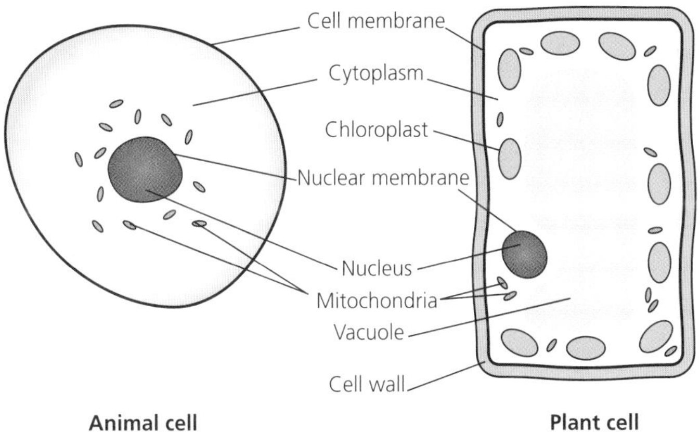

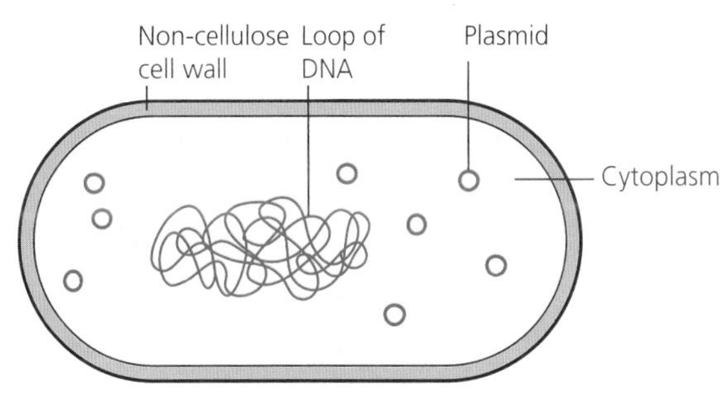

Figure 1.1 Animal, plant and bacterial cells

Using a microscope to examine plant and animal cells

REVISED

Making a slide of plant cells

- Peel a small section of onion tissue and place on the centre of a microscope slide.
- Add water using a drop pipette to the onion tissue to stop it drying out.
- Gently lower a coverslip onto the onion tissue. The coverslip will help protect the lens should the lens make contact with the slide, and it also prevents the cells from drying out.
- Set the slide onto the stage of the microscope and examine using low power first and then high power.

Exam tip

The coverslip should be lowered one end first on to the onion tissue, to avoid trapping air bubbles (seen as black rings if air is trapped between the slide and the coverslip).

Making a slide of animal cells

- Using your nail or an inter-tooth brush, gently scrape the inside of your cheek.
- Smear the material gathered onto the centre of a microscope slide.
- Carefully lower a coverslip on top, as described above.
- Observe using a light microscope; first at low power, then using high power.

Exam tips

- When looking at onion cells under a microscope you will probably see the cell walls, cytoplasm, nuclei (if stained, with iodine for example) and possibly the vacuole. You are unlikely to see cell membranes and will not see chloroplasts or mitochondria.
- Plant cells are much more **regularly shaped** than animal cells – they are usually much **larger** as well.
- When using a microscope you should always use **low power first**. You can see more at low power – there is a **greater field of view** – so it is easier to find what you are looking for. It is also **easier to focus** at low power than high power.

Now test yourself

TESTED

1. State the function of the cell membrane in cells.
2. Name **one** structure that is present in bacterial cells but not in plant or animal cells.
3. Give **one** reason for using a coverslip when making slides of plant or animal cells.

Multicelled organisms and specialisation

REVISED

In multicelled organisms cells and regions of the body become specialised, as summarised in Table 1.2.

Table 1.2 Specialisation in multicelled organisms

Structure	Description
Cell	Basic building block of living organisms, e.g. animal cell
Tissue	Groups of cells with similar structures and functions, e.g. skin
Organ	Groups of different tissues working together to form a structure with a particular function, e.g. brain
Organ system	Organs organised into organ systems, e.g. the nervous system
Organism	Different organ systems make up the organism, e.g. human

Exam tip

In terms of increasing complexity, you should remember the order:
cell → tissue → organ → organ system → organism

Exam practice

1 (a) Figure 1.2 represents a plant cell.

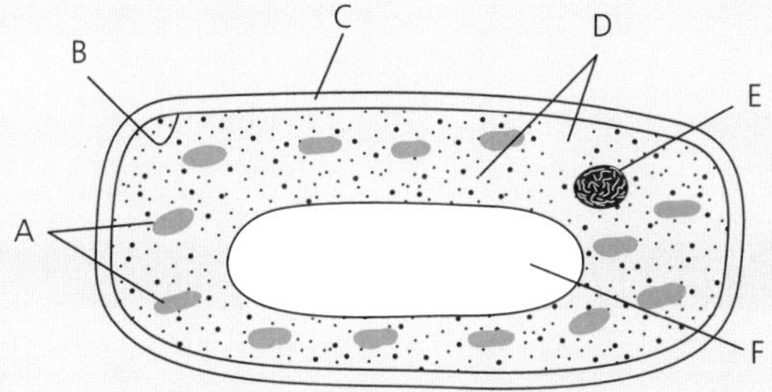

Figure 1.2

(i) Name the parts labelled **A**, **B** and **C**. [3]
(ii) Give the letters of **three** structures that are also present in animal cells. [3]

(b) Complete the sentences below.
The .... is the part of the cell that contains chromosomes. Chemical reactions take place in the .... [2]

(c) State **two** structures that are found in bacterial cells but are not in plant or animal cells. [2]

2 (a) A thin piece of onion epidermis (skin) is placed onto a microscope slide. What else must you do before the slide is ready for observation using a microscope? [3]

(b) Suggest why onion cells do not contain chloroplasts. [1]

Answers online

ONLINE

2 Photosynthesis and plants

Photosynthesis

REVISED

In **photosynthesis** plants make food (sugars and starch) using **light energy**. The light is trapped by **chlorophyll** in **chloroplasts** in plant leaves.

The word equation for photosynthesis is:

carbon dioxide + water → glucose + oxygen

H The balanced chemical equation is:

$$6CO_2 + 6H_2O \rightarrow C_6H_{12}O_6 + 6O_2$$

As photosynthesis requires (light) energy to work, it is an **endothermic reaction.**

Photosynthesis – a process in plants in which light energy is trapped by chlorophyll to produce food.

Endothermic reactions – reactions that require energy to be absorbed (taken in) to work.

Photosynthesis experiments

REVISED

The glucose produced during photosynthesis is usually converted into **starch** for storage. One way of showing that photosynthesis has taken place is by showing that starch is present in a leaf.

This can be done using a **starch test**, as described in Table 2.1.

Starch test – a test to show whether or not starch is present in a plant leaf.

Table 2.1 Carrying out the test for starch

Step	Method	Reason
1	Put the leaf in boiling water	This kills the leaf and stops further reactions
2	Boil the leaf in ethanol (alcohol) – this must be done in a water bath with the very hot/boiling water poured from an electrical kettle, as alcohol is flammable (Bunsen burners should not be used)	This removes chlorophyll (green colour) from the leaf
3	Dip the leaf in boiling water again	This makes the leaf soft and less brittle (boiling in ethanol makes the leaf rigid)
4	Spread the leaf on a white tile and add iodine	If starch is present, the iodine will turn from yellow-brown to blue-black

Exam tips

- The chlorophyll is removed as it makes it easier to see any colour change with iodine.
- If asked to give the **colour change** of a positive starch test make sure you give the actual colour *change* and not just the final colour – if starch is present the colour change is yellow-brown to blue-black.

Before carrying out investigations into photosynthesis, it is usually necessary to **destarch** the plant. This involves leaving the plant in **darkness** (e.g. a dark cupboard) for **48 hours**. This is necessary to make sure that any starch produced is only produced during the investigation.

Exam tip

If plants were not destarched in these investigations the investigation would not be **valid**, as it would be impossible to say whether any starch present was produced during the investigation or was there before the investigation started.

You should carry out investigations that can show that light, carbon dioxide and chlorophyll are necessary for photosynthesis.

Showing that light is needed for photosynthesis

- Destarch a plant.
- Partially cover a leaf on a plant with foil.
- Put the plant in bright light for at least 6 hours.
- Test the leaf for starch (Figure 2.1).

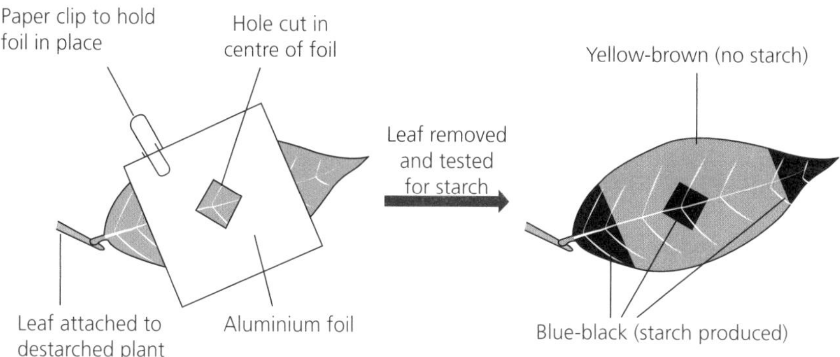

Figure 2.1 Experiment to show that light is required for photosynthesis to occur

The results show that starch is only produced in the parts of the leaf that received light – showing that light is necessary for photosynthesis.

Showing that chlorophyll is needed for photosynthesis

Some plants have leaves that are part green and part white. These leaves are described as **variegated** leaves.

- Destarch a variegated plant.
- Put the plant in bright light for at least 6 hours.
- Test the leaf for starch.

The starch test will show that starch is only produced in those parts of the leaf that had chlorophyll (were green).

Showing that carbon dioxide is needed for photosynthesis

- Destarch a plant.
- Set up as shown in Figure 2.2 – the sodium hydroxide absorbs carbon dioxide from the air inside the plastic bag.
- Leave the plant in bright light for at least 6 hours.
- Test one of the leaves for starch.

A negative starch test will show that carbon dioxide is necessary for photosynthesis.

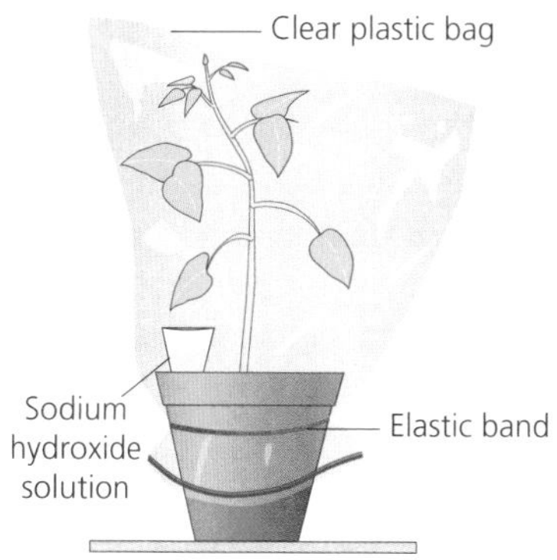

Figure 2.2 Experiment to show that carbon dioxide is needed for photosynthesis

Showing that oxygen is produced

Using apparatus similar to that in Figure 2.3, it is possible to demonstrate that oxygen is produced in photosynthesis.

The rate of photosynthesis affects the rate at which the bubbles of oxygen are given off, and this can be used to compare photosynthesis rates in different conditions. For example, by moving the position of the lamp it is possible to investigate the effect of light intensity on photosynthesis.

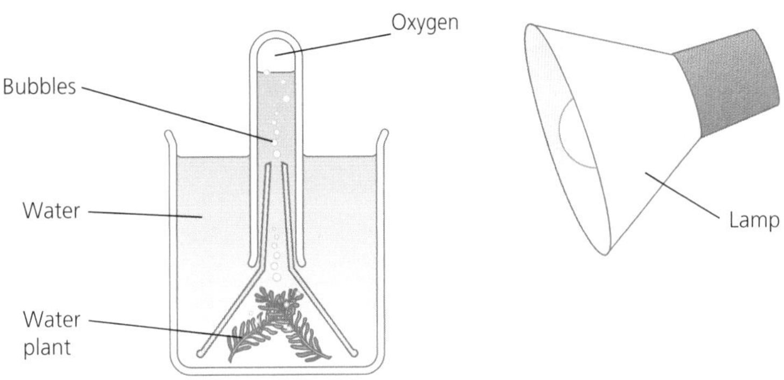

Figure 2.3 Experiment to show that oxygen is produced during photosynthesis

Now test yourself

TESTED

1. State why leaves are dipped in boiling water after the chlorophyll is removed during the starch test.
2. Name the chemical used in photosynthesis investigations to absorb carbon dioxide.
3. What is a variegated leaf?

Prescribed practical B1

Investigate the need for light and chlorophyll in photosynthesis by testing a leaf for starch

Limiting factors in photosynthesis

REVISED

The environmental factors **light**, **carbon dioxide** and **temperature** all affect the rate of photosynthesis. If all three are present in sufficient quantities, the rate of photosynthesis will be at its optimum.

If any of these factors are at sub-optimal levels, the rate of photosynthesis will be reduced. A **limiting factor** is a factor that limits the rate of photosynthesis due to that factor being present at a sub-optimal level.

Limiting factor – an environmental factor that limits the rate of photosynthesis due to that factor being present in too small an amount.

Exam tip

Light (to provide energy) and **carbon dioxide** (a raw material) are necessary for photosynthesis, so normally the more of these are present the faster the photosynthesis reaction takes place (up to a maximum). As **temperature** affects the rate of all reactions (e.g. through the speed of molecules diffusing and its effect on enzyme activity), it will affect the rate of the photosynthesis reaction.

H Higher tier candidates need to be able to interpret data on limiting factors – typically in the form of graphs or tables. Figure 2.4 is a graph providing data on all three limiting factors. In this graph, lines A, B and C represent the rates of photosynthesis in three different regimes (A – low carbon dioxide and low temperature; B – high carbon dioxide and low temperature; and C – high carbon dioxide and high temperature).

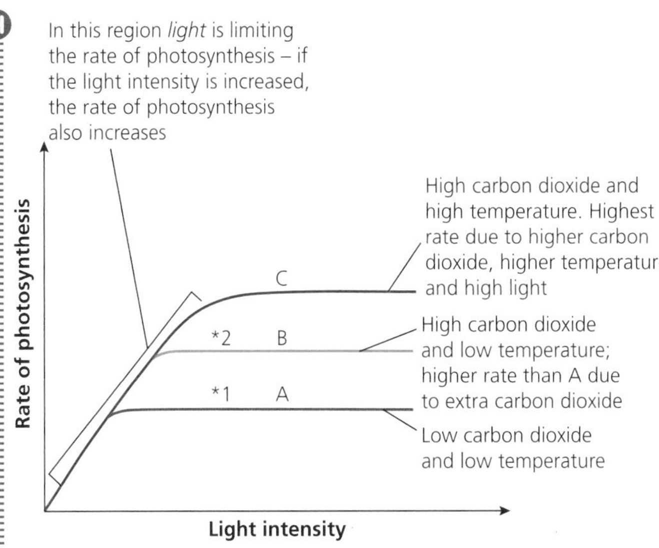

Figure 2.4 Limiting factors in photosynthesis

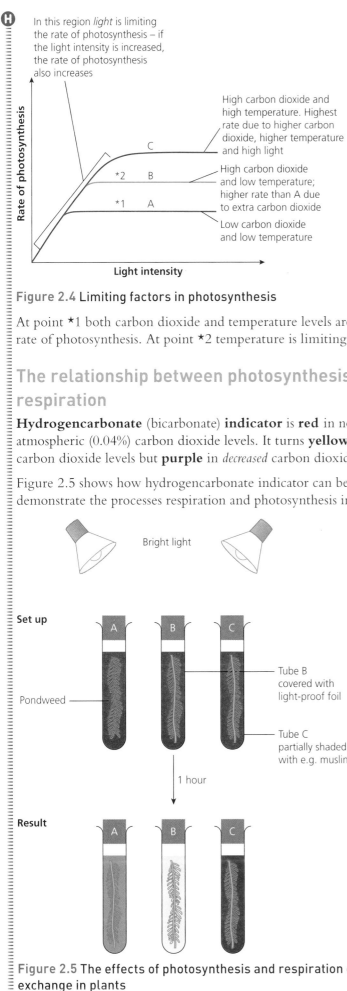

At point ⋆1 both carbon dioxide and temperature levels are limiting the rate of photosynthesis. At point ⋆2 temperature is limiting the rate.

The relationship between photosynthesis and respiration

Hydrogencarbonate (bicarbonate) **indicator** is **red** in normal atmospheric (0.04%) carbon dioxide levels. It turns **yellow** in *increased* carbon dioxide levels but **purple** in *decreased* carbon dioxide levels.

Figure 2.5 shows how hydrogencarbonate indicator can be used to demonstrate the processes respiration and photosynthesis in plants.

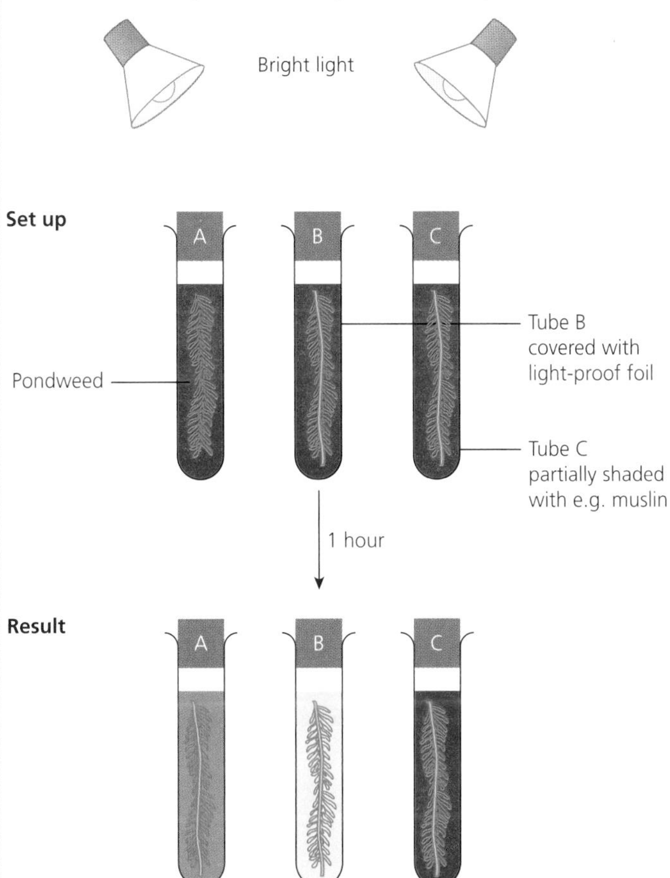

Figure 2.5 The effects of photosynthesis and respiration on gas exchange in plants

H Interpretation of results

- In **boiling tube A** the rate of photosynthesis exceeded the rate of respiration, resulting in the concentration of carbon dioxide in the tubes *decreasing* and turning the indicator **purple**.
- In **boiling tube B** there was only respiration taking place (the foil prevented photosynthesis). The carbon dioxide concentration in the boiling tube *increased*, turning the indicator **yellow**.
- In **boiling tube C** both photosynthesis and respiration were taking place but the rate of photosynthesis was much reduced (due to the partial shading) – the rates of photosynthesis and respiration were equal and the amount of carbon dioxide used in photosynthesis was the same as the amount released in respiration. There was *no change* in carbon dioxide levels and the indicator remained **red**. The point at which the rates of photosynthesis and respiration are equal is referred to as the **compensation point**.

Example

Describe how to use hydrogencarbonate indicator to find the amount of light required to reach the compensation point (the point at which the rates of photosynthesis are equal) in pondweed.

Answer

1. Add a section of pondweed to hydrogencarbonate indicator in a boiling tube.
2. Add a bung.
3. Leave the pondweed for 20–30 minutes in a particular light intensity and check the colour of the hydrogencarbonate indicator.
4. Increase or decrease the light intensity by moving a lamp or using a dimmer switch and repeat step 3.
5. Repeat step 4 until the indicator remains red.

Leaf structure

REVISED

Leaves are plant organs in which **photosynthesis** occurs. Figure 2.6 shows a cross-section of a mesophytic leaf.

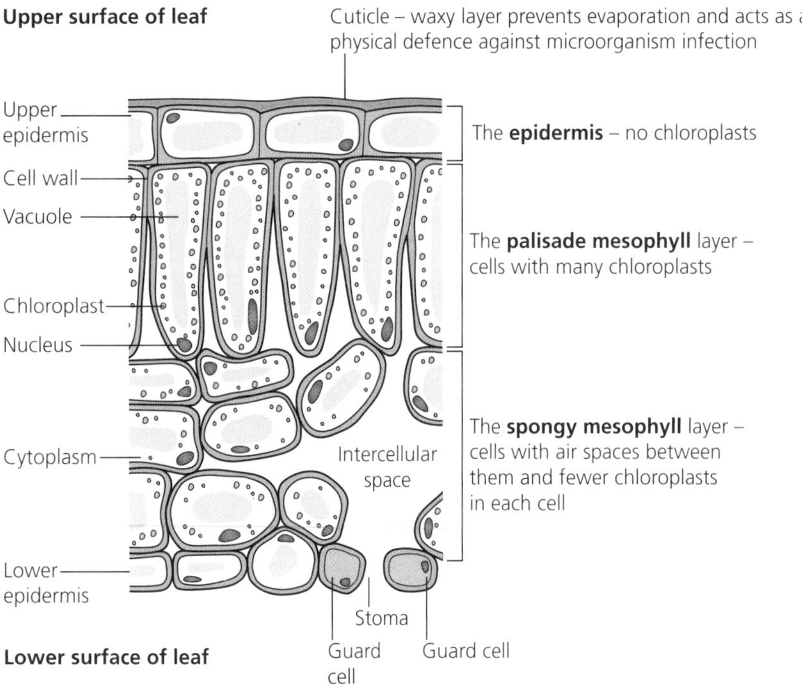

Figure 2.6 Cross-section of a mesophytic leaf

Exam tip

A mesophytic leaf is a typical unspecialised leaf – the plants have reasonable supplies of water, but not so much that they don't need to have some adaptations to reduce water loss.

Leaves are highly adapted for **light absorption** and **gas exchange**, as summarised in Figure 2.7.

Adaptations for light absorption

- Large surface area
- Thin, transparent cuticle
- Presence of many tightly packed palisade mesophyll cells, end-on to the upper surface, with many chloroplasts rich in chlorophyll

Adaptations for gas exchange

- The spongy mesophyll cells have a large surface area for gas exchange
- The intercellular spaces in the spongy mesophyll allow carbon dioxide to enter and oxygen to leave the photosynthesising cells, which are mainly concentrated in the palisade layer
- Stomata that allow carbon dioxide and oxygen to enter the leaf; the guard cells can open and close the stomatal pore – in many plants stomata are open during the day and closed at night

Figure 2.7 Leaf adaptations for light absorption and gas exchange

Now test yourself

TESTED

4 What is the function of the waxy cuticle in leaves?
5 Why do the epidermal cells at the top of the leaf have no chloroplasts?
6 What is the function of the intercellular spaces in leaves?

Exam practice

1 (a) Give the word equation for photosynthesis. [2]
(b) (i) Name the chemical that absorbs light in plants. [1]
(ii) Name the cellular structures that contain this chemical. [1]
(c) Photosynthesis is an endothermic reaction. What is meant by the term 'endothermic reaction'? [1]
2 (a) (i) Describe how you would destarch a plant. [1]
(ii) Why is it necessary to destarch a plant in a photosynthesis investigation? [1]
(b) Give **one** safety precaution necessary when carrying out a starch test. [1]
(c) Explain fully why leaves are boiled in ethanol when testing for starch. [2]
3 Explain the changes in carbon dioxide level between the periods **A** and **B** in Figure 2.8. [4]

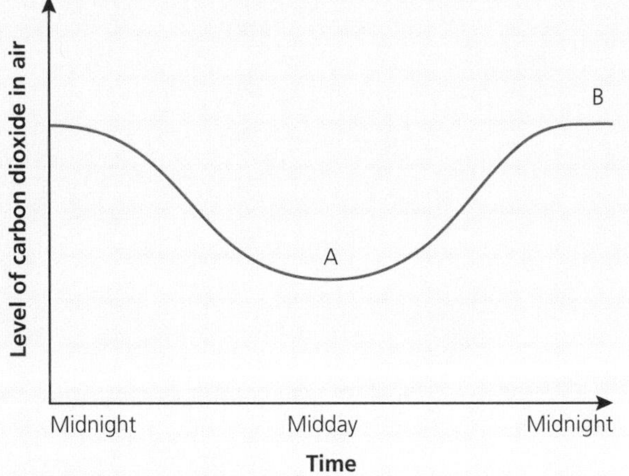

Figure 2.8

4 Figure 2.9 shows a cross-section through a leaf.

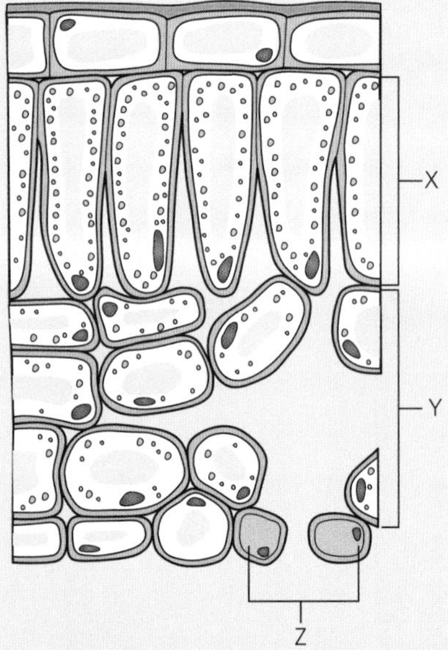

Figure 2.9

(a) (i) Identify the layer **X**. [1]

(ii) Using the diagram, give **three** reasons why more light will be absorbed in layer **X** than in layer **Y**. [3]

(b) (i) Name the cells labelled **Z**. [1]

(ii) Give the function of these cells. [1]

Answers online

ONLINE

3 Food and energy

Food tests

REVISED

Food tests can be used to identify the food types present in food.

Table 3.1 Food tests

Food type	Test	Method	Result (if food type present)
Starch	Starch test	Add iodine solution.	Iodine turns from yellow-brown to blue-black.
Sugar	Benedict's test	Add Benedict's solution and **heat** in a water bath.	The solution changes from blue to a brick-red precipitate.
Protein (amino acids)	Biuret test	Add sodium hydroxide, then a few drops of copper sulfate and shake.	The solution turns from blue to purple/lilac.
Fat	Ethanol test	Shake the fat with ethanol (alcohol), then add an equal amount of water.	The colourless ethanol changes to a cloudy white emulsion.

Exam tips

- The **Benedict's test** is the only food test that requires **heating**.
- The **Benedict's test** is the only food test in the table that is **partially quantitative** – i.e. it can give an estimate of the amount of sugar present. The reagent will change from blue to green or orange or brick red, depending on how much sugar is present.
- Most foods contain more than one food type – e.g. bacon contains protein and fat.

Now test yourself

TESTED

1 Name the reagent used to test for protein.
2 Name the food group that gives a positive test with ethanol.

Biological molecules

REVISED

Carbohydrates, proteins and fats are very important biological molecules. Each of them contains the elements **carbon**, **hydrogen** and **oxygen**, and protein also contains **nitrogen**.

Carbohydrates

'Simple' carbohydrates are sugars such as glucose and lactose – they are described as 'simple' as they are made up of one (e.g. glucose) or two (e.g. lactose) basic sugar units. They are good sources of **energy**. **Glucose** is the sugar that is normally used in respiration. **Lactose** is the sugar and energy source in milk. Sugars are **'fast-release' energy stores**, which means that they can be quickly metabolised to release energy. Examples of foods rich in sugars are some fizzy drinks and cakes.

Carbohydrate – biological molecule formed of sugar sub-units. Carbohydrates differ by having different numbers or types of sugar sub-unit.

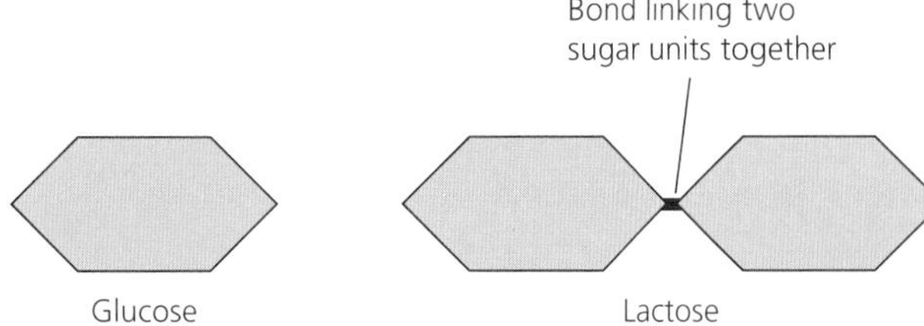

Figure 3.1 Glucose and lactose ('simple sugars')

Complex carbohydrates include starch, glycogen and cellulose. They are made up of many **glucose** (sugar) units linked together, as shown in Figure 3.2. Starch and glycogen are very important **storage molecules** – **starch** is the main storage molecule in plants and **glycogen** (stored in the liver and muscles) is an important storage molecule in animals. **Cellulose** is a **structural carbohydrate**, providing support in plant cell walls.

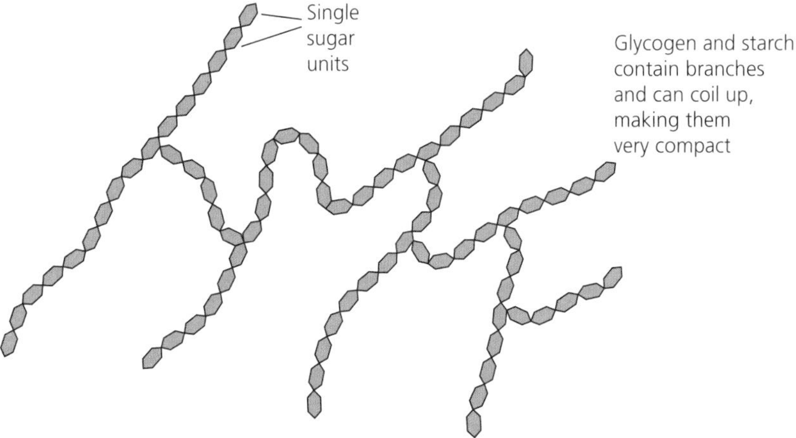

Figure 3.2 Starch and glycogen (complex storage molecules)

Example

Using Figure 3.2, explain how starch and glycogen are adapted as storage molecules.

Answer

They are made up of many glucose molecules that can be used for respiration. They can be packed (coiled) into small spaces, allowing many glucose units to be contained in a very small volume.

Proteins

Proteins are long chains of **amino acids** bonded together. As there are 20 different types of amino acid, there are many different arrangements in which the amino acids can be linked together (Figure 3.3). Proteins can be **structural** (e.g. in muscle) or **functional** (e.g. as enzymes or antibodies). Sources of protein include lean meat, lentils and fish.

Protein – a biological molecule formed of sub-units of amino acids. Proteins can differ by containing different types or numbers of amino acids or by these being arranged in different sequences.

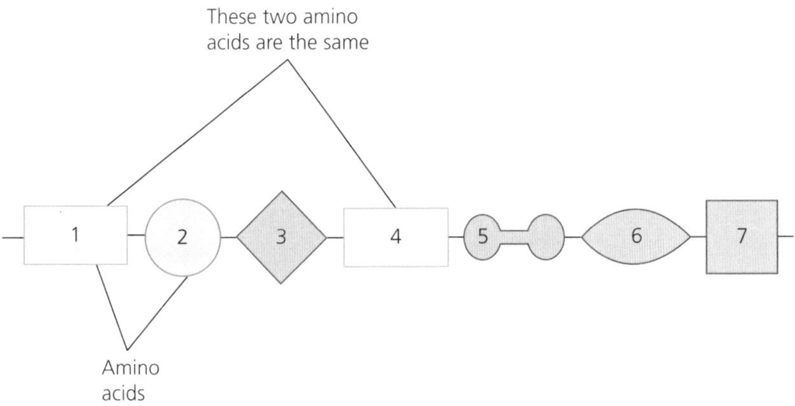

Figure 3.3 Proteins are formed of long chains of amino acids. The section of protein shown in the diagram has seven amino acids – six of these are different but the first and fourth are the same amino acid

Fats

The basic sub-unit of a **fat** consists of one molecule of **glycerol** and three **fatty acid** molecules, as shown in Figure 3.4. Fats are **high in energy** (1 gram of fat contains approximately twice as much energy as 1 gram of carbohydrate or protein) so are excellent **storage** molecules. Oils such as olive and rapeseed oil are fats. Rich sources of fat are sausages, streaky bacon, butter and lard.

Fat – the basic unit of a fat is glycerol and three fatty acids.

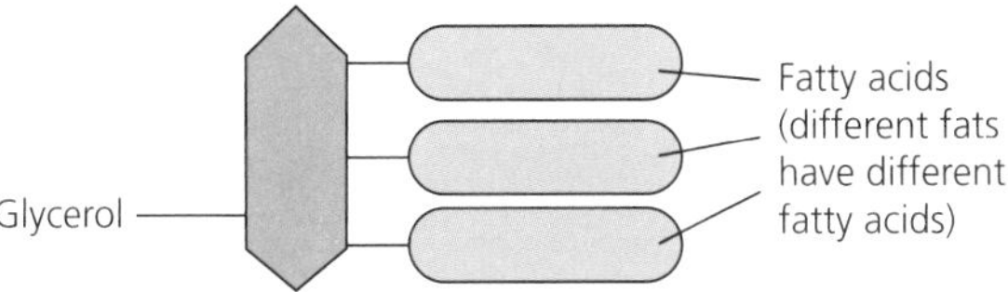

Figure 3.4 Fats

Exam tip

You need to be aware that too much fat in the diet can lead to health problems – see Chapter 13.

H Higher tier candidates need to be aware that fats are also called **lipids**.

Now test yourself

TESTED

3 Name **one** complex carbohydrate that has a structural role.
4 Explain precisely what fats are formed from.

Prescribed practical B2

Investigate the energy content of food by burning food samples

The apparatus shown in Figure 3.5 can be used to investigate the energy in food or to compare the energy in different foods. The difference in temperature of the water between the start of the investigation and when the food has completely burned gives an indication of the amount of energy in the food.

→

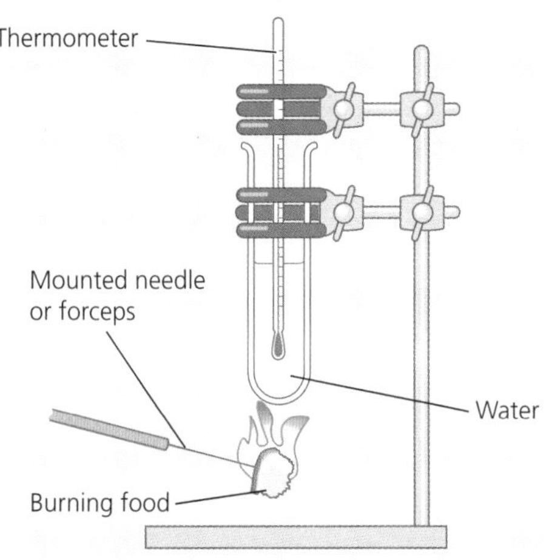

Figure 3.5 Measuring the energy content of food

The energy released when burning food is calculated using the following equation:

energy released in joules (J) = mass of water (g) × rise in temp (°C) × 4.2

Exam tips

- If the same mass of food is burned each time, the result of this equation can be used to compare foods.
- As 1 g of water = 1 cm^3 of water the volume of water can be substituted into the equation instead of mass.
- Many exam questions ask about **validity** and why all the energy in a food is not used to heat the water (i.e. why some is lost – Figure 3.6).

To make sure results are valid (a fair test) when comparing different foods:
- use the same amount of each food
- hold the burning food the same distance from the boiling tube.

Some (heat) energy will be:
- lost to the air
- lost to heat the glass
- left in the burned food remains.

Figure 3.6 Key things you must know in food-burning investigations

Exam practice

1 (a) Copy and complete Table 3.2, which is about food tests. [3]

Table 3.2

Reagent	Initial colour	End colour if food present
Iodine	Yellow-brown	
Biuret		Purple (lilac)
	Clear	White emulsion

(b) Name the food test that requires heating. [1]

2 (a) Name the sub-units of protein. [1]

(b) There are only 20 different amino acids, yet there are thousands of different proteins in the body. Explain this statement. [1]

(c) (i) Give **one** example of a functional protein. [1]

(ii) Name **one** structure in the body that you would expect to contain structural proteins. [1]

3 The amount of energy in food can be calculated by burning it under a boiling tube of a known volume of water, measuring the temperature increase and then applying the formula below:

energy (J) = volume of water(cm^3) x temperature rise (°C) x 4.2

25 cm^3 of water was used in each boiling tube and 1 g of each food sample was burned. Table 3.3 shows the results for three foods.

→

Table 3.3

Food	Temperature of water/°C			Energy per gram /J
	Before burning	After burning	Difference	
A	16	43	27	2835
B	17	56	39	4095
C	17	64		

(a) Calculate the energy per gram for food **C**. [2]

(b) Apart from information provided above, give **one** other variable that should have been controlled in this investigation. [1]

(c) The energy values in the table for foods **A** and **B** are probably an underestimation of their real values. Suggest **one** reason for this. [1]

Answers online

ONLINE

4 Enzymes and digestion

Enzymes

REVISED

Enzymes are **proteins** that act as **biological catalysts**, speeding up the rates of reactions in the body. The enzymes themselves are not used up in the reaction. Enzymes can both build up and break down molecules.

Enzyme – an enzyme is a biological catalyst that speeds up reactions without being used in the reaction itself.

How enzymes work

In enzyme action the substrate fits snugly into the active site of the enzyme. This tight fit enables the enzyme to catalyse the reaction and split the substrate into its products, as shown in Figure 4.1.

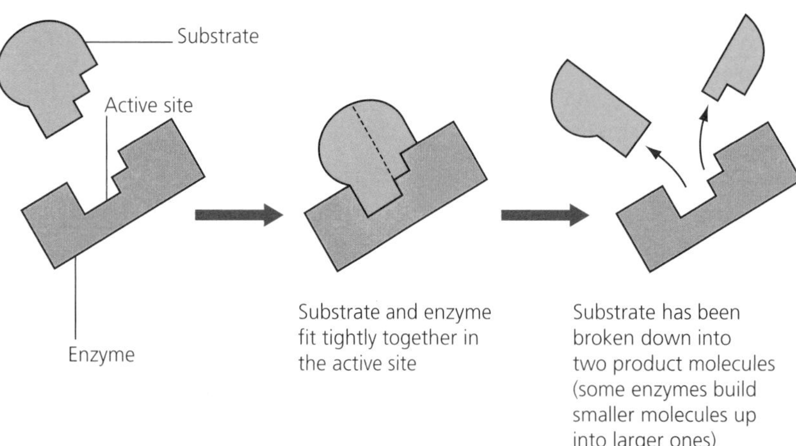

Figure 4.1 How enzymes work

Exam tip

The shape of the active site and the substrate are complementary (mirror images) to each other – they are not the same!

The action of enzymes as described in Figure 4.1 is referred to as the **'lock and key'** model due to the importance of the tight fit between the enzyme's **active site** and the **substrate**. This tight fit explains the process of **enzyme specificity** – each enzyme is specific in that it will only work on one (or a very small range of) substrates.

Factors affecting enzyme action

Temperature, pH and enzyme concentration all affect the action of enzymes. The effect of each of these factors is shown in Figures 4.2–4.4. The maximum rate of enzyme activity is described as the **optimum**.

Temperature

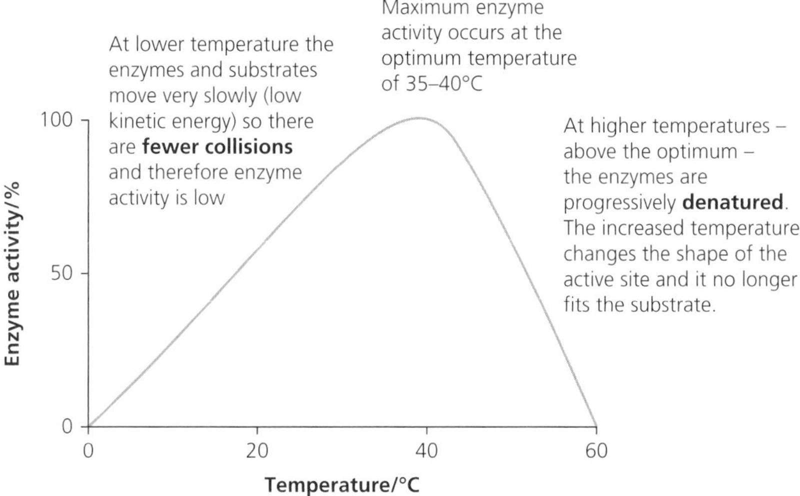

Maximum enzyme activity occurs at the optimum temperature of 35–40°C
At lower temperature the enzymes and substrates move very slowly (low kinetic energy) so there are **fewer collisions** and therefore enzyme activity is low
At higher temperatures – above the optimum – the enzymes are progressively **denatured**. The increased temperature changes the shape of the active site and it no longer fits the substrate.
100
50
0
Amylase acts in neutral/ slightly alkaline conditions in the mouth
As more enzymes become available there is more activity
Rate levels off as enzyme concentration increases because the number of substrate molecules becomes limiting
Enzyme activity/%
0
20
40
60
Temperature/°C

Figure 4.2 The effect of temperature on enzyme activity

> **Exam tip**
>
> The changes in enzyme active site shape that take place when enzymes are heated above their optimum temperatures are **irreversible**.

pH

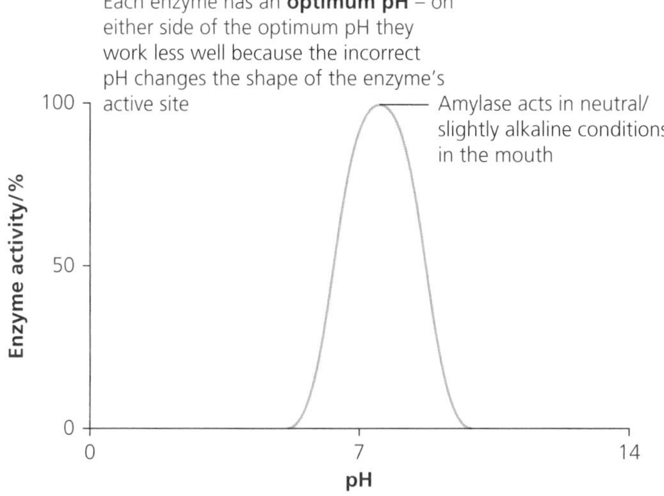

Figure 4.3 The effect of pH on enzyme activity

Enzyme concentration

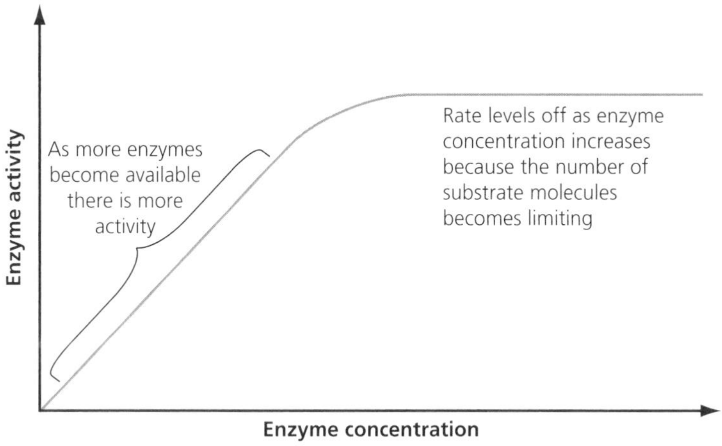

Figure 4.4 The effect of enzyme concentration on enzyme activity

Now test yourself

TESTED

1 Enzymes are biological catalysts. Explain the term 'biological catalyst'.
2 'Enzymes become denatured at high temperatures.' Explain this statement.
3 State the term that describes the temperature at which enzymes work at their maximum rate.

Example

You should be able to interpret graphs and tables showing enzyme activity in different temperatures, pH or enzyme concentrations.

Use Figure 4.5 to answer the questions which follow.

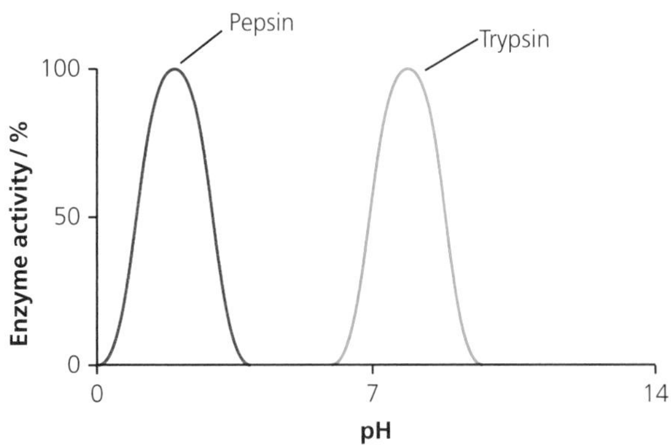

Figure 4.5 **The effect of pH on pepsin and trypsin**

1 Describe and explain the activity of pepsin in different pHs.
2 Trypsin is a protease found in the upper parts of the small intestine. Use Figure 4.5 to suggest the pH of that part of the small intestine.

Answers

1 Pepsin has a pH optimum of around 2.5, and its activity decreases either side of this optimum. Its activity decreases as the enzyme's active site becomes denatured when the pH is not the same as the optimum pH of the enzyme.
2 The pH is probably around 8, as the pH optimum of enzymes in the body reflect the pH of the environment in which they are found.

Exam tip

Describe and **explain** – if you are asked to *describe* a feature such as enzyme activity, as in this worked example, you should answer in terms of what is taking place. *Explanation* refers to *why* it taking place. In summary, describe = what and explain = why.

H Inhibitors

Some molecules or substances can fit loosely into an enzyme's active site – they may not be an exact fit, but they can fit well enough to stop the normal substrate from fitting. These are referred to as **inhibitors**. Inhibitors are not broken down by the enzyme.

Exam tip

Inhibitor molecules **inhibit (reduce)** normal enzyme activity.

Prescribed practical B3

Investigate the effect of temperature on the action of an enzyme

Digestion and absorption in the digestive system

REVISED

The two main processes taking place in the digestive system are **digestion** and **absorption**.

Digestion – the breaking down of large, insoluble food molecules into small, soluble molecules that can be absorbed.

Absorption – the process in which small, soluble food molecules are transferred from the gut to the blood system (in the ileum).

Enzymes in the digestive system

In the digestive system, enzymes are needed to **break down** (digest) **large, insoluble** food molecules into **small, soluble** ones that can be **absorbed** into the bloodstream.

Most of the digestion of food takes place in the stomach and the first part of the ileum.

There are three main groups of enzyme involved in the digestive system, as shown in Table 4.1.

Table 4.1 The main groups of enzymes in the digestive system

Enzyme	Food digested (substrate)	Products of digestion
Carbohydrase (amylase)	Starch	Glucose (and other sugars)
Protease	Protein	Amino acids
Lipase	Fat	Glycerol and fatty acids

Now test yourself

TESTED

4 Define the term 'digestion'.

Commercial enzymes

REVISED

Many commercial processes now use enzymes. For example, they are used in **biological washing powders**. These enzymes are very effective at breaking down a wide range of stains, but they are also **thermostable** – they can work effectively at a wide range of temperatures.

Commercial enzymes are widely used in the food industry, for example, amylases are used to break down starch into sweeter and more soluble sugars in many products.

Exam tip

Many detergents (washing powders) contain both lipases and proteases – this is because some of the most difficult stains to remove are fats and proteins.

Exam practice

1 (a) Give the function of enzymes. [2]

(b) Copy and complete Table 4.2. [3]

Table 4.2

Enzyme	Food digested (substrate)	Products of digestion
Carbohydrase (amylase)	Starch	
	Protein	Amino acids
Lipase		Glycerol and fatty acids

(c) Explain what is meant by the term 'enzyme specificity'. [2]

→

2 (a) Figure 4.6 shows the effect of temperature on an enzyme.

(i) Give the optimum temperature of this enzyme. [1]

(ii) Explain the sub-optimum enzyme activity at point **W**. [2]

(iii) Explain the shape of the graph at point **X**. [2]

H (b) Explain fully how an enzyme inhibitor affects enzyme activity. [3]

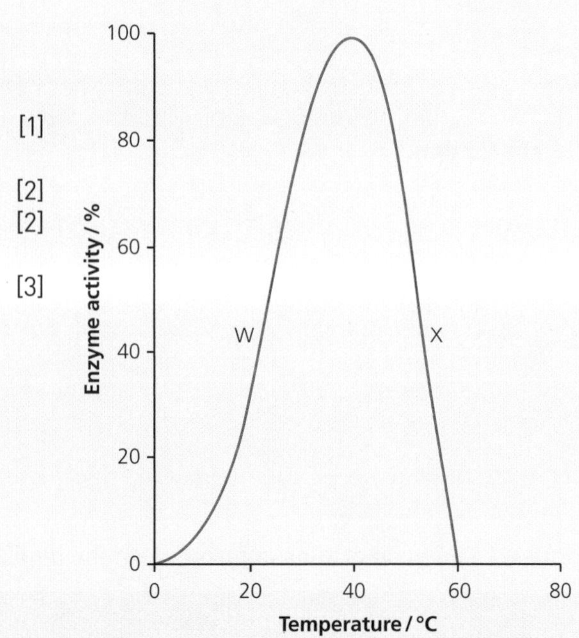

Figure 4.6

Answers online

ONLINE

5 The respiratory system and cell respiration

Respiration

REVISED

Respiration (cell respiration) is a biological process that **continually releases energy** from food (usually glucose). The energy released is used for heat, movement, growth, reproduction and

H active transport (uptake).

The word equation for **aerobic respiration** is:

glucose + oxygen → carbon dioxide + water + energy

H Respiration is an **exothermic reaction** (it releases energy to the surroundings) and the sites of respiration are the **mitochondria**, which are found in the cytoplasm of animal and plant cells.

The balanced chemical equation for aerobic respiration is:

$C_6H_{12}O_6 + 6O_2 \rightarrow 6CO_2 + 6H_2O$ + energy

Respiration – the release of energy from food.

Aerobic respiration – respiration in the presence of oxygen.

Exam tip

Energy is **released** from food in respiration – it is **not** produced!

Aerobic and anaerobic respiration

The type of respiration described in the equations above is **aerobic respiration** – respiration in the presence of oxygen.

Anaerobic respiration is respiration without oxygen.

We can respire anaerobically in our muscles for short periods of time. The equation for anaerobic respiration in **mammalian muscle** is:

glucose → lactic acid + energy

Yeast can also respire anaerobically and the equation is:

glucose → alcohol + carbon dioxide + energy

Anaerobic respiration – respiration in the absence of oxygen.

Exam tip

Anaerobic respiration in muscles is only likely to take place during **strenuous exercise** – it is only then that we are respiring at a rate at which we cannot supply enough oxygen to our muscles to respire aerobically.

Exam tips

- **Anaerobic respiration** releases much **less energy** than aerobic respiration – this is why we cannot live in anaerobic conditions for any length of time – we need more energy to stay alive.
- Anaerobic respiration in yeast is a very useful commercial process – it is important in beer and wine making and also baking (the carbon dioxide released causes bread and cakes to rise).

Table 5.1 summarises the similarities and differences between aerobic and anaerobic respiration.

Table 5.1 Similarities and differences between aerobic and anaerobic respiration

Similarities	Differences
• They both produce energy • They both use glucose as an energy source	• Aerobic respiration produces more energy than anaerobic respiration • Oxygen is not used in anaerobic respiration • Water is not produced in anaerobic respiration • Lactic acid is produced in anaerobic respiration in mammalian muscle and alcohol is produced in yeast – carbon dioxide is produced in yeast but not in mammalian muscle

Figure 5.1 shows how anaerobic respiration can be demonstrated in yeast.

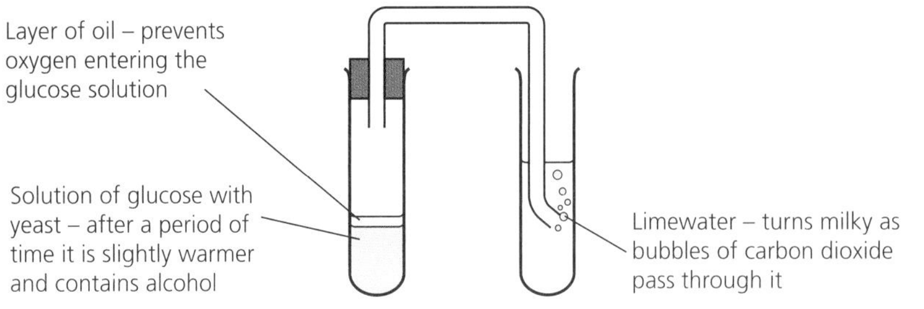

Layer of oil – prevents oxygen entering the glucose solution
Solution of glucose with yeast – after a period of time it is slightly warmer and contains alcohol
Limewater – turns milky as bubbles of carbon dioxide pass through it

Figure 5.1 Demonstrating anaerobic respiration in yeast

This apparatus can be used to investigate how different factors, such as temperature and the type of sugars added, affect the rate of anaerobic respiration in yeast. You should be able to work out the controlled variables when investigating different independent variables.

Exam tip

You need to be able to describe the differences between aerobic and anaerobic respiration, and also the differences between anaerobic respiration in mammalian muscle and in yeast.

Now test yourself

TESTED

1 Give the word equation for anaerobic respiration in mammalian muscle.

The effect of exercise on the depth and rate of breathing

Breathing is a process which brings air rich in oxygen into the lungs and thus supplies the oxygen the body needs for respiration. It also removes carbon dioxide produced during respiration from the body.

Breathing – the process in which air rich in oxygen is taken into the lungs and air rich in carbon dioxide is removed.

When someone is active, he or she will need to respire more to produce the extra energy required. Extra respiration requires extra oxygen (Figure 5.2).

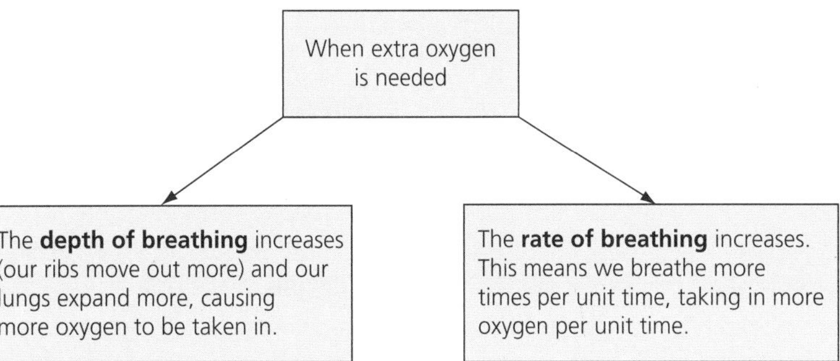

When extra oxygen is needed
The **depth of breathing** increases (our ribs move out more) and our lungs expand more, causing more oxygen to be taken in.
The **rate of breathing** increases. This means we breathe more times per unit time, taking in more oxygen per unit time.

Figure 5.2 The effect of exercise on the depth and rate of breathing

Example

Describe and explain what happens to a person's depth of breathing during exercise.

Answer

The depth of breathing increases. This means that more oxygen is taken in during each breath. This extra oxygen is used in respiration to release the extra energy required during exercise. The increased depth of breathing also removes the extra carbon dioxide produced.

Exam tip

Do not confuse the terms 'respiration' and 'breathing'. Respiration is the release of energy from food, and breathing is the process in which oxygen is taken into the lungs and carbon dioxide removed.

Respiratory surfaces

REVISED

Respiratory surfaces are parts of a living organism in which **respiratory gases** (oxygen and carbon dioxide) are **exchanged** between the atmosphere and cells (or blood).

Respiratory surface – the parts of living organisms across which respiratory gases can be exchanged between the environment (atmosphere) and the organism's cells.

Respiratory surfaces in humans (animals)

Figure 5.3 shows that lungs consist of many microscopic air sacs called alveoli. Gas exchange takes place between the alveoli and the blood.

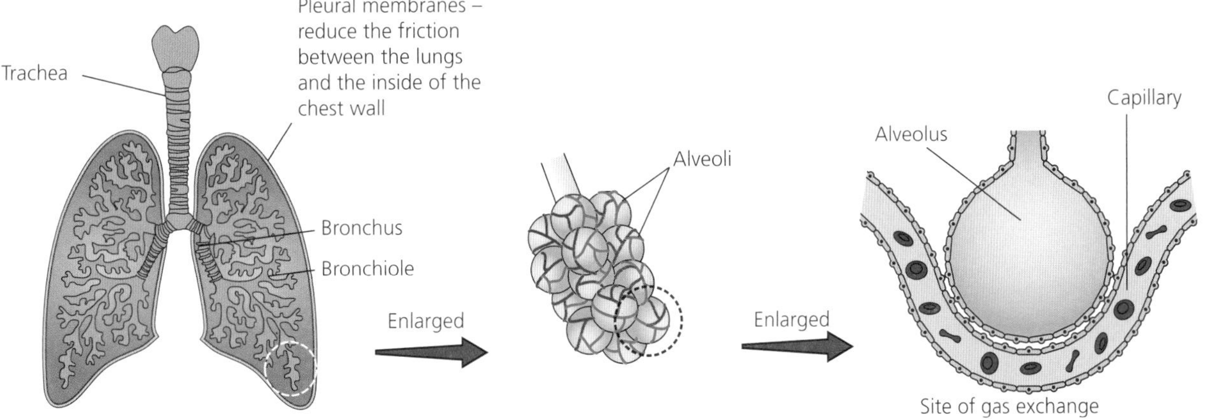

Figure 5.3 Alveoli, the site of gas exchange in humans

Figure 5.4 shows the gas exchange that takes place between the alveoli and the blood and also summarises adaptations that maximise the exchange of gases.

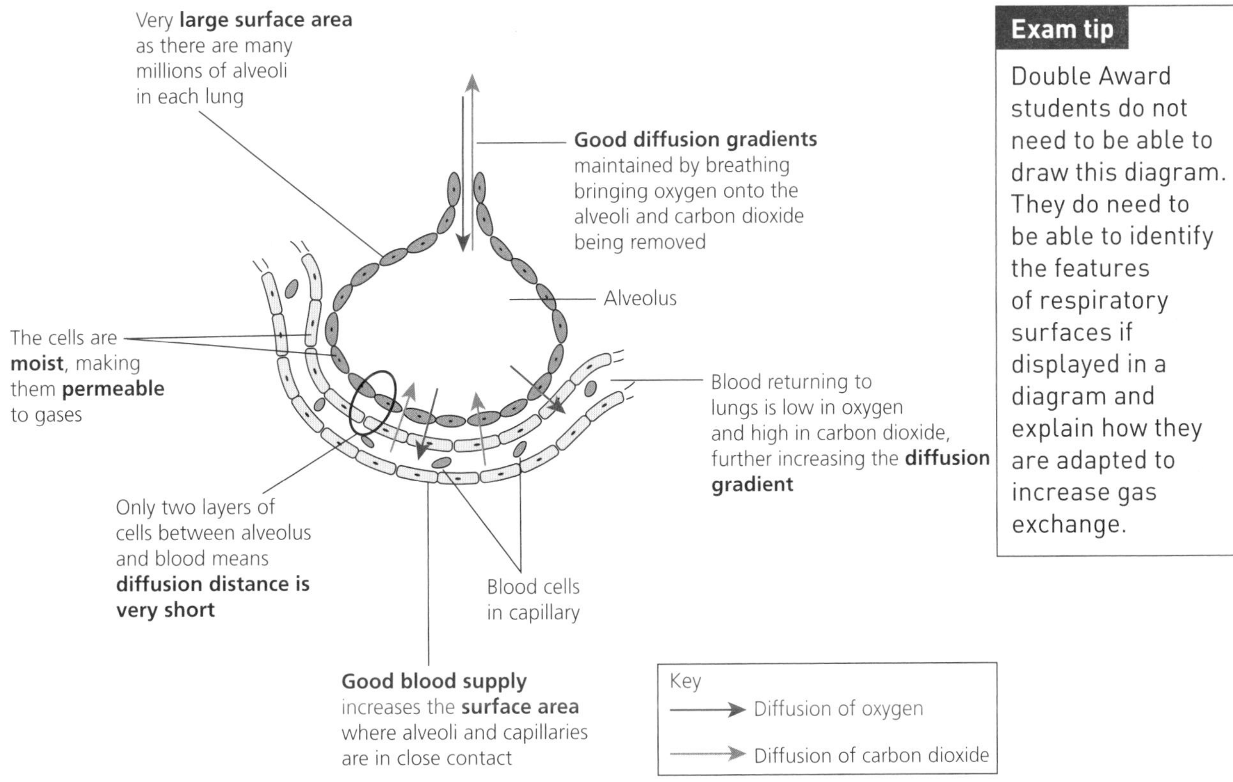

Figure 5.4 Adaptations of respiratory surfaces

Exam tip

Double Award students do not need to be able to draw this diagram. They do need to be able to identify the features of respiratory surfaces if displayed in a diagram and explain how they are adapted to increase gas exchange.

Exam tip

In animals, respiratory surfaces usually have **large surface areas**, are **thin** (have a short diffusion distance), are **moist**, are **permeable**, have a **good blood supply** and have **steep concentration (diffusion) gradients**.

Respiratory surfaces in plants

The main respiratory surfaces in plants are the **cells surrounding the intercellular air spaces** in leaves. Plant respiratory surfaces also are adapted by having:

- a large surface area
- thin exchange surfaces
- moist and permeable walls
- a diffusion gradient caused by respiration (and photosynthesis during the day) in cells, leading to the diffusing gases being in lower or higher concentrations than in the intercellular air spaces.

Exam tip

Plants do *not* have an active **breathing system** to further increase concentration gradients or a **blood system** to carry diffusing gases to and from the gas exchange surfaces.

Exam practice

1 (a) Give the word equation for aerobic respiration. [2]
(b) In terms of products, give **two** differences between anaerobic respiration in mammalian muscle and in yeast. [2]

2 Describe and explain the effect of exercise on the rate of breathing. [3]

3 (a) Respiratory surfaces in plants and animals normally have a large surface area and are thin (involve short diffusion distances).
(i) What is the advantage of respiratory surfaces having a large surface area? [1]
(ii) How is a large surface area achieved in the lungs? [1]
(iii) How is a short diffusion distance achieved between the alveoli and capillaries in the lungs? [2]
(b) Explain fully why a diffusion gradient exists between the intercellular air spaces in leaves and the spongy mesophyll cells, allowing oxygen to diffuse in. [2]

Answers online

ONLINE

6 Coordination and control

The nervous system

REVISED

We are able to respond to the environment around us. Anything that we respond to is called a **stimulus**.

In animals, **stimuli** (e.g. sound, smell, visual stimuli) affect **receptors** in the body. There are many types of receptor, each responding to a particular type of stimulus. If a receptor is stimulated, it may cause an **effector** such as a **muscle** or **gland** to produce a **response**.

Nerve cells, or **neurones**, link the receptors to a coordinator. The coordinator is the **central nervous system (CNS)**, consisting of the brain and spinal cord. In effect, the role of the CNS is to make sure the correct receptor is linked to the correct effector. A neurone carries information in the form of small electrical charges called **nerve impulses**.

Neurone – a nerve cell.

Central nervous system (CNS) – the part of the nervous system that links receptors and effectors.

The role of the nervous system in animals is summarised in Figure 6.1.

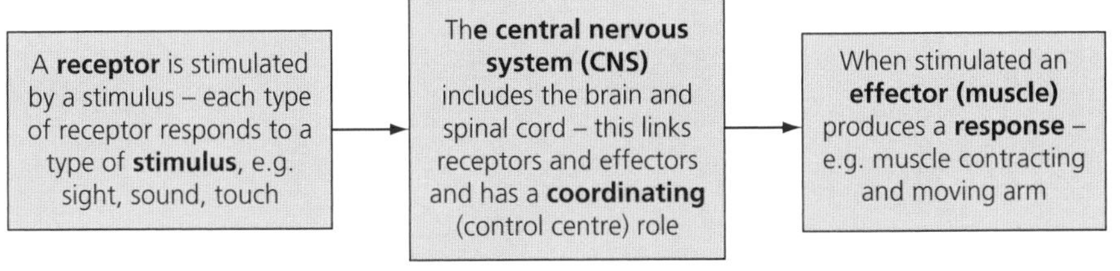

Figure 6.1 The nervous system

Exam tip

Remember that the **CNS** includes both the brain and the spinal cord.

Voluntary and reflex actions

Voluntary and **reflex** actions are the two main types of nervous action, as summarised in Table 6.1.

Table 6.1 Voluntary and reflex actions

	Voluntary	Reflex
Conscious control (brain and thinking time) involved	Yes	No
Speed of action	Variable – usually much slower	Fast

Exam tip

Reflexes are **automatic** and often **protective** – such as the withdrawal of a hand from a hot object.

Now test yourself

TESTED

1 Name the **two** parts of the central nervous system.
2 Give **two** differences between reflex and voluntary actions.

H The reflex arc

The pathway of neurones in a reflex action is described as a **reflex arc**. The reflex arc for the reflex that occurs when someone puts his or her hand on a hot object is shown in Figure 6.2.

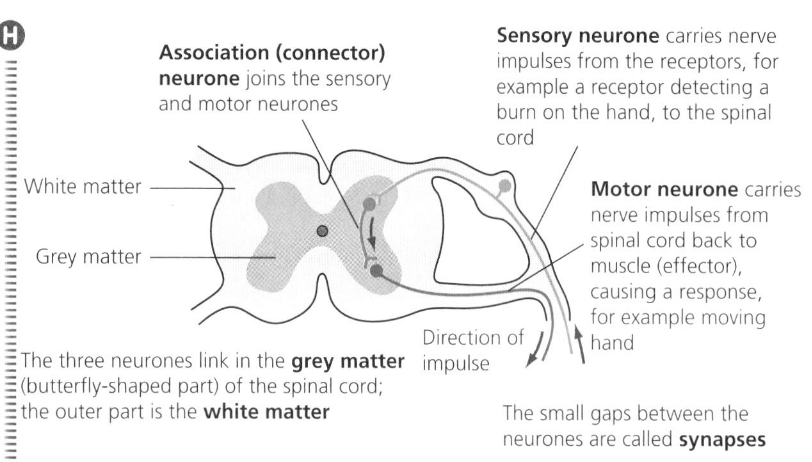
Association (connector) neurone joins the sensory and motor neurones
Sensory neurone carries nerve impulses from the receptors, for example a receptor detecting a burn on the hand, to the spinal cord
White matter
Grey matter
Motor neurone carries nerve impulses from spinal cord back to muscle (effector), causing a response, for example moving hand
Direction of impulse
The three neurones link in the grey matter (butterfly-shaped part) of the spinal cord; the outer part is the white matter
The small gaps between the neurones are called synapses

Figure 6.2 A reflex arc

Example

You need to be able to draw or complete and label diagrams of a reflex arc.

Students were asked to add the three neurones to an outline of the spinal cord. One student's answer is shown in Figure 6.3, but there are a number of mistakes. Can you spot them?

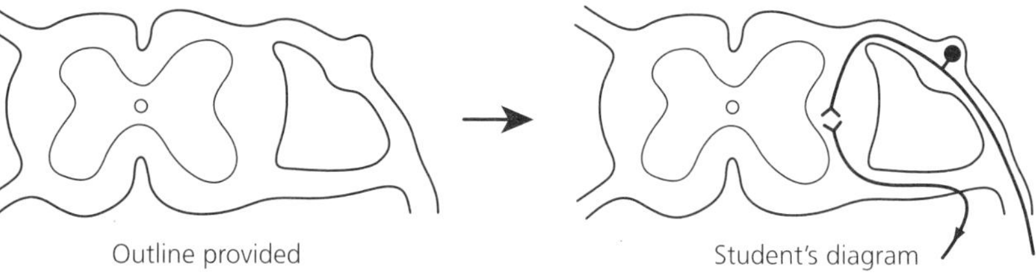
Outline provided
Student's diagram

Figure 6.3 Drawing a reflex arc

Answer

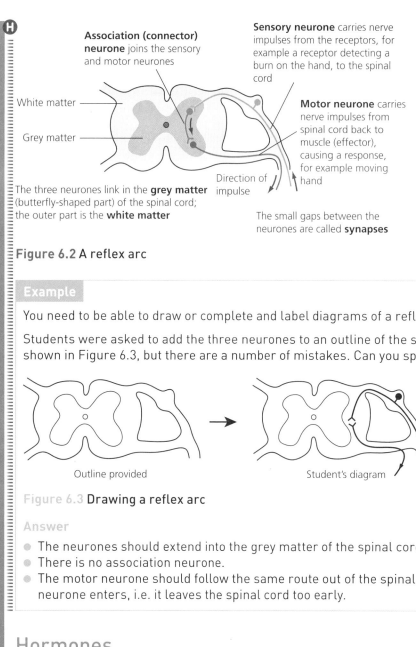

- The neurones should extend into the grey matter of the spinal cord.
- There is no association neurone.
- The motor neurone should follow the same route out of the spinal cord through which the sensory neurone enters, i.e. it leaves the spinal cord too early.

Hormones

REVISED

Hormones are chemical messengers, produced by glands, that travel in the blood to bring about a response in a target organ.

The main differences between the hormones and the nervous system are summarised in Table 6.2.

Hormone – a chemical produced by a gland that travels in the blood to bring about a response in a target organ.

Table 6.2 The main differences between hormone and nervous communication

	Nervous system	Hormones
Method of communication	Electrical impulses along neurones	Chemicals in blood
Speed of action	Fast-acting	Usually slow-acting

Insulin

Insulin is a hormone that **lowers blood glucose** concentrations (Figure 6.4). It is important that the amount of glucose (sugar) in the blood is kept at the optimum concentration.

Insulin – the hormone that lowers blood glucose concentrations.

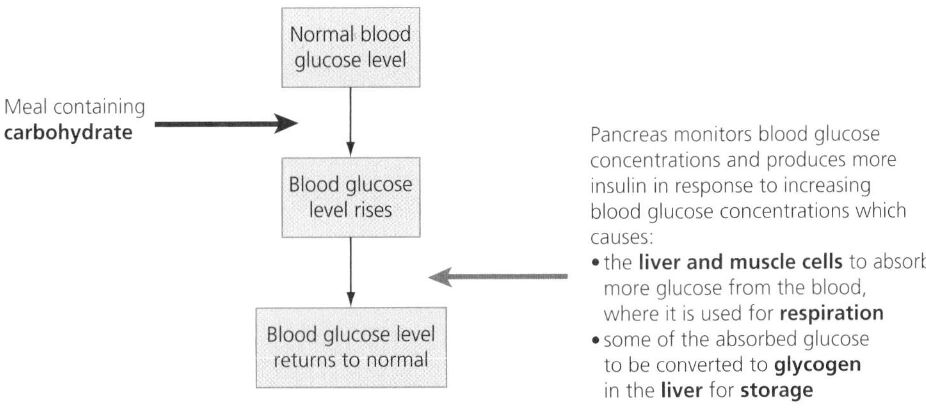

Figure 6.4 The action of insulin

Exam tips

- In an exam you need to state that insulin **lowers/reduces** blood glucose concentrations – it is not enough to say that it controls blood glucose concentrations.
- The action of insulin highlights the definition of a hormone. Insulin is a **chemical messenger**, produced by the pancreas (a **gland**) that travels in the **blood** to bring about a **response** (converting glucose to glycogen and/or causing glucose to move from the blood to cells for respiration) in the liver and/or muscles (**target organs**).

Example

1 Explain why the body needs glucose in the blood.
2 Suggest why the concentration of insulin in the blood is usually at its lowest in the middle of the night.

Answer

1 The glucose is transported to body cells to provide energy in respiration.
2 Glucose concentrations will also be low during the night. By this time the glucose from the last meal of the day has been used up in respiration or converted to glycogen for storage.

Diabetes

Diabetes is a lifelong condition in which the body does not produce enough insulin (or the insulin produced does not work). Therefore, people with diabetes have very high (and dangerous) blood glucose concentrations unless treated.

The **symptoms** (signs that show something is wrong) of diabetes include:

- high blood glucose concentrations
- glucose in the urine
- lethargy
- thirst.

Exam tip

Diabetes is a condition in which the blood glucose control mechanism fails.

There are two types of diabetes: type 1 and type 2. The main differences between type 1 and type 2 diabetes are summarised in Table 6.3.

Table 6.3 The main differences between type 1 and type 2 diabetes

	Type 1	Type 2
Main effect	Insulin is not produced by the pancreas	Insulin is produced but stops working properly or the pancreas does not produce enough insulin
Treatment	Insulin injections for life (plus controlled diet and exercise)	Usually controlled by diet initially but later requires medication and/or insulin injections
Preventative measures	None – not caused by lifestyle	Take exercise, reduce sugar intake, avoid obesity
Age of first occurrence	Often in childhood	Usually as an adult

Long-term effects and future trends

People who have had diabetes for a long time and whose blood glucose concentration is not tightly controlled are at risk of developing **long-term complications**. These include:

- **eye damage** (and blindness)
- **heart disease** and **strokes** (circulatory diseases)
- **kidney damage**.

Exam tip

Type 2 diabetes is linked to **lifestyle** but type 1 diabetes is ***not*** caused by lifestyle.

The number of people who suffer diabetes is increasing rapidly, and the cost of treatment is becoming very high. The large increase in the number of people with type 2 diabetes is linked to poor diet and a lack of exercise.

The excretory system and osmoregulation

REVISED

The kidney has two main roles in the body. It is important in the **excretion** of waste products such as urea, and **osmoregulation** – controlling the water balance in the body (Figure 6.5).

Excretion – the removal of waste products from the body, e.g. carbon dioxide during breathing and urea in the kidneys.

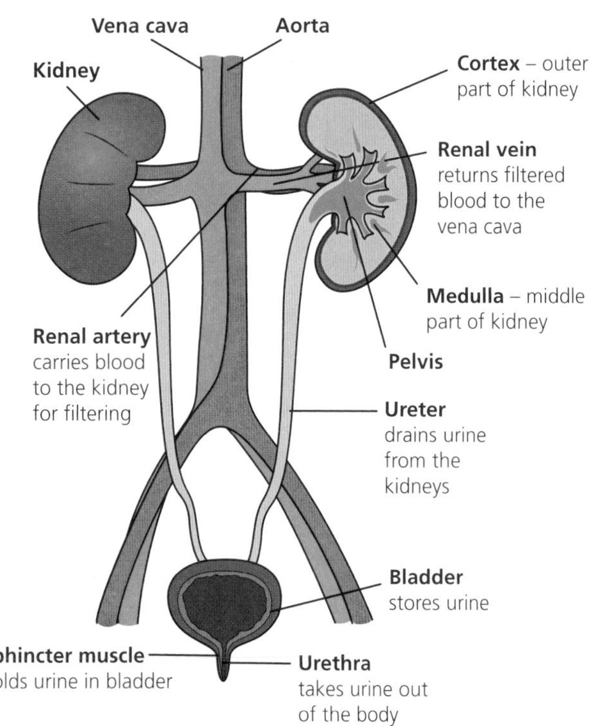

Figure 6.5 The excretory system

The kidney **filters** the blood and excretes waste products. Within the kidney most of the blood is filtered out of the blood vessels but only the useful material, such as glucose and water, is **reabsorbed** back into the blood. Most of this takes place in the **cortex**. The waste materials (those not reabsorbed, such as urea) are drained into special structures called collecting ducts, which pass through the medulla and into the **pelvis** (base) of the kidney before emptying into the ureters for excretion.

Homeostasis and osmoregulation

Homeostasis means maintaining a **constant internal environment** in the body for the proper functioning of cells and enzymes in response to internal and external change.

Osmoregulation means maintaining the water balance at a constant level in the body.

Table 6.4 shows some of the different ways in which we can gain and lose water.

Table 6.4 Taking in and losing water

Gain water	Lose water
• Drinking liquids • In food • Water produced in body cells as a waste product in respiration	• Evaporation of sweat • Breathing out water vapour • In urine

The kidney is the organ that controls water balance, and it does this by controlling the amount of water that is reabsorbed back into the blood during the filtering process.

Figure 6.6 shows the role of the kidney in osmoregulation.

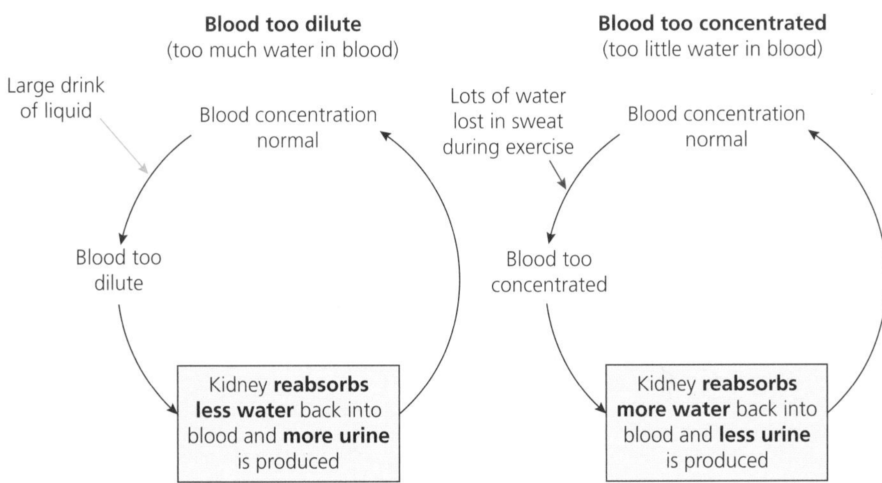

Figure 6.6 The role of the kidney in osmoregulation

Figure 6.6 shows that the kidney can control water balance by reabsorbing more or less water back into the blood (following filtration), depending on the concentration of the blood.

H The amount of water reabsorbed is controlled by **antidiuretic hormone** (**ADH**). If the blood is too concentrated, more ADH is produced, and this results in more water reabsorption and less urine produced. (If the blood is too dilute, less ADH is produced, and this results in less water reabsorption with more urine produced.)

Exam tips

- Osmoregulation effectively means keeping the concentration of the blood correct.
- The control of **blood glucose** is another example of **homeostatic** control.

Exam tip

The only bullet point in Table 6.5 that can be 'fine-tuned' to control water balance is the amount of water lost in the urine.

Now test yourself

TESTED

3 Name the **two** main functions of the kidney.
4 Name the **three** ways in which the body gains water.
H 5 Name the hormone that controls the amount of water reabsorbed in the kidneys.

Plant hormones

REVISED

Hormones are also important in coordination in plants. **Phototropism** is the growth response involving plants bending in the direction of light. When a plant experiences uneven light (unidirectional or unilateral light), more of the hormone **auxin** passes to the shaded side. The auxin causes increased elongation of cells on the shaded side, resulting in uneven growth and the plant bending towards the light.

Auxin – the hormone responsible for phototropism in plants.

Exam tip

By bending in the direction of the light, the plant gets **more light** and so undergoes **more photosynthesis** and **more growth**.

H Figure 6.7 shows that the auxin is produced in the tip of the plant but more passes down the shaded (non-illuminated side) than the illuminated side. This means that the cells in this region get more auxin and therefore grow more (compared with the cells in the side getting most light).

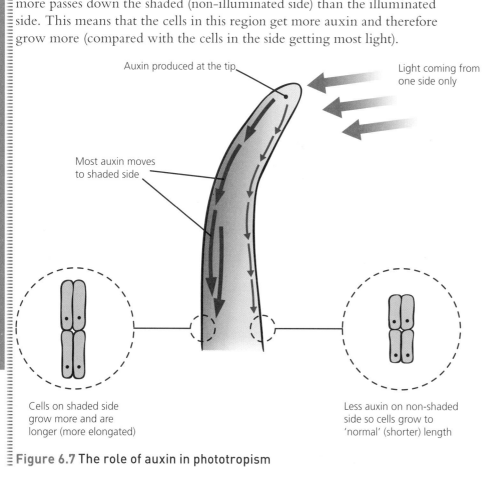

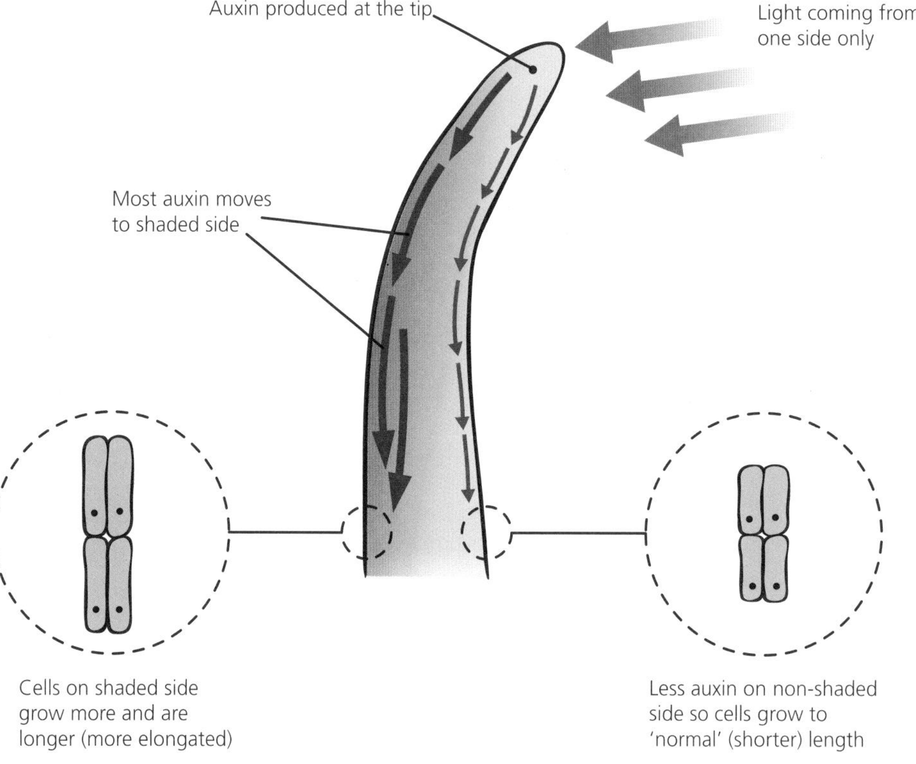
Auxin produced at the tip
Light coming from one side only
Most auxin moves to shaded side
Cells on shaded side grow more and are longer (more elongated)
Less auxin on non-shaded side so cells grow to 'normal' (shorter) length

Figure 6.7 The role of auxin in phototropism

Exam practice

1 (a) Figure 6.8 shows a reflex arc.

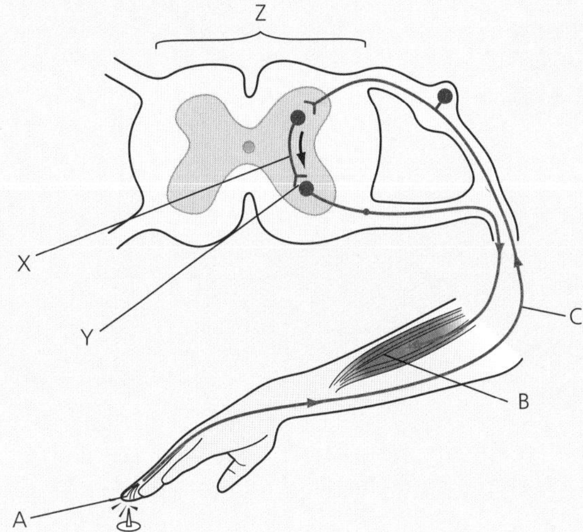

Figure 6.8

(i) Name neurone **X**. [1]
(ii) Give the name for the small gap **Y**. [1]
(iii) What is structure **Z** called? [1]
(iv) Which part of the diagram (**A**, **B** or **C**) is the receptor? [1]
(v) Which part (**A**, **B** or **C**) is the effector? [1]

(b) One feature of reflex actions is that the response is rapid. In terms of neurone pathways, explain why reflex actions are usually very fast. [2]

2 (a) Complete the sentence below about hormones.
A hormone is a chemical .... that travels in the .... to a target organ, where it acts. [2]

(b) Figure 6.9 shows how a person's blood glucose concentration changes over a period of time.

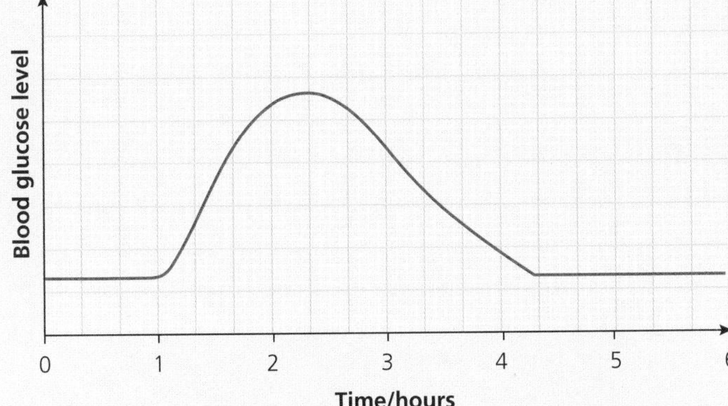

Figure 6.9

(i) Name the organ that produces insulin. [1]
(ii) Suggest what caused the blood glucose concentration to rise after 1 hour. [1]
(iii) Describe the role of insulin in causing the blood glucose concentration to decrease. [2]

→

3 (a) Table 6.5 shows the numbers of people being treated for type 1 and type 2 diabetes in a hospital over a 5-year period.

Table 6.5

Year	Number of people treated with diabetes	
	Type 1	Type 2
2005	14	94
2006	15	99
2007	17	105
2008	17	121
2009	19	133
2010	20	141

(i) Calculate the percentage increase in type 2 diabetes between 2005 and 2010. [2]

(ii) The table shows that the number of patients with type 2 diabetes is increasing over time. Describe **one** other trend shown by the information in the table. [1]

(iii) Suggest **two** reasons for the increase in number of people with type 2 diabetes. [2]

(b) Give **two** long-term complications of diabetes. [2]

4 (a) Name **one** substance that is filtered out of the blood in the kidney but not reabsorbed back. [1]

(b) (i) Define the term 'osmoregulation'. [1]

(ii) Explain why only small amounts of urine are produced on a very hot day. [2]

H (iii) Explain the role of ADH in kidney function. [2]

5 (a) Figure 6.10 shows some seedlings that receive light from one side only.

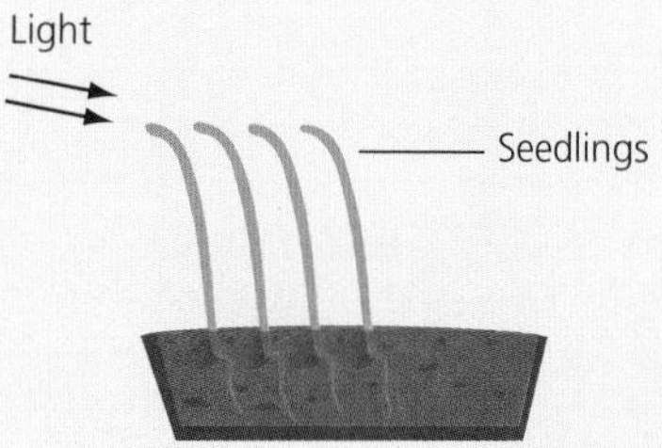

Figure 6.10

(i) Name this growth response in plants. [1]

(ii) Explain how this response benefits the seedlings. [2]

H (b) Figure 6.11 shows a plant shoot bending towards a light source. Use the diagram and your knowledge to explain why this plant shoot is bending to the right. [3]

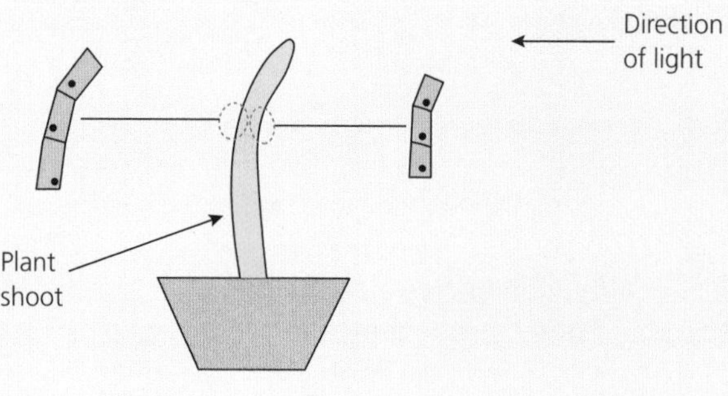

Figure 6.11

Answers online

ONLINE

7 Ecological relationships and energy flow

Ecology deals with the distribution of living organisms and the relationships between them.

Some key terms that you need to know are: **habitat**, **population**, **community**, **biodiversity**, **environment** and **ecosystem**.

The environment can be subdivided into:

- **abiotic** (physical, or non-living) factors, such as temperature and light
- **biotic** (living) factors, such as the effect of predators or competitors.

Habitat – the place where an organism lives and breeds.

Population – the number of organisms of a **single species** in a given habitat/area.

Community – all the populations (from **all species**) in a particular area.

Biodiversity – a measure of the **range** of different **species** of organisms living in an area.

Environment – the surroundings in which an organism lives.

Ecosystem – an area in which a community of organisms interact with each other and their physical surroundings.

Sampling – a process used to give a good estimate of the number, or percentage cover, of an organism (or organisms) in a particular area.

Fieldwork

REVISED

Fieldwork normally involves **sampling** the numbers or distribution of organisms in an area. Additionally, fieldwork often involves the measurement of environmental factors, such as light intensity.

Sampling

With most fieldwork it is impossible to count all the individuals of a species in an area, so the species is sampled. This is usually done using **quadrats**.

When using quadrats, the abundance of individuals can be estimated by recording:

- **number** – this is used when the number of individuals is easy to determine, e.g. limpets on a shore or thistles (which stand out above the grass)
- **percentage cover** – this is used when it is difficult to see where one individual stops and another one starts, e.g. many grass and moss plants. Percentage cover is usually rounded to the nearest 10%.

Random sampling is carried out when the area to be sampled is uniform.

When random sampling:

- enough samples (quadrats) should be used to give a **representative** sample of the area
- the quadrats should be positioned using **random numbers** to **avoid bias** (Figure 7.1).

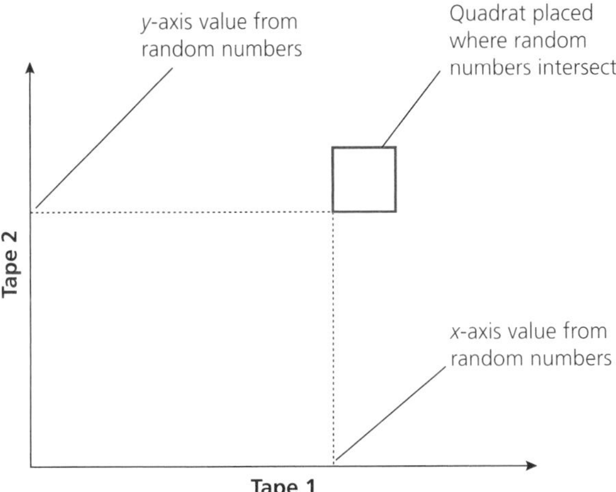

Figure 7.1 Random sampling

A **belt transect** is used if **zonation is evident** (there is a clear transition from one area into another) – examples include sampling from a grassland into woodland or on a rocky shore from the low tide line to the high tide line. Here the sampling is not random but along a **transect line**.

In a belt transect, the quadrats are typically placed end to end along a transect tape stretching from one end of the area to be sampled to the other – alternatively, if the transect is very long, the quadrats could be placed every 5 m or at other set intervals along the tape.

Exam tip

Using quadrats to count the number of each species in an area measures **biodiversity** (the number of species in an area) as well as the number of individuals for each species.

Example

You are asked to estimate the number of thistles in a field measuring 100 m × 50 m. You are also provided with a 50 cm × 50 cm quadrat. Describe how you could sample the field to estimate the number of thistles.

Answer

- Use a random number to identify, e.g. 20, places where the quadrat will be placed.
- Count the number of thistles in the quadrat at each position.
- Calculate the average number of thistles in the 20 quadrats.
- The overall area is 5000 m^2 and the area of the quadrat is 0.25 m^2 – so the estimated number of thistles in the field is the average in 20 quadrats × 20000 (5000 × 4).

Measuring environmental factors

Table 7.1 shows how many of the environmental factors that affect plant and animal numbers and distributions can be measured.

Table 7.1 Measuring environmental factors

Factor	Means of measurement	Comment
Wind speed	Using an **anemometer**	Wind speed can be very important in exposed habitats, such as a rocky shore or sand dune system.
Water	Weighing soil mass then drying in an oven until completely dry (constant mass), then reweighing; the percentage moisture is the difference divided by the initial mass × 100	Most plant species are restricted to soils of a particular moisture range, e.g. rushes are usually found in wetter soils.
pH	Using **soil test kits** or **pH probes** or **sensors**	Heathers are found in acid soils (soils with a low pH), but most plants prefer soils around neutral pH (pH 7).
Light	Using a **light meter**	Many woodland plants are adapted to growing in moderate or low light levels.
Temperature	Using a **thermometer**	Temperature is important in plant and animal distribution on a global scale (rather than at a local level).

Now test yourself

TESTED

1 Define the term 'community'.
2 When sampling organisms in a habitat, when would you use a belt transect?
3 What apparatus would you use to measure soil pH in a habitat?

Prescribed practical B4

Use quadrats to investigate the abundance of plants and/or animals in a habitat

Competition

REVISED

Living things compete with each other for resources.

Plants compete for:

- water
- light
- space to grow
- minerals.

Animals compete for:

- water
- food
- territory (space to live)
- mates.

Predators will also affect the distribution and number of animals present.

Competition – the term used to describe the 'battle' between living organisms for the same resource or resources.

Exam tip

Competition affects both the **distribution** of living organisms and also the **number** of any one species in a particular place.

Food chains and food webs

REVISED

A **food chain** describes the order in which energy passes through living organisms – i.e. a feeding sequence (Figure 7.2).

The **Sun** is the initial **source of energy** for all food chains.

A food chain always follows the order:

producer → primary consumer → secondary consumer → tertiary consumer

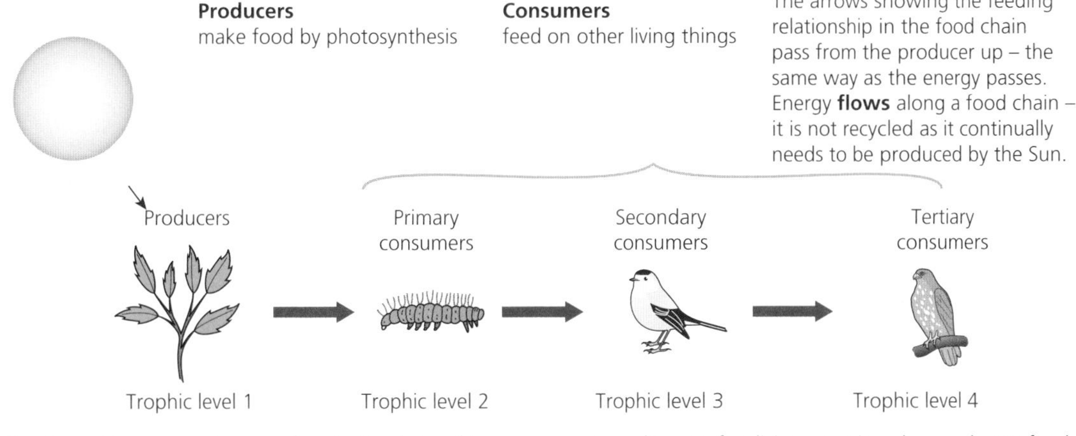

Producers
make food by photosynthesis
Consumers
feed on other living things
The arrows showing the feeding relationship in the food chain pass from the producer up – the same way as the energy passes. Energy **flows** along a food chain – it is not recycled as it continually needs to be produced by the Sun.
Producers
Primary consumers
Secondary consumers
Tertiary consumers
Trophic level 1
Trophic level 2
Trophic level 3
Trophic level 4
The **Sun** is the source of energy for all other living things. **Trophic level** refers to the 'feeding' level and begins with producers as level 1.
In reality very few living organisms have only one food source as shown in the food chain – usually there are many interlinked organisms as in a **food web**.

Figure 7.2 A food chain – a chain of living organisms through which energy passes

Table 7.2 Food chain terms

	Producer	Primary consumer	Secondary consumer	Tertiary consumer
Description	A plant that makes food by photosynthesis	An animal that feeds on a plant	An animal that feeds on a primary consumer	An animal that feeds on a secondary consumer
Example	Grass	A leaf-eating insect	A beetle	A bird that eats beetles
Trophic level	1	2	3	4

Exam tips

- The **arrows** in a food chain show the direction of the **transfer of substances** (e.g. carbon and nitrogen) through a food chain as well as showing what eats what (consumption).
- Animals need plants both as an initial **food source** and also to provide **oxygen** (for respiration) through photosynthesis.

Food webs show how a number of food chains are interlinked. They are more realistic because very few consumers feed on only one thing.

Energy flow

REVISED

Plants use energy from the Sun to produce food. As this 'food' passes through consumers, the energy initially trapped by plants passes through the consumers. Food chains and webs therefore also show the direction of **energy flow**.

Energy flow – the transfer of energy between organisms in a food chain.

Now test yourself

TESTED

4 What is meant by the term 'energy flow'?

Nutrient cycles

REVISED

In ecosystems, carbon, nitrogen and other elements are recycled as a result of many processes.

Decomposition

Decomposition is a key process in the recycling of **carbon** and **nitrogen**. **Decay** involves the initial part of the breakdown of plants and animals (and their products) – this involves earthworms and many types of insect.

Bacteria and **fungi** are the microorganisms responsible for the **decomposition** of decayed material to form mineral nutrients:

- **Saprophytic** (decomposing) fungi and bacteria secrete (release) digestive enzymes onto the decaying material.
- These enzymes break down the decaying organic material (**extracellular digestion**).
- The digested soluble products of digestion are absorbed into the fungi and bacteria.

Decay and decomposition form **humus** – the part of the soil in which plants grow and from which they obtain minerals.

Decomposition and humus formation take place more quickly when environmental conditions are optimum. These include:

- **adequate moisture** (decomposition cannot take place in totally dry conditions)
- a **warm temperature**
- presence of **oxygen**.

Exam tips

- A warm temperature and oxygen provide:
  - maximum rates of **aerobic respiration** in saprophytic microorganisms
  - optimum **enzyme activity** in these organisms.
- You need to be able to describe how you could investigate the factors that affect decomposition rate. You can do this by placing some plant material, such as leaves, in different environmental conditions, including the presence or absence of oxygen.

The carbon cycle

Figure 7.3 summarises the carbon cycle.

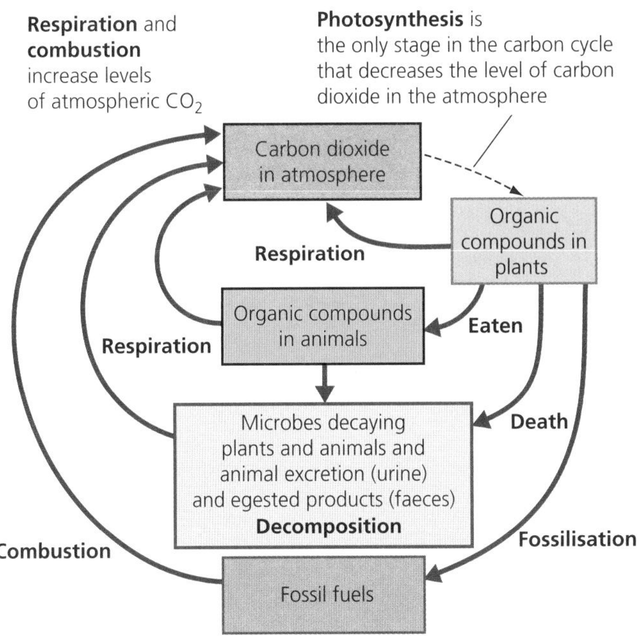

Figure 7.3 The carbon cycle

> **Exam tips**
>
> - The processes **respiration** and **combustion** *increase* levels of atmospheric carbon dioxide.
> - **Photosynthesis** is the only process that *reduces* atmospheric carbon dioxide.

H The nitrogen cycle

Plants obtain nitrogen in the form of **nitrates** from the soil. Nitrates are used to make **amino acids**, which are built up to form **proteins**.

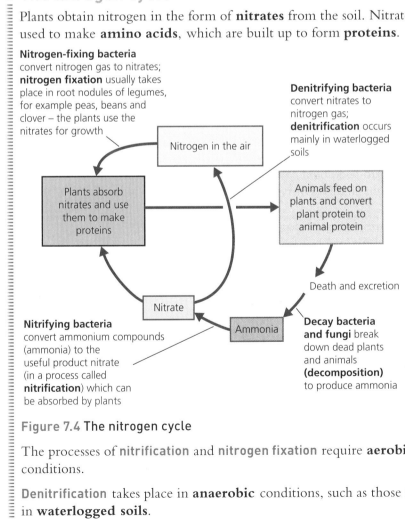

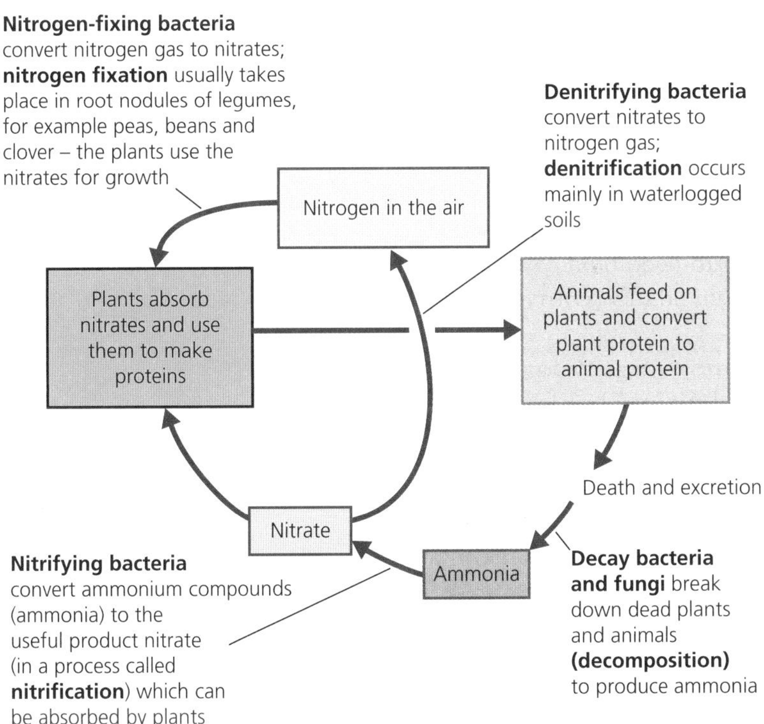

Nitrogen-fixing bacteria convert nitrogen gas to nitrates; **nitrogen fixation** usually takes place in root nodules of legumes, for example peas, beans and clover – the plants use the nitrates for growth
Nitrogen in the air
Denitrifying bacteria convert nitrates to nitrogen gas; **denitrification** occurs mainly in waterlogged soils
Plants absorb nitrates and use them to make proteins
Animals feed on plants and convert plant protein to animal protein
Death and excretion
Nitrate
Ammonia
Nitrifying bacteria convert ammonium compounds (ammonia) to the useful product nitrate (in a process called **nitrification**) which can be absorbed by plants
Decay bacteria and fungi break down dead plants and animals **(decomposition)** to produce ammonia

Figure 7.4 The nitrogen cycle

The processes of **nitrification** and **nitrogen fixation** require **aerobic** conditions.

Denitrification takes place in **anaerobic** conditions, such as those found in **waterlogged soils**.

> **Nitrification** – the process that describes the conversion of ammonium compounds (ammonia) to nitrate (in the nitrogen cycle).
>
> **Nitrogen fixation** – the process that describes the conversion of nitrogen gas to nitrate (in the nitrogen cycle).
>
> **Denitrification** – the process that describes the conversion of nitrate to nitrogen gas (in the nitrogen cycle).

Now test yourself

TESTED

5 State **three** environmental conditions necessary for decomposition to take place at a rapid rate.
6 In the carbon cycle, name the only process that reduces atmospheric carbon dioxide levels.
H 7 Explain what is meant by the term 'nitrification'.

Root hair cells as the site for mineral absorption

Root hair cells are specialised cells in the epidermal layer of roots that are adapted by having an extended shape, which gives a **large surface area** for the uptake of minerals (e.g. **nitrates** for making **protein**) and water from the soil.

H Figure 7.5 shows that the minerals are taken into the root **against the concentration gradient** – this is described as **active uptake**. Active uptake requires **energy** from **respiration** to move the minerals against the concentration gradient.

H **Active uptake (transport)** – the movement of particles from an area of low concentration to an area of high concentration using energy from respiration.

H
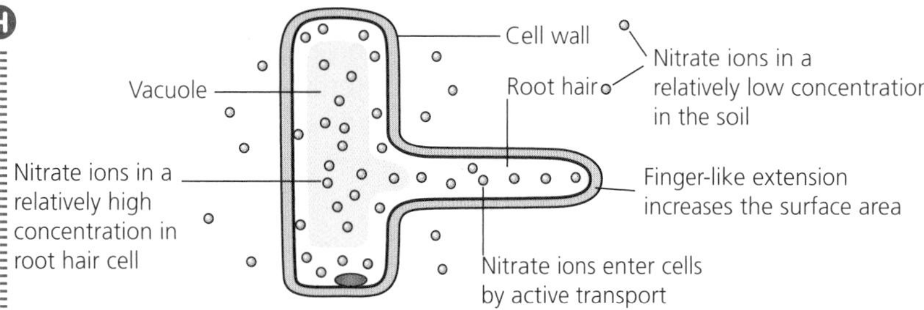

Figure 7.5 Taking minerals into the root

H Eutrophication

REVISED

Eutrophication is a form of water pollution caused by the water becoming enriched with minerals. This occurs due to:

- **fertiliser** runoff into waterways
- minerals in **sewage** entering waterways.

Eutrophication – a type of water pollution that is triggered by too many minerals/nutrients entering the water.

Exam tip

Fertiliser is more likely to leach into waterways when:

- too much is used on the land
- it is sprayed during rainfall or onto wet or sloping ground.

The sequence of events that occur in eutrophication are summarised in Figure 7.6.

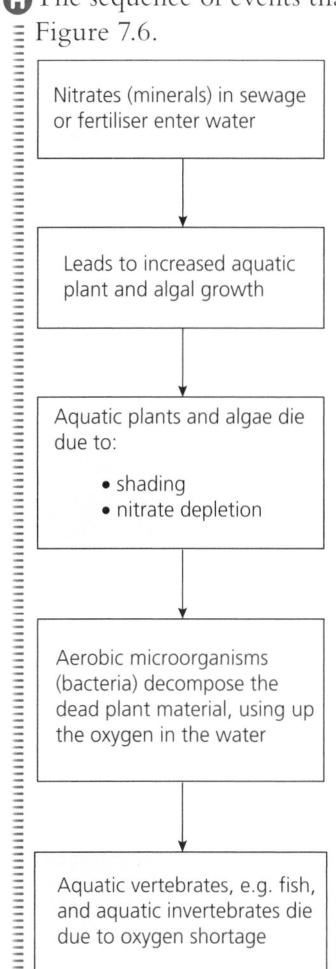

Figure 7.6 Eutrophication

Now test yourself

TESTED

8 Define the term 'eutrophication'.

Exam practice

1 (a) Define the term 'population'. [1]
(b) Describe how you would estimate the number of daisies in a school playing field. [3]

2 (a) What do the arrows in a food chain represent? [1]
(b) Define the term 'primary consumer'. [1]
(c) Which trophic level is a secondary consumer? [1]

3 Figure 7.7 shows the role coal (a fossil fuel) plays in the carbon cycle.

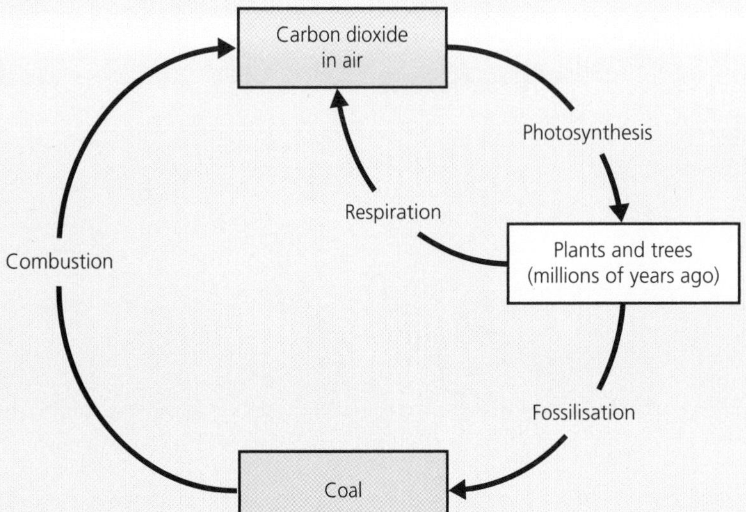

Figure 7.7

(a) Name the process that lowers carbon dioxide in the atmosphere. [1]
(b) Name the process that does not affect carbon dioxide levels. [1]

Answers online

ONLINE

8 Osmosis and plant transport

Osmosis

REVISED

Osmosis is the movement of water from a dilute solution to a more concentrated solution through a selectively permeable membrane (Figure 8.1).

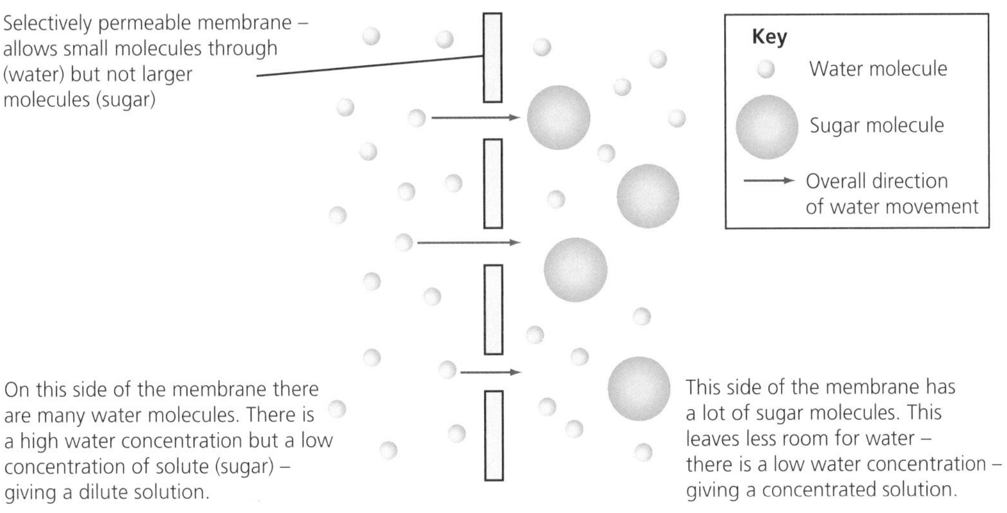

Selectively permeable membrane – allows small molecules through (water) but not larger molecules (sugar)
Key
Water molecule
Sugar molecule
Overall direction of water movement
On this side of the membrane there are many water molecules. There is a high water concentration but a low concentration of solute (sugar) – giving a dilute solution.
This side of the membrane has a lot of sugar molecules. This leaves less room for water – there is a low water concentration – giving a concentrated solution.

Figure 8.1 Osmosis

Osmosis – the movement of water from a dilute solution to a more concentrated solution through a selectively permeable membrane.

Exam tip

Figure 8.1 shows the two things required for osmosis; two different concentrations of solutions (or water and a solution) separated by a selectively permeable membrane.

Example

1 The apparatus shown in Figure 8.2 was set up in a classroom to demonstrate the process of osmosis. Describe and explain what happens after 24 hours.

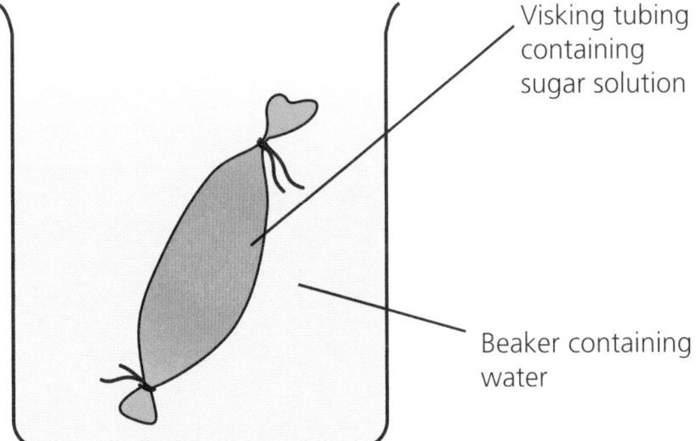

Visking tubing containing sugar solution
Beaker containing water

Figure 8.2 **Using Visking tubing**

2 Five potato cylinders each measuring 40 mm were placed in concentrated salt solution for 24 hours. When they were re-measured, their average length was 37 mm. Explain this result.

Answers

1

- There are more water molecules outside than inside the tubing.
- Water would move into the tubing and the tubing would expand.
- The Visking tubing is selectively permeable.

2
- There was more water in the cells of the potato than in the concentrated salt solution (the salt solution is more concentrated than the potato).
- Water moved from the potato into the concentrated salt solution
- by osmosis
- through a selectively permeable membrane.

Osmosis and plants

Normally a plant cell is more concentrated than its surroundings:
- **Water enters** the cell by **osmosis**.
- The vacuole expands, pushing the cell membrane against the cell wall.
- This causes the **turgor** necessary for support.
- The **cell wall** stops the membrane expanding too far to cause damage and therefore limits the water intake.

If a plant cell is surrounded by a more concentrated solution (this very seldom happens in nature) the cell **loses water** by **osmosis**.

The cell loses turgor and the membrane pulls away from the cell wall as the vacuole shrinks. This is called **plasmolysis**.

Turgor – the state of a plant cell when it has gained enough water by osmosis for the cell membrane to push against the cell wall, making the cell firm.

Plasmolysis – a plant cell is plasmolysed when it has lost water by osmosis and its membrane separates from the cell wall.

Why do plants need water?

Plants use water:
- for **support** (turgor)
- for **transpiration** – the movement of water up through a plant, its **evaporation** from leaf cells followed by **diffusion** out of the **stomata**
- for **transport** – as the water moves up through the plant it carries **minerals**
- as a raw material in **photosynthesis**.

Now test yourself

TESTED

1 Define the term 'osmosis'.
2 Describe the appearance of a plasmolysed plant cell when viewed under the microscope.

Prescribed practical B5

Investigate the process of osmosis by measuring the change in length or mass of plant tissue or model cells, using Visking tubing

Transpiration

REVISED

Transpiration is the evaporation of water from mesophyll cells followed by diffusion through the leaf air spaces and stomata.

Transpiration – the evaporation of water from mesophyll cells, followed by diffusion through the leaf air spaces and stomata.

Measuring transpiration

The bubble potometer

The rate of water loss can be measured or compared in different conditions using a **potometer**, as shown in Figure 8.3. This apparatus measures the rate of **water uptake** by a cut shoot. It does not accurately measure the exact amount of transpiration (water loss through the leaves), as some of the water entering the leaves is used and does not evaporate. However, a potometer is an excellent method of **comparing transpiration rates** in different conditions.

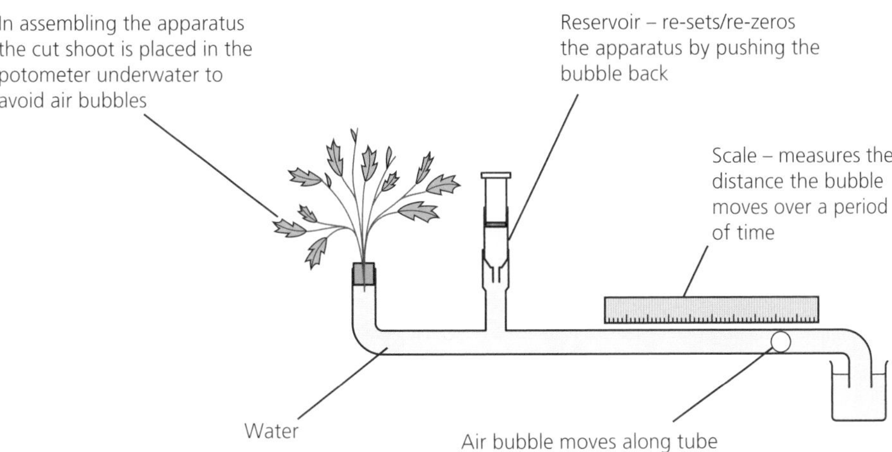
In assembling the apparatus the cut shoot is placed in the potometer underwater to avoid air bubbles
Reservoir – re-sets/re-zeros the apparatus by pushing the bubble back
Scale – measures the distance the bubble moves over a period of time
Water
Air bubble moves along tube

Figure 8.3 The potometer

The weight potometer

We can compare rates of transpiration by measuring the loss in mass of a pot plant (or shoot in a flask) in different conditions. Normally the plant is placed on a top-pan balance for at least 24 hours and the mass is recorded at intervals (Figure 8.4).

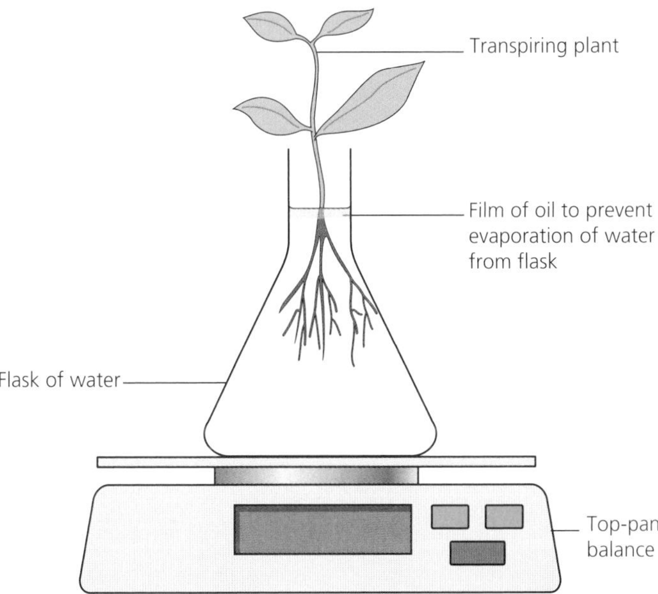
Transpiring plant
Film of oil to prevent evaporation of water from flask
Flask of water
Top-pan balance

Figure 8.4 The weight potometer

Exam tips

- Using the weighing method, it is important that water can only escape by transpiration (through the leaves) – the compost around the shoot must be covered by polythene to stop the evaporation of soil water.
- As with the bubble potometer, the weighing method can only be used to *compare* rates of transpiration, not give absolute values.
- It is possible that the pot plant could grow enough to partially offset the loss in mass due to transpiration.

The washing line method

This method can **compare water loss** in different conditions. Leaves are detached from a plant/tree, numbered and weighed before being attached to a string, as shown in Figure 8.5. After a period of time the leaves are reweighed and the loss of leaf mass compared between the two conditions.

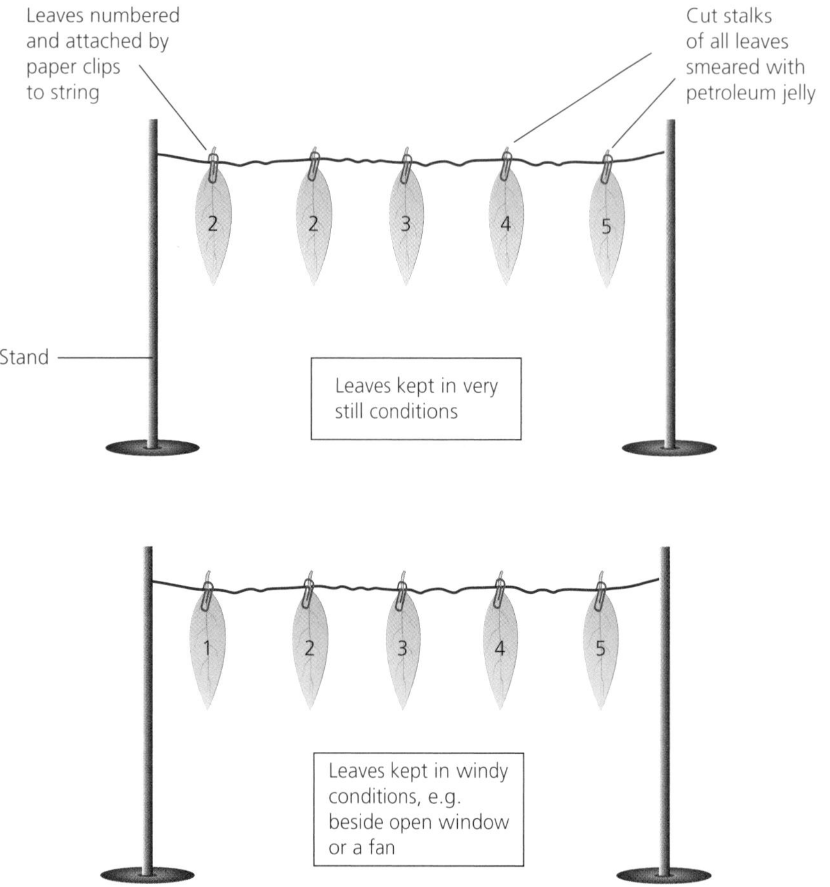

Figure 8.5 Comparing transpiration rates using the 'washing line' method

Factors affecting the rate of transpiration

As the following environmental factors affect the rate of evaporation of water from leaves, they also affect the transpiration rate.

- **Temperature**: in warmer conditions water evaporates faster.
- **Wind speed**: evaporation is faster in higher wind speeds as the wind rapidly removes the evaporating water away from the stomata and leaf surface, thus maintaining a steep gradient of moisture.
- **Humidity**: humid conditions restrict evaporation, as there is a decrease in moisture gradient between the leaf surface and the surrounding air.
- **Light/darkness**: many plants close their stomata in darkness (during the night) to reduce water loss.

The **surface area of leaves** (or number of leaves) affects the rate at which transpiration takes place – the greater the surface area the greater the number of stomata, and the faster evaporation takes place.

Exam tip

In this type of investigation it is important that the leaves used are from the same species and that they are only taken from the tree/plant just before starting the investigation.

Exam tip

A common exam question asks you to state how you would use a potometer to measure the rate of transpiration in different conditions.

Marks are likely to be available for:

- describing how you create the different conditions, e.g. using a polythene bag around the plant to produce humid conditions
- calculating rate – measuring both the bubble movement/mass loss and the time period
- controlling variables.

Now test yourself

3 Describe and explain the effect of humidity on the rate of transpiration in plants.
4 Describe and explain the effect of leaf surface area on the rate of transpiration in plants.

TESTED

Prescribed practical B6

Use a potometer (bubble and weight potometer) to investigate the factors affecting the rate of water uptake by a plant and washing line method to investigate the factors affecting the rate of water loss from plant leaves

Exam practice

1 You are given two solutions (**A** and **B**). One solution is 5% sucrose and the other is 10% sucrose, but they are not labelled. You are also provided with Visking tubing and a top-pan balance and any other standard laboratory equipment you might require.
Plan an investigation that will allow you to identify the sugar solutions. [5]

2 (a) The results in Table 8.1 were obtained from an investigation using a weight potometer to compare the rate of transpiration in windy and still conditions. A fan was used to create windy conditions.

Table 8.1

	Still conditions	Windy conditions
Mass of pot plant at start/g	450	490
Mass of pot plant after 24 hours/g	437	392
Change in mass/g	13	98
% change of mass	2.9	

(i) Calculate the percentage change in mass in windy conditions. [2]

(ii) Why is it important to calculate percentage change of mass rather than just use change in mass? [1]

(iii) Give **two** variables you would have to keep constant in this experiment. [2]

(iv) Describe and explain the results of the investigation. [3]

(b) Describe how turgor occurs. [3]

3 (a) An investigation into osmosis was carried out with carrot cylinders. Three test tubes (**A**, **B** and **C**) were set up and different solutions were added, as described in Table 8.2. A carrot cylinder of 50 mm length was placed in each of the test tubes and the tubes were left for 2 hours before the cylinders were surface-dried and remeasured.

→

Table 8.2

Test tube	Solution	Length of carrot cylinder/mm	
		at start	after 2 hours
A	Concentrated sugar solution	50	46
B	Dilute sugar solution	50	50
C	Water	50	52

(i) Describe and explain the results for test tubes **A** and **B**. [4]

(ii) Give **two** reasons why it would be more accurate to measure change in mass (rather than change in cylinder length) in this experiment. [2]

(iii) Give **one** factor that should be kept constant in this experiment. [1]

(b) Give **two** functions of water in plants. [2]

Answers online

ONLINE

9 The circulatory system

The circulatory system has two main functions:

- **transport** – blood cells, food (glucose/amino acids), carbon dioxide, urea
- **protection** against disease (Figure 9.1).

Blood

REVISED

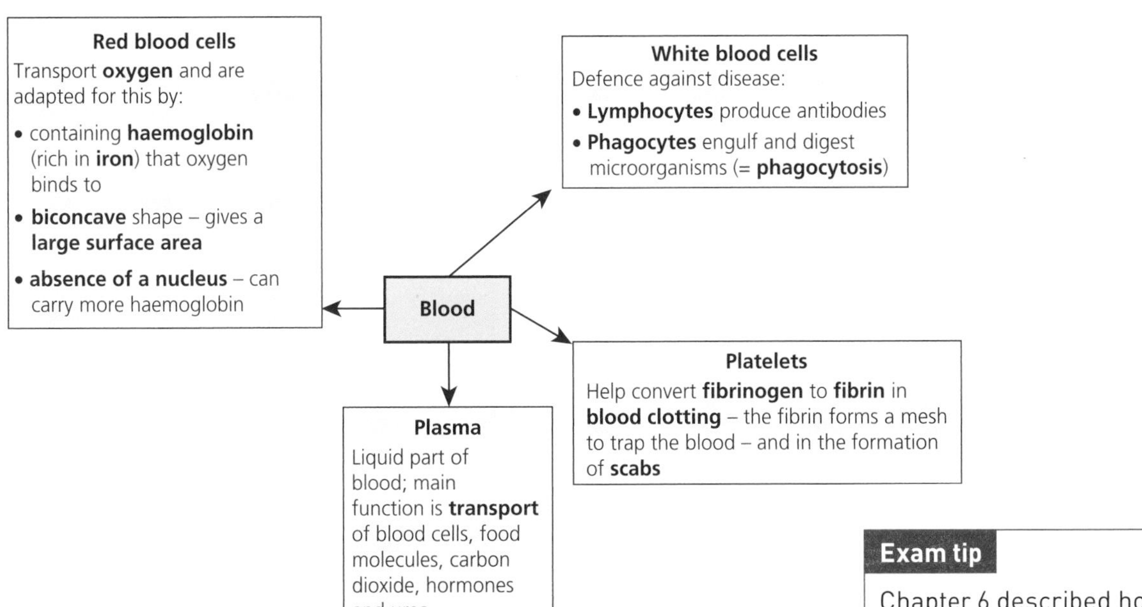

Red blood cells
Transport oxygen and are adapted for this by:
• containing haemoglobin (rich in iron) that oxygen binds to
• biconcave shape – gives a large surface area
• absence of a nucleus – can carry more haemoglobin
White blood cells
Defence against disease:
• Lymphocytes produce antibodies
• Phagocytes engulf and digest microorganisms (= phagocytosis)
Blood
Platelets
Help convert fibrinogen to fibrin in blood clotting – the fibrin forms a mesh to trap the blood – and in the formation of scabs
Plasma
Liquid part of blood; main function is transport of blood cells, food molecules, carbon dioxide, hormones and urea

Figure 9.1 The main components of blood

H Salts and other chemicals in the plasma keep its concentration stable and at a concentration similar to that of the blood cells. This is important, because if **red blood cells** are placed in water they will take in water by **osmosis** and burst in a process called **cell lysis**.

Exam tip

Chapter 6 described how the **kidney** keeps the blood at the correct concentration – it does this by controlling the amount of **water reabsorbed** back into the blood.

Blood vessels

REVISED

The structures and functions of the three types of blood vessel (**arteries**, **veins** and **capillaries**) are described in Table 9.1.

Table 9.1 The structures and functions of the main blood vessels

Vessel	Direction of blood flow	Thickness of wall	Blood pressure	Valves	Lumen diameter
Artery	Away from the heart	Thick – contains muscle for strength, as blood pressure is high, and elastic fibres that allow arteries to expand and recoil as blood pulses through	High	None	Relatively small
Vein	Back to the heart	Thinner than artery – less muscle and fewer elastic fibres	Low	Yes – to prevent backflow of blood	Relatively large
Capillary	From arteries to veins	One cell thick to allow exchange between the blood and body cells	Low	None	Very small

Figure 9.2 represents the three types of blood vessel in cross-section.

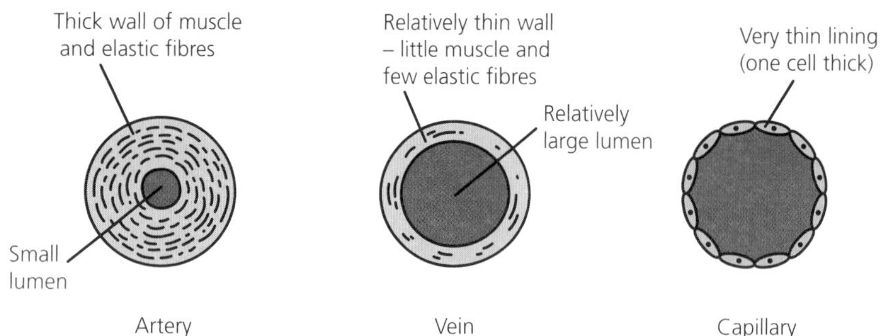

Figure 9.2 An artery, vein and capillary in cross-section (not to scale)

Figure 9.3 shows the circulatory system.

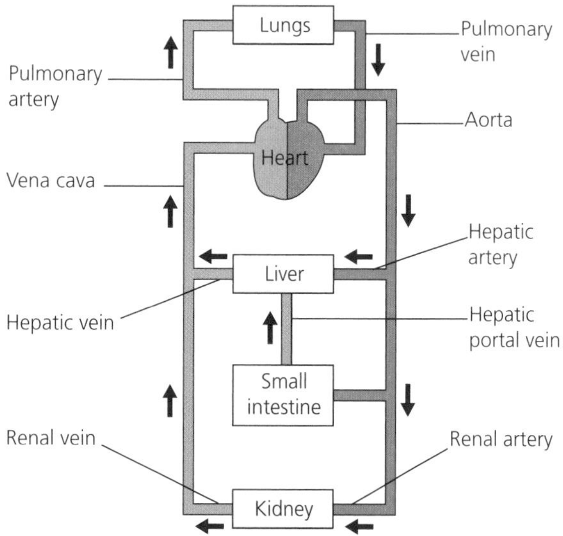

Figure 9.3 The circulatory system

Artery – a blood vessel that carries blood under high pressure away from the heart.

Vein – a blood vessel that carries blood back to the heart.

Capillary – a very thin blood vessel through which the exchange of material between blood and cells takes place.

Exam tip

Veins have a **larger lumen** than arteries to reduce friction as they carry blood under a much lower pressure.

Exam tip

Figure 9.3 shows that the pulmonary artery and the pulmonary vein are exceptions to the usual rule, in that the pulmonary artery carries deoxygenated blood and the pulmonary vein carries oxygenated blood.

Now test yourself

TESTED

1 What is a vein?
2 Give **two** differences between the structure of a vein and an artery.

You should be aware of how the composition of blood changes before and after the main organs it passes through. Two examples are given in the worked example below.

Example

1 Describe differences in the composition of the blood in:
   (a) the hepatic portal vein and the hepatic vein
   (b) the renal artery and the renal vein.

Answer

(a) The hepatic vein will have fewer dissolved food molecules (e.g. glucose) than the hepatic portal vein. Glucose is converted to glycogen for storage in the liver. The hepatic portal vein will also have more oxygen (and less carbon dioxide) than the hepatic vein due to respiration by liver cells.

(b) The renal artery will have more waste products (e.g. urea) and probably more water in the blood (due to the excretory and osmoregulatory roles of the kidney) than the renal vein. The renal artery will also have more oxygen (and less carbon dioxide) than the renal vein due to respiration by kidney cells.

The heart

REVISED

The heart is the organ that pumps blood round the body. Figure 9.4 shows that the body has a **double circulation** – the blood travels through the heart twice for each complete circuit of the body.

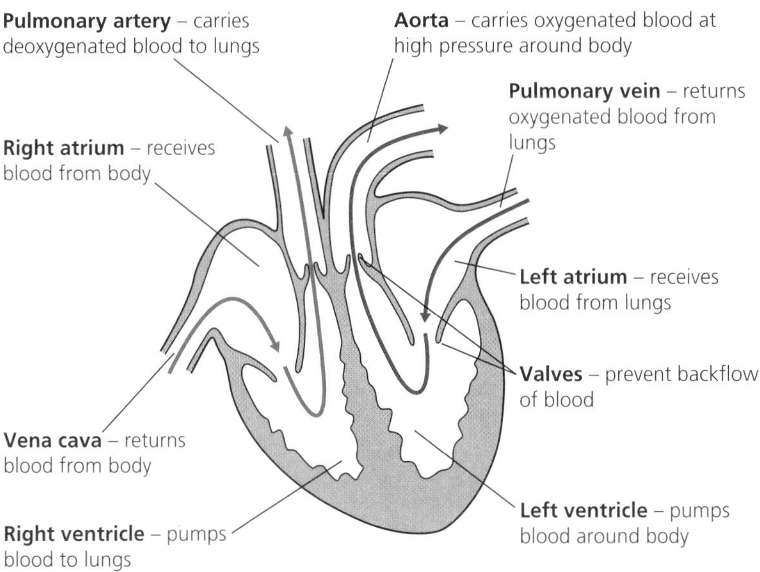

Figure 9.4 The heart

- The **ventricles** are thicker than the **atria** as they are the chambers that pump the blood.
- The **left ventricle** has a thicker muscular wall than the right ventricle as it pumps blood round the body – not just to the lungs.
- The **valves** prevent backflow and make sure that the heart acts as a unidirectional (one-way) pump.
- The **coronary blood vessels** supply the heart muscle with blood.

Ventricles – the lower chambers in the heart, which pump blood to the lungs and around the body.

Atria – the upper chambers in the heart, which receive blood from the body and the lungs.

Now test yourself

TESTED

3 Name the heart chamber that has the thickest walls.
4 Name the blood vessel that carries blood from the right ventricle to the lungs.

Exercise and the circulatory system

REVISED

Regular **exercise** benefits the circulatory system in a number of ways. Exercise helps to:

- strengthen the heart muscle
- increase the **cardiac output** (pump more blood per minute) even when not exercising.

You should also investigate the effect of exercise on **pulse rate**.

Exam tip

Exam questions often use graphs or tables to show how the pulse (or heart rate) changes during and after exercise. The **recovery time** is the length of time it takes for someone's pulse rate to return to normal after exercise.

Cardiac output – the volume (amount) of blood pumped by the heart per minute.

Exam tip

Heart rate and **pulse rate** will be the same – heart rate is how often the heart beats and the pulse rate how often a 'pulse' or surge of blood passes round the body – they are the same, as each beat causes a new pulse.

Example

Figure 9.5 shows the effect of exercise on the heart and recovery rate of two girls.

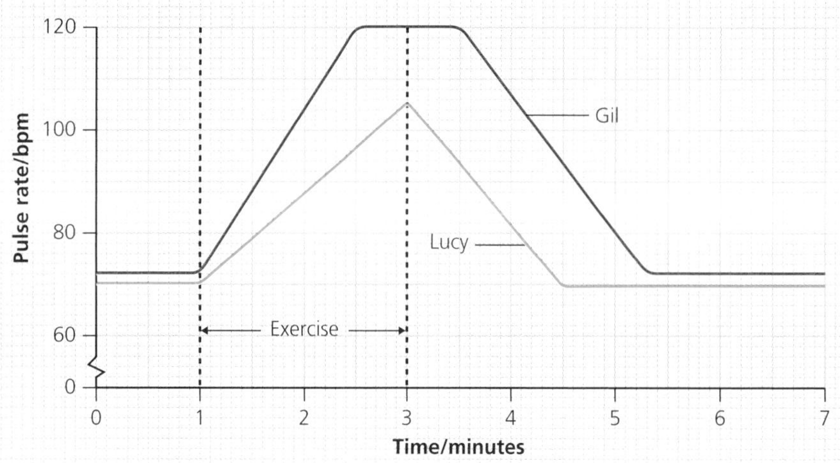

Figure 9.5 **The effect of exercise on the heart**

From the graph state:

(a) when exercise started
(b) the maximum increase in pulse rate for Lucy
(c) the maximum percentage pulse rate increase for Lucy
(d) Lucy's recovery time
(e) **two** things that suggest that Gil is less fit than Lucy.

Answers

(a) 1 minute
(b) 105 – 70 = 35 bpm
(c) (35/70) × 100 = 50%
(d) 3 min to 4.5 min = 1.5 min
(e) Any two from: Gil's resting rate is higher/her heart rate increases more during exercise/her recovery time is longer.

Exam practice

1 (a) Red blood cells are rich in haemoglobin.
(i) What is the function of haemoglobin? [1]
(ii) Apart from containing haemoglobin, describe and explain **two** other ways in which red blood cells are adapted for their function. [4]
(b) Describe fully the function of platelets. [2]

2 (a) Figure 9.6 represents a cross-section through the aorta.

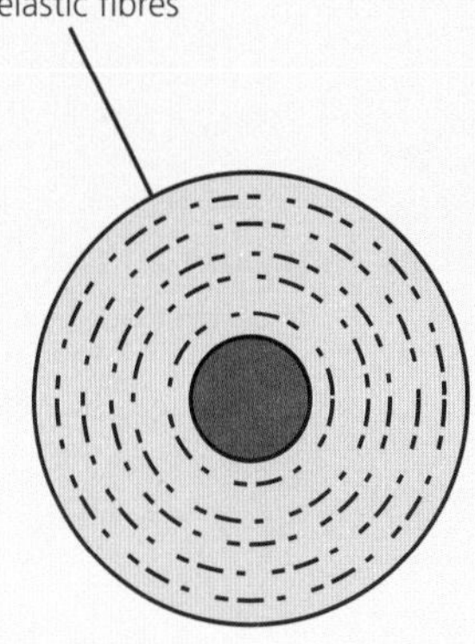

Figure 9.6

(i) What is the function of the muscle in the wall of the artery? [1]
(ii) What is the function of the elastic fibres? [2]
(b) Suggest why the renal artery has less elastic fibre than the aorta. [1]
(c) Name the artery that carries deoxygenated blood. [1]

3 (a) Describe the passage of blood from the right atrium until it reaches the left ventricle. You should name any other heart chambers, organs and blood vessels involved. [3]
(b) Name the blood vessels that supply the heart muscle with oxygen and glucose for respiration. [1]

4 (a) Table 9.2 shows the pulse rate of two boys before, during and after exercise.
The boys started exercising after 2 minutes and stopped after 5 minutes.
(i) Calculate the maximum increase in Jack's heart rate during exercise compared with rest. [2]
(ii) How long did it take Sean's pulse rate to return to its resting value after exercise stopped? [1]
(iii) Give **two** pieces of evidence that suggest that Jack is fitter than Sean. [2]

Table 9.2

Boy	Time/min									
	1	2	3	4	5	6	7	8	9	10
Jack	68	68	84	103	108	90	81	69	68	68
Sean	76	76	112	134	141	133	115	110	86	76

(b) Give **two** benefits to the heart of regular exercise. [2]

Answers online

ONLINE

10 DNA, cell division and genetics

The genome, chromosomes, genes and DNA

REVISED

1 The genetic material (**DNA**, deoxyribonucleic acid) is contained in **chromosomes** in the **nucleus** of the cell.
2 Chromosomes occur as **functional pairs**, except in gametes (sex cells).
3 **Genes** are short sections of chromosomes that control specific characteristics.
4 **Genes** are therefore **short lengths of DNA**.
5 All the DNA in an individual is referred to as the **genome**.

The structure of DNA

Figure 10.1 shows the structure of DNA.

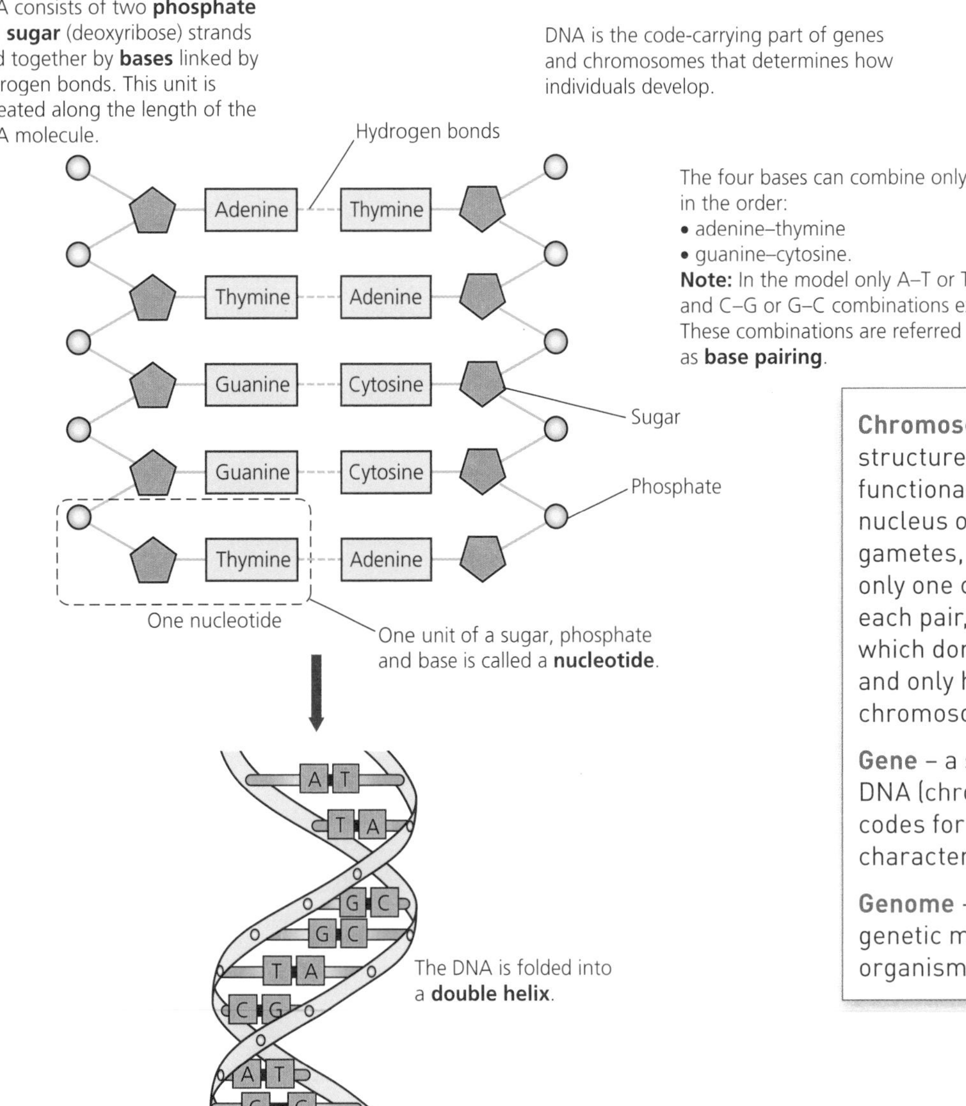

DNA consists of two **phosphate** and **sugar** (deoxyribose) strands held together by **bases** linked by hydrogen bonds. This unit is repeated along the length of the DNA molecule.
DNA is the code-carrying part of genes and chromosomes that determines how individuals develop.
Hydrogen bonds
Adenine
Thymine
Thymine
Adenine
Guanine
Cytosine
Guanine
Cytosine
Thymine
Adenine
The four bases can combine only in the order:
• adenine–thymine
• guanine–cytosine.
Note: In the model only A–T or T–A and C–G or G–C combinations exist. These combinations are referred to as **base pairing**.
Sugar
Phosphate
One nucleotide
One unit of a sugar, phosphate and base is called a **nucleotide**.
The DNA is folded into a **double helix**.

Figure 10.1 The structure of DNA

Chromosomes – genetic structures that occur in functional pairs in the nucleus of cells (except gametes, where there is only one chromosome from each pair, and bacteria, which don't have a nucleus and only have a single chromosome).

Gene – a short section of DNA (chromosome) that codes for a particular characteristic.

Genome – the entire genetic material found in an organism.

The **sequence** of bases (along the length of each chromosome) in each individual is **unique**.

How does DNA work?

DNA works by coding for different **amino acids**, which then combine to form **proteins**, as shown in Figure 10.2.

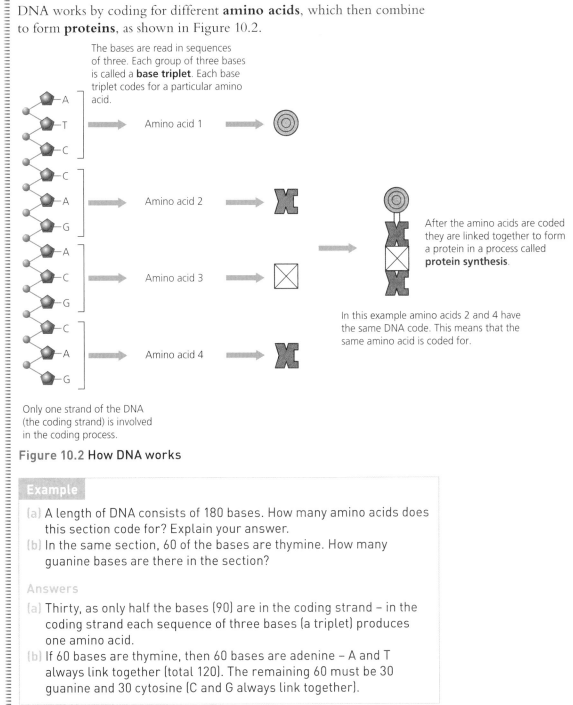

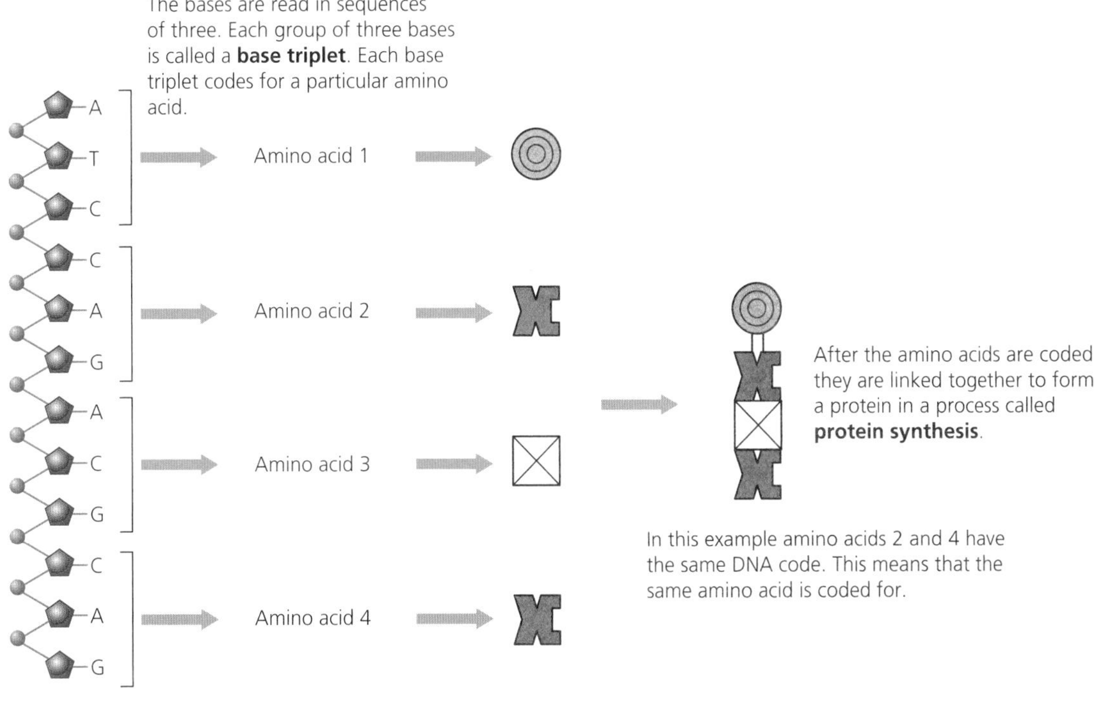

The bases are read in sequences of three. Each group of three bases is called a **base triplet**. Each base triplet codes for a particular amino acid.

Amino acid 1
Amino acid 2
Amino acid 3
Amino acid 4
After the amino acids are coded they are linked together to form a protein in a process called **protein synthesis**.
In this example amino acids 2 and 4 have the same DNA code. This means that the same amino acid is coded for.
Only one strand of the DNA (the coding strand) is involved in the coding process.

Figure 10.2 How DNA works

Example

(a) A length of DNA consists of 180 bases. How many amino acids does this section code for? Explain your answer.
(b) In the same section, 60 of the bases are thymine. How many guanine bases are there in the section?

Answers

(a) Thirty, as only half the bases (90) are in the coding strand – in the coding strand each sequence of three bases (a triplet) produces one amino acid.
(b) If 60 bases are thymine, then 60 bases are adenine – A and T always link together (total 120). The remaining 60 must be 30 guanine and 30 cytosine (C and G always link together).

Cell division

REVISED

Mitosis and **meiosis** are two different types of cell division.

Exam tip

Make sure you can spell mitosis and meiosis – it is very easy to mix them up!

Mitosis

Cell division by mitosis is one part of the **cell cycle**, which also includes the new 'daughter' cells **growing** before dividing again.

Mitosis:

- takes place throughout the body
- is important for growth, replacing worn out cells and repairing damaged tissue
- ensures that new 'daughter' cells have exactly the same chromosome arrangement as each other and as the parent cell.

Mitosis produces **clones of cells** – they are all identical.

Mitosis – a type of cell division that produces cells genetically identical to the parent cell and to each other.

Meiosis

Meiosis:

- occurs in **sex organs** (testes and ovaries) only
- produces **gametes**
- is **reduction division**, as it produces gametes with half the number of chromosomes (**haploid number**) as in other cells (**diploid number**) – this ensures that when gametes fuse in fertilisation the normal diploid number is restored
- one cell produces **four genetically different**, **haploid** cells in two divisions.

Meiosis – a type of cell division that produces cells (gametes) that have half the normal chromosome number.

H Either chromosome in a pair of chromosomes can combine with either chromosome from another pair in gamete formation (and so on for all 23 pairs in humans). This ensures that there are millions of possible chromosome arrangements in the gametes of one person – this **independent assortment** is a major cause of **variation** in individuals.

Independent assortment – a process that takes place during meiosis, in which chromosomes are reassorted in the formation of gametes.

The differences between mitosis and meiosis are summarised in Figure 10.3.

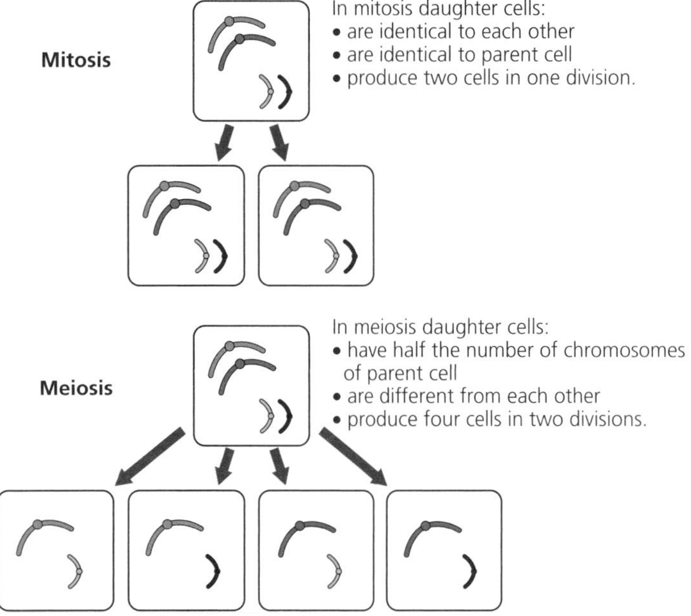

Mitosis
In mitosis daughter cells:
• are identical to each other
• are identical to parent cell
• produce two cells in one division.
Meiosis
In meiosis daughter cells:
• have half the number of chromosomes of parent cell
• are different from each other
• produce four cells in two divisions.

Figure 10.3 The differences between mitosis and meiosis (only two pairs of chromosomes are shown)

Exam tip

Half for **h**aploid

Now test yourself

TESTED

1 Name the structures in the cell that contain DNA.
2 What term is used to describe the shape of DNA?
3 Name the type of cell division that reduces the number of chromosomes in a cell by half.

Genetics

REVISED

The science of **genetics** explains how characteristics pass from parents to offspring.

Each gene carries the code for a particular characteristic, such as eye colour. As chromosomes occur in pairs, each chromosome in a pair carries the same gene, but the gene for eye colour may have different forms (called **alleles**) in the two chromosomes (one allele may be for brown eyes and one for blue eyes). This is shown in Figure 10.4.

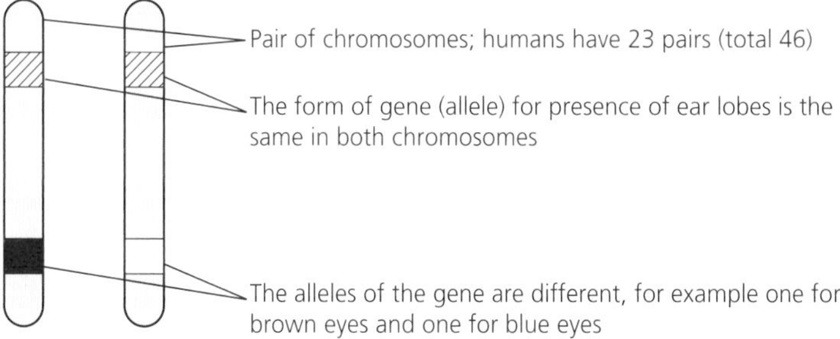
Pair of chromosomes; humans have 23 pairs (total 46)
The form of gene (allele) for presence of ear lobes is the same in both chromosomes
The alleles of the gene are different, for example one for brown eyes and one for blue eyes

Figure 10.4 Arrangement of alleles in a chromosome pair

Some of the key genetic terms are defined in Table 10.1.

Table 10.1 Some important genetic terms

Term	Definition	Example
Gene	A short section of chromosome that codes for a particular characteristic	Gene for eye colour
Allele	A particular form of a gene	Brown eyes and blue eyes are different alleles of the eye colour gene
Homozygous	Describes the situation when both alleles of a gene are the same	Both alleles are for brown eyes
Heterozygous	Describes the situation when the two alleles of a gene are different	One allele is for brown eyes and the other is for blue eyes (Figure 10.4)

Genetic crosses

Figure 10.5 shows how to set out a genetic cross when you are asked to work out the offspring produced from two heterozygous parents using the example of seed shape in peas.

In pea plants, seeds can be either round or wrinkled – this is controlled by a single gene that has alleles for round and wrinkled seeds. The allele for round seeds is dominant to the recessive allele for wrinkled seeds – the alleles are given the symbols **R** (for round) and **r** (for wrinkled). A cross involving one characteristic (e.g seed shape) is a **monohybrid cross**.

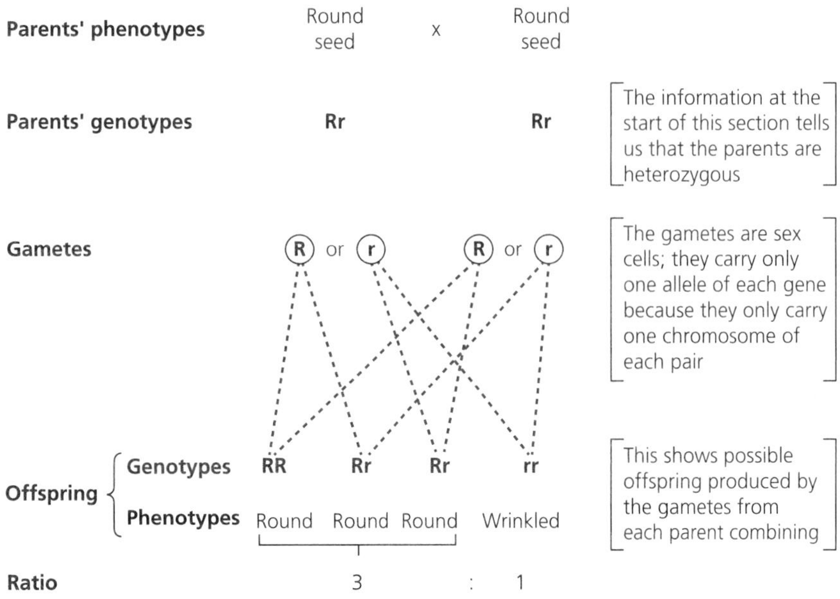

Figure 10.5 A genetic cross showing how two heterozygous parents produce offspring in a 3:1 ratio

Exam tips

- The gametes produced by one parent can combine only with the gametes of another parent – different gametes in the same individual cannot combine.
- You only get **two different types** of **gamete** in one individual if it is **heterozygous**.
- Ratios are accurate only when large numbers of offspring are involved. For example, if there were only two seeds produced in the genetic cross in Figure 10.5, the ratio could not be 3:1.
- In genetic crosses you could be asked to predict the probability or percentage chance of the offspring having a particular genotype or phenotype. If the chance is 1 in 4 (the chance of having an offspring pea producing wrinkled seeds in the cross above), then the **probability** could be written as 1 in 4; 1:3; 1/4, 25%, and the **percentage chance** as 25%.
- Sometimes the offspring are referred to as the **F_1 generation**, and if the parents are described as **pure breeding** this means they are homozygous.

Some other important terms used in the genetic cross described above are defined in Table 10.2.

Table 10.2 Other important genetic terms

Term	Definition	Example
Genotype	Paired symbols showing the allele arrangement in an individual	The parents in Figure 10.5 have the genotype **Rr**
Phenotype	Outward appearance of an individual	The parents in Figure 10.5 have a round seed phenotype
Dominant	In the heterozygous condition the dominant allele overrides the non-dominant (recessive) allele	The parents in Figure 10.5 both produce round seeds, even though they are heterozygous and have an allele for producing wrinkled seeds
Recessive	The recessive allele is dominated by the dominant allele – it only shows itself in the phenotype if there are two recessive alleles	Only one-quarter of the offspring in the cross produce wrinkled seeds, as only one-quarter have no dominant **R** allele present

Figure 10.6 shows how a **Punnett square** can be used in setting out genetic crosses. In this example, using seed shape as before, a heterozygote **(Rr)** pea is crossed with a homozygous recessive **(rr)** pea.

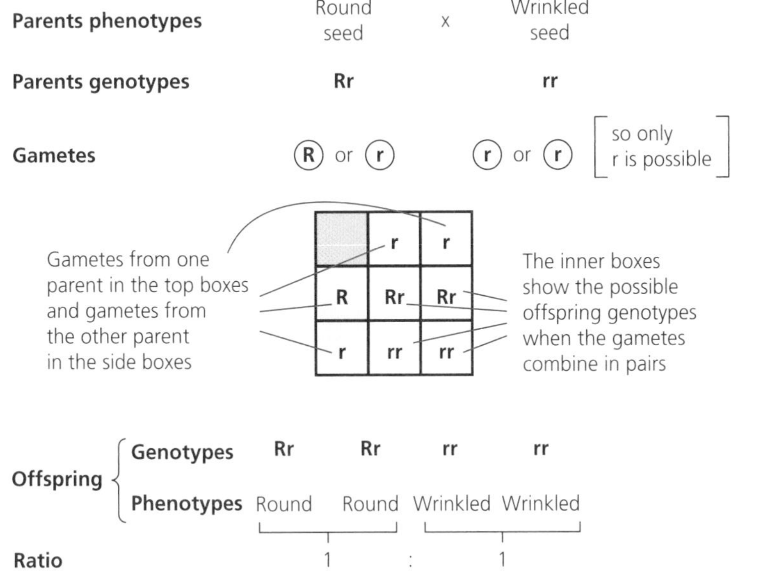

Figure 10.6 Using a Punnett square

Example

Brown eyes are dominant to blue eyes. Using the symbols **B** for brown and **b** for blue, use a Punnett square to show how brown-eyed parents can have children with blue eyes.

Answer

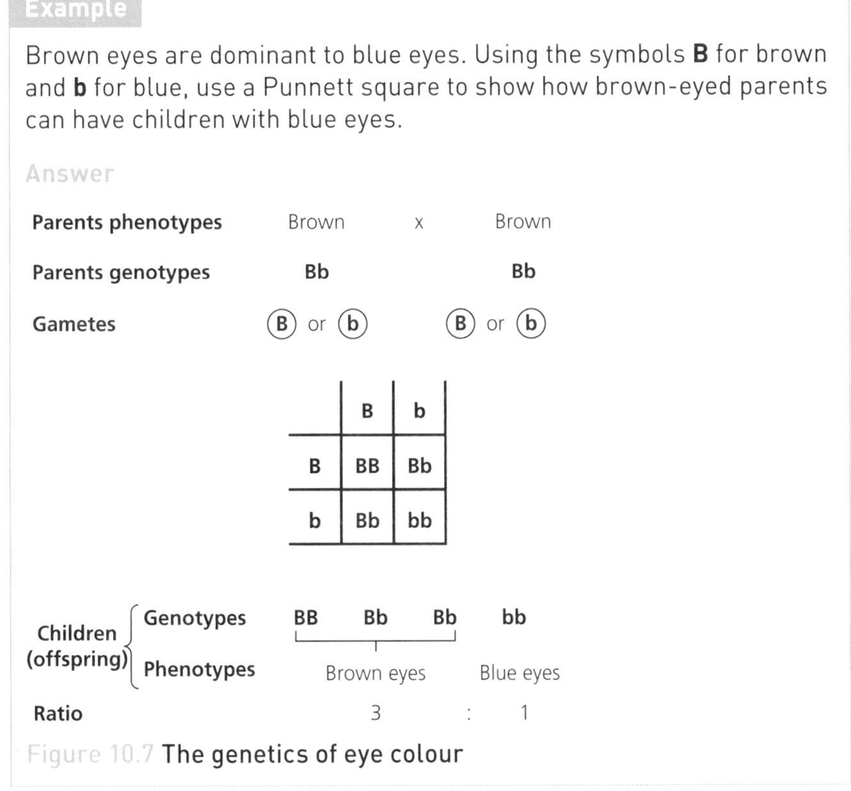

Figure 10.7 The genetics of eye colour

Exam tips

- It is usually easier to work out crosses using Punnett squares.
- If the offspring ratio is **3:1** in a genetic cross, then **both parents** must be **heterozygous**.
- If the offspring ratio is **1:1** in a genetic cross, then one parent is **heterozygous** and the other is **homozygous recessive**.

H The test (back cross) to determine an unknown genotype

Individuals that are homozygous dominant or heterozygous have different genotypes but the same phenotype. The test cross can be used to determine the **genotype** of an individual of **dominant phenotype** but **unknown genotype**, as shown in Figure 10.8.

In the example of the pea, plants producing round seeds could be homozygous (**RR**) or heterozygous (**Rr**). To identify the unknown genotype of the plant it is crossed with a homozygous recessive plant (one with genotype **rr**).

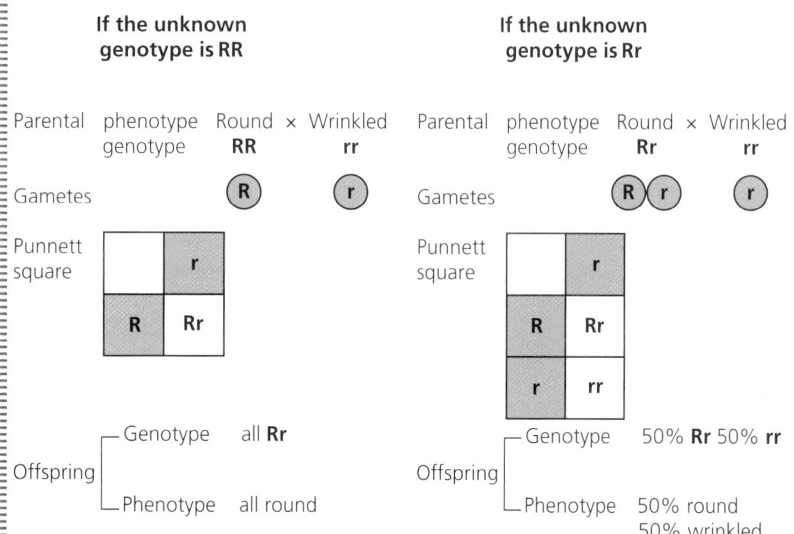

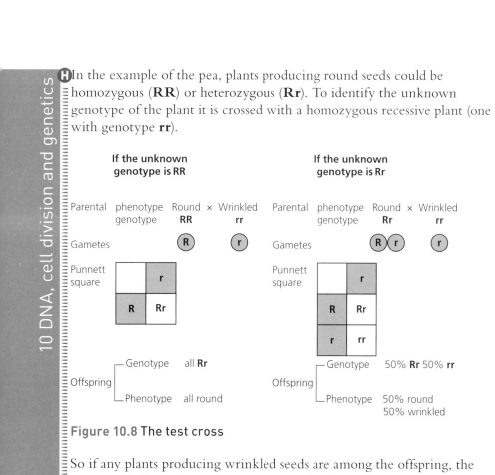

Figure 10.8 The test cross

So if any plants producing wrinkled seeds are among the offspring, the parent with the unknown genotype must be heterozygous (**Rr**).

Higher tier candidates should be able to interpret **pedigree diagrams** – see the Exam practice questions at the end of this chapter.

Exam tip

If both alleles from a parent are identical it is not necessary to use two identical gametes in the genetic cross – see Figure 10.8.

Sex determination in humans

Humans have 22 pairs of normal chromosomes and one pair of sex chromosomes. The male sex chromosomes are XY and females have two XX chromosomes. As the sex chromosomes (and alleles) act in the same way as in other genetic crosses, Figure 10.9 shows that equal numbers of boys and girls are produced.

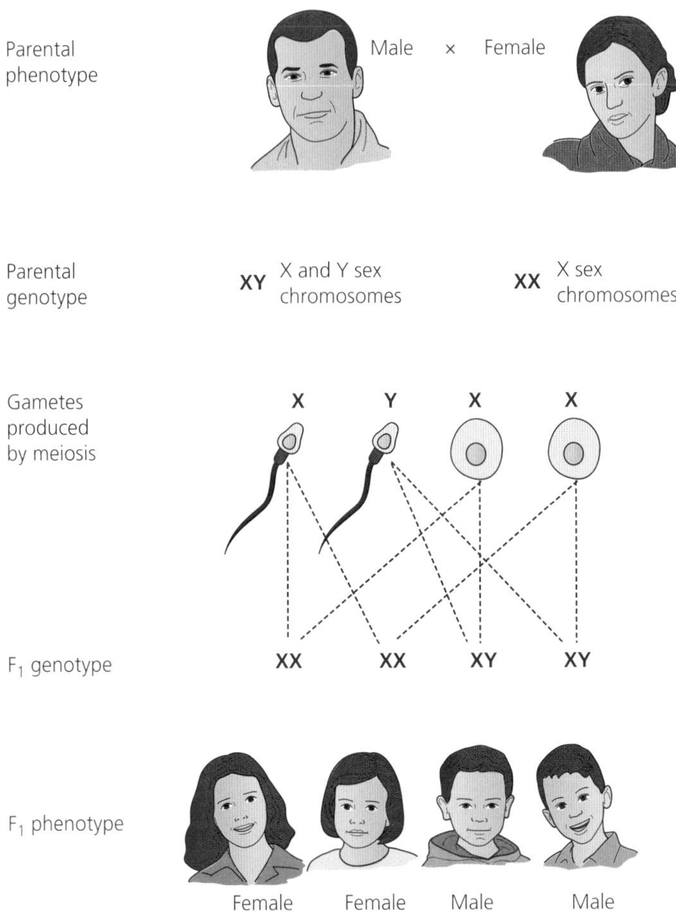

Figure 10.9 How equal numbers of boys and girls are produced

Now test yourself

TESTED

4 In terms of alleles, explain what is meant by the term 'homozygous'.
5 Define the term 'recessive'.
6 Give the sex chromosomes present in males.

Sex linkage

As well as determining sex, sex chromosomes can carry genes and alleles that control other characteristics. As the Y chromosome does not contain any alleles, any recessive alleles present on the X chromosome in males cannot be masked by a dominant allele and therefore show in the phenotype.

In females where there are two X chromosomes, the recessive condition can be masked by a dominant allele (as in other chromosomes). Examples of sex-linked conditions are red-green colour blindness and haemophilia, as shown in Figure 10.10.

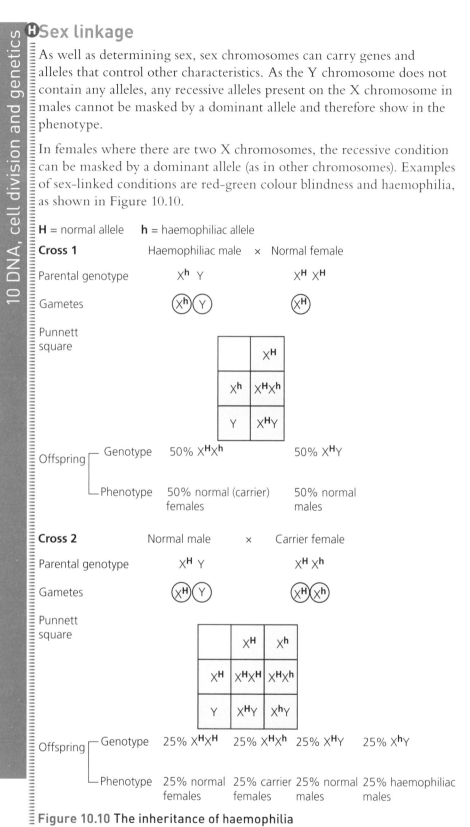

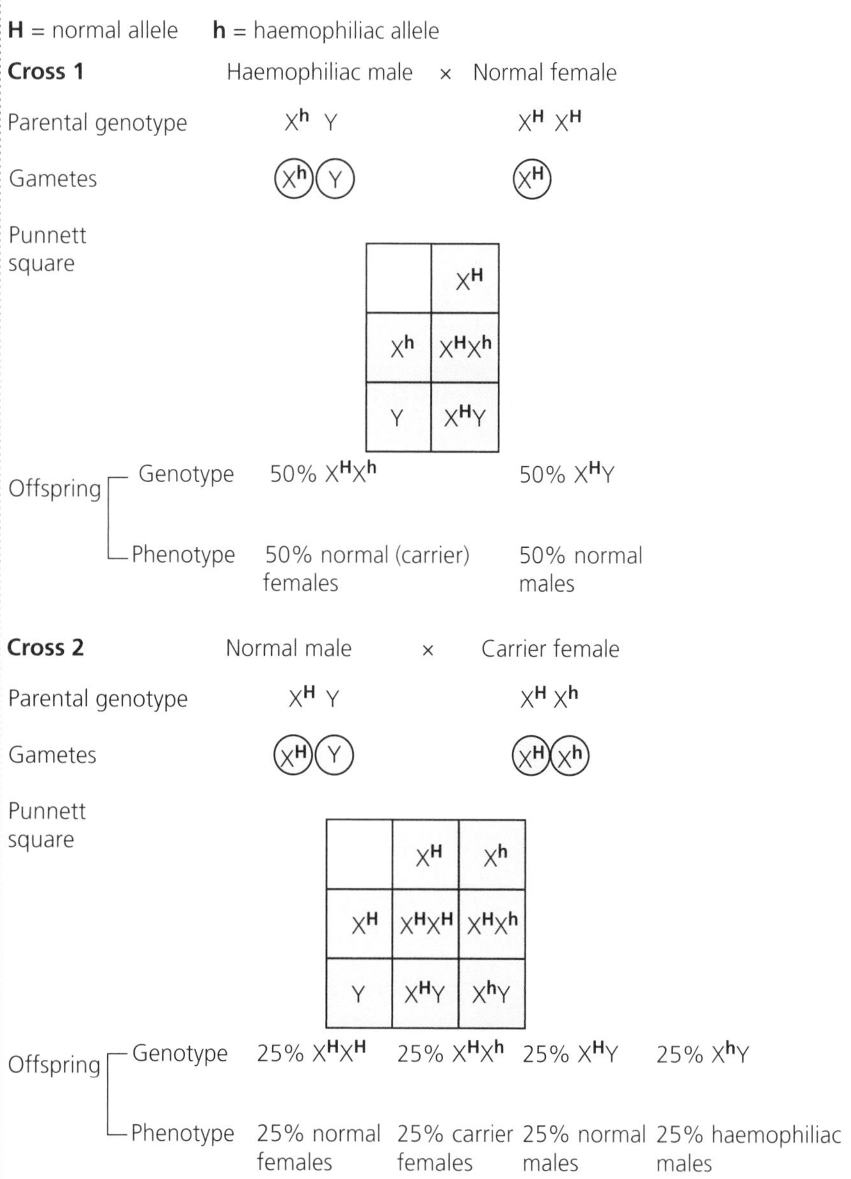

Figure 10.10 The inheritance of haemophilia

These crosses show why haemophilia is usually found only in males. Very occasionally, females may inherit the condition.

Exam tips

- When carrying out crosses involving sex linkage it is important to use symbols that show *both* the type of sex chromosome (X or Y) and any allele carried, e.g. X^h Y.
- In genetic conditions, individuals who are heterozygous for the condition (i.e. they 'carry' the harmful allele but don't show the condition) are referred to as **carriers**.

Genetic conditions

REVISED

Genetic conditions are conditions caused by a fault with genes or chromosomes (a genetic fault). Some genetic conditions (but not all) are **inherited**; inherited conditions are passed down from parent to child.

Four genetic conditions, each with a different cause, are described in Table 10.3.

Table 10.3 Genetic conditions

Genetic condition	Explanation
Haemophilia	This condition is caused by a problem with the blood clotting mechanism. Sufferers are at risk of excessive bleeding, even from very small wounds or bruising. It is a sex-linked inherited condition caused by a **recessive allele on the X chromosome**. Most people with haemophilia are males with the genotype **X^hY**.
Cystic fibrosis	Individuals with cystic fibrosis have frequent and serious lung infections and problems with food digestion. It is caused by a **recessive allele**, so affected individuals must be **homozygous recessive**.
Huntington's disease	Individuals with Huntington's disease have progressive brain deterioration, which usually becomes apparent in middle age. It is fatal and there is no cure. It is caused by the presence of a **dominant allele**.
Down's Syndrome	This condition is caused by the presence of an extra chromosome so that affected individuals have **47 chromosomes** rather than 46. Humans normally have 23 chromosomes in each gamete (sperm or egg). Occasionally gametes are formed with 24 chromosomes, so if one of these gametes is involved in fertilisation with a 'normal' gamete then the child produced will have 47 chromosomes. Individuals with Down's Syndrome have easily identified facial features particular to the condition, reduced muscle tone and reduced cognitive development.

Exam tip

Haemophilia, **cystic fibrosis** and **Huntington's disease** are **inherited** (the faulty alleles are passed from parent to child). **Down's Syndrome** is caused by a **mistake during gamete formation** – it is not inherited as such (it is not the passing of a faulty gene or chromosome from parent to child). Individuals with Down's Syndrome can be identified by counting the number of chromosomes in a **karyotype**.

Genetic screening

REVISED

Genetic screening can be used to identify the presence of genetic conditions. For example, it can be used to test for the presence of Down's Syndrome (Figure 10.11) and other conditions, including cystic fibrosis, in a foetus.

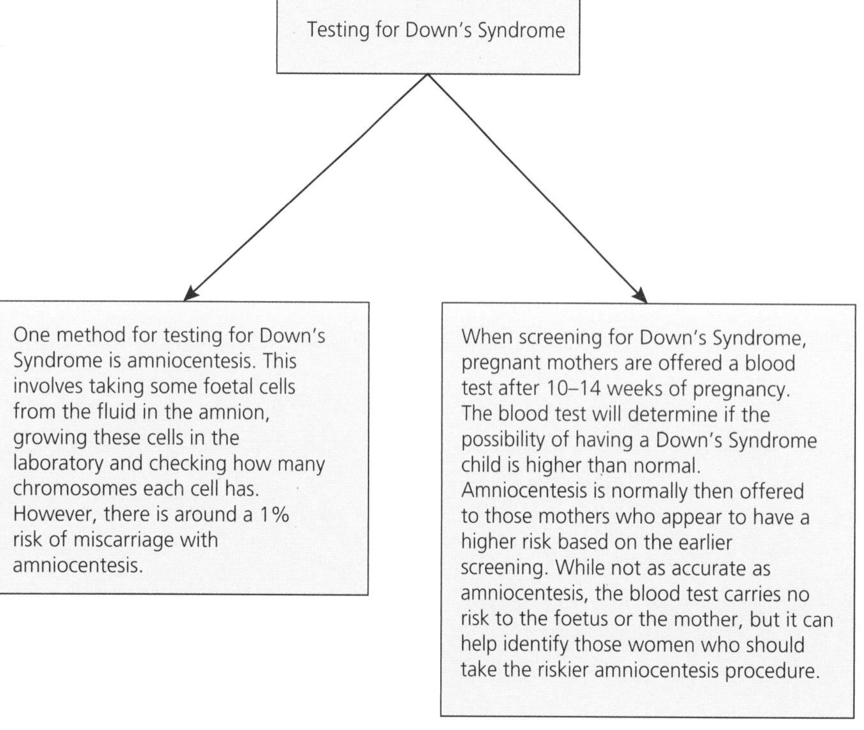

Testing for Down's Syndrome
One method for testing for Down's Syndrome is amniocentesis. This involves taking some foetal cells from the fluid in the amnion, growing these cells in the laboratory and checking how many chromosomes each cell has. However, there is around a 1% risk of miscarriage with amniocentesis.
When screening for Down's Syndrome, pregnant mothers are offered a blood test after 10–14 weeks of pregnancy. The blood test will determine if the possibility of having a Down's Syndrome child is higher than normal. Amniocentesis is normally then offered to those mothers who appear to have a higher risk based on the earlier screening. While not as accurate as amniocentesis, the blood test carries no risk to the foetus or the mother, but it can help identify those women who should take the riskier amniocentesis procedure.

Figure 10.11 Testing for Down's Syndrome

Genetic screening – a process used to test a foetus, or a person, for the presence of harmful alleles or other genetic abnormalities.

Exam tip

The **blood test** for Down's Syndrome is **less precise** than amniocentesis testing, but it poses **no risk**.

Genetic screening – ethical and moral issues

If a foetus is diagnosed with a genetic condition the potential parents have some very difficult decisions to make, and this creates a **real dilemma** for many.

Is abortion the best thing to do?

Many parents will argue yes, as:

- it prevents having a child that could have a poor quality of life
- a lot of time may need to be spent caring for the child with the abnormality at the possible expense of time with their other children.

Many parents will argue no, as:

- the unborn child doesn't have a say
- they argue that it is not morally right to 'kill' a foetus
- abortion is banned in some religions and in some countries.

Some other issues arising from genetic screening are listed below:

- Who decides on who should be screened?
- Is there an acceptable risk associated with genetic screening? For example, amniocentesis for Down's Syndrome screening has a small risk of miscarriage.
- Costs of screening compared with the costs of treating individuals with a genetic condition – should cost be a factor?

Should information from genetic screening be made public? Again, there are arguments for and against.

For making genetic information **public**:

- It could help with medical research.

Against making genetic information **public**:

- Possible discrimination – **insurance companies** may not give life insurance or it could be more expensive.

Now test yourself

TESTED

7 How many chromosomes are in a cell in someone who has Down's Syndrome?
8 Give **one** disadvantage of having an amniocentesis test.
9 Give **one** argument against making someone's genetic information available to the general public.

Genetic engineering

REVISED

Genetic engineering is the **modification of the genome** of an organism to introduce **desirable characteristics**. This usually involves adding a human gene to the DNA of another organism, e.g. bacteria – the other organism makes the product that the human DNA codes for.

Genetic engineering – the deliberate modification of the DNA in an organism to introduce desirable characteristics.

Bacteria are genetically engineered to make **human insulin** (used in the treatment of diabetes) — see Figure 10.12.

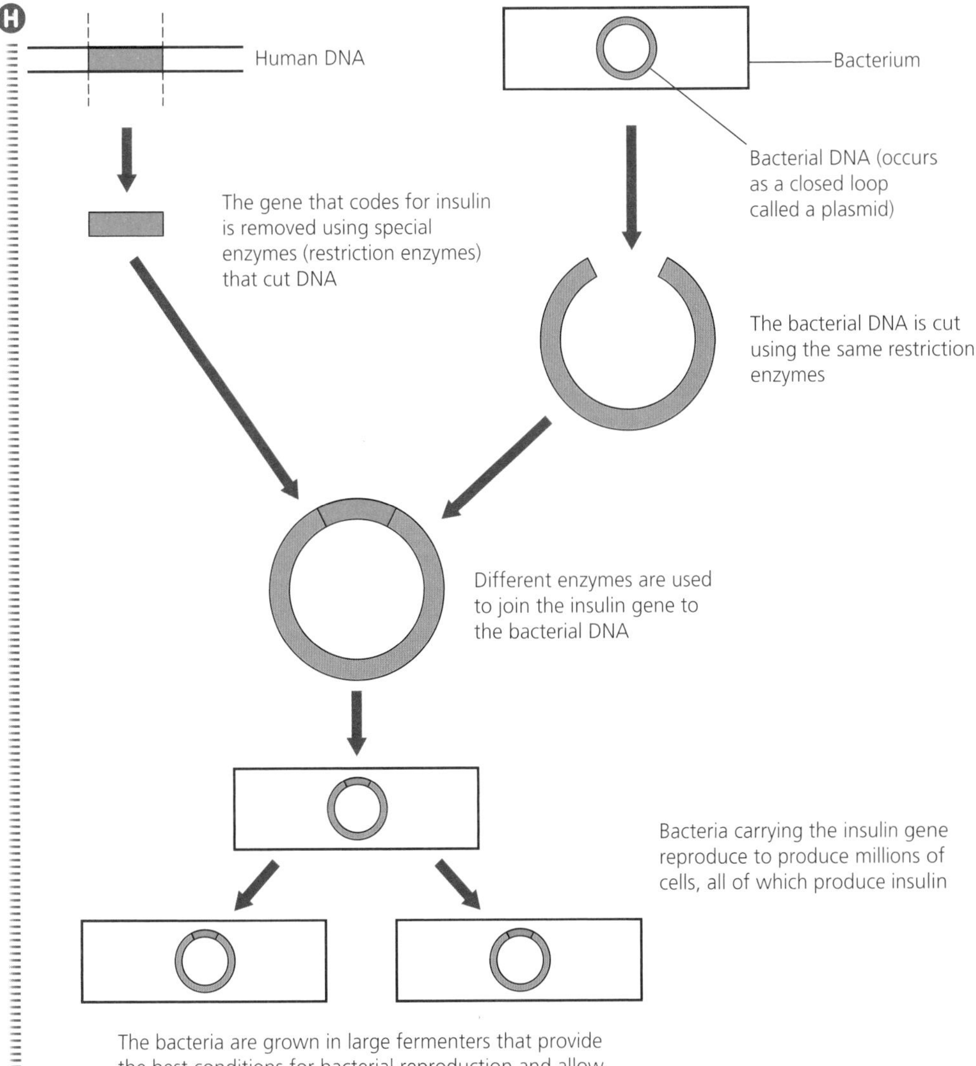

Human DNA
Bacterium
Bacterial DNA (occurs as a closed loop called a plasmid)
The gene that codes for insulin is removed using special enzymes (restriction enzymes) that cut DNA
The bacterial DNA is cut using the same restriction enzymes
Different enzymes are used to join the insulin gene to the bacterial DNA
Bacteria carrying the insulin gene reproduce to produce millions of cells, all of which produce insulin
The bacteria are grown in large fermenters that provide the best conditions for bacterial reproduction and allow easy extraction of insulin

Figure 10.12 Genetic engineering – making human insulin

Special enzymes (**restriction enzymes**) cut the human gene in such a way as to leave overlapping strands of DNA. The same enzymes cut the bacterial plasmid (DNA) in the same way to leave complementary **sticky ends**. The sticky ends make it easy for the human and bacterial DNA to join through **base pairing** (Figure 10.13).

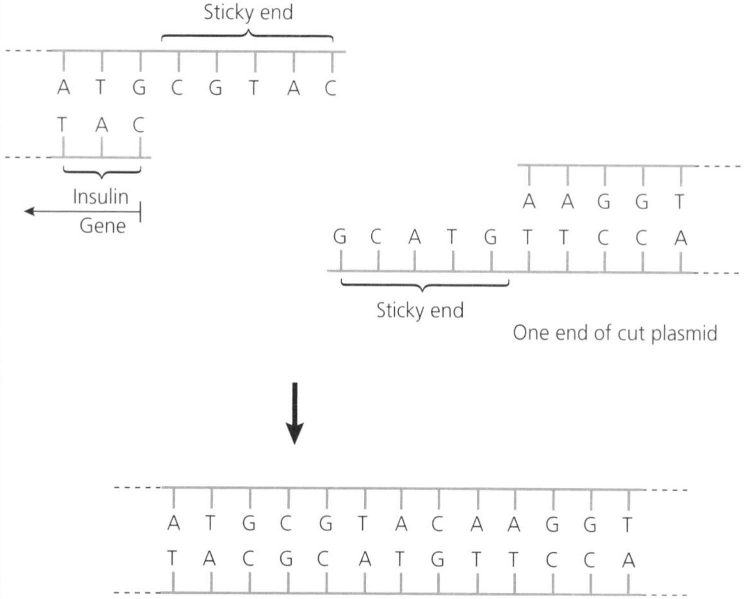

Figure 10.13 Sticky ends allow pairing to take place and links human insulin DNA into the bacterial plasmid

Exam tip

In genetic engineering, enzymes are needed to both cut out the human insulin gene and also to cut a gap in the plasmid to allow the human gene to fit.

Commercially, the genetically engineered bacteria are cultured in **large fermenters** that provide the optimum conditions for growth and reproduction (and for producing insulin).

Following the production of insulin it is **extracted**, **purified** and **packaged** in a process called **downstreaming**.

Advantages of producing insulin by genetic engineering include the following:

- Before genetic engineering, the **amount of insulin available was limited** by the amount that could be extracted from dead animals in slaughterhouses. This restricted the amount that could be obtained and made the insulin relatively expensive.
- **Human insulin** is slightly **different in structure** from the insulin of other animals – so insulin obtained from dead animals might not be as effective and could cause allergies.
- There was the **risk of spreading viruses** when transferring the insulin from animals.
- Many people with diabetes are ethically opposed to injecting animal insulin into their body.

Many other products, including the human growth hormone, are now made by genetic engineering.

Exam practice

1 (a) Define the term 'genome'. [1]

(b) (i) In terms of DNA structure, explain what is meant by the 'base pairing rule'. [2]

(ii) Explain what is meant by the 'unique nature of an individual's DNA'. [2]

(c) Explain what is meant by the 'base triplet hypothesis'. [2]

2 (a) Explain the roles of mitosis and meiosis in maintaining constancy of chromosome number in a species. [4]

(b) Explain the role of meiosis in providing variation. [2]

3 Flowers can be red or white in a certain type of plant. When two red flowers were crossed and their offspring counted the results in Table 10.4 were obtained.

Table 10.4

	Red flowers	White flowers
Number of offspring	148	53

(a) (i) What genetic ratio do the offspring results approximate to? [1]

(ii) Explain why the offspring numbers do not fit the ratio exactly. [1]

(b) Use a Punnett square to explain the outcome of this cross. [3]

4 Huntington's disease is a medical condition caused by the presence of a single allele. Huntington's disease is caused by a non-sex-linked dominant allele.
Figure 10.14 shows the inheritance of Huntington's disease in a family through three generations.

(a) How many male grandchildren do individuals 1 and 2 have? [1]

(b) What is the evidence that suggests that the allele for Huntington's disease is dominant and not recessive? [1]

(c) What is the probability that the next child of parents 7 and 8 will be a boy with Huntington's disease? Explain your answer. [3]

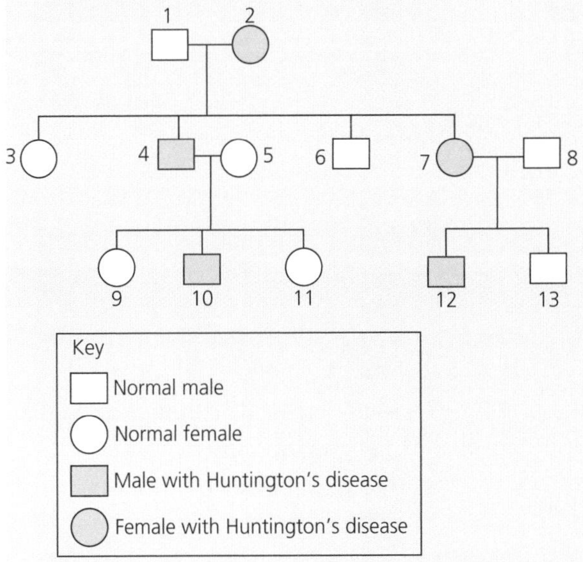

Key
Normal male
Normal female
Male with Huntington's disease
Female with Huntington's disease

Figure 10.14

5 (a) Place the following structures into order of size starting with the smallest.
human gene **bacterium** **plasmid** [1]

(b) Describe the process of adding a human gene into a bacterial plasmid during genetic engineering. [4]

(c) Give **two** advantages of making insulin for people with diabetes by genetic engineering. [2]

Answers online

ONLINE

11 Reproduction, fertility and contraception

Reproduction

REVISED

Living organisms need to be able to reproduce. Like most other animals, humans carry out **sexual reproduction**, which involves the joining together of two **gametes** – the sperm and the egg (ovum).

The male reproductive system

Figure 11.1 shows the male reproductive system and describes the role of each part of the system.

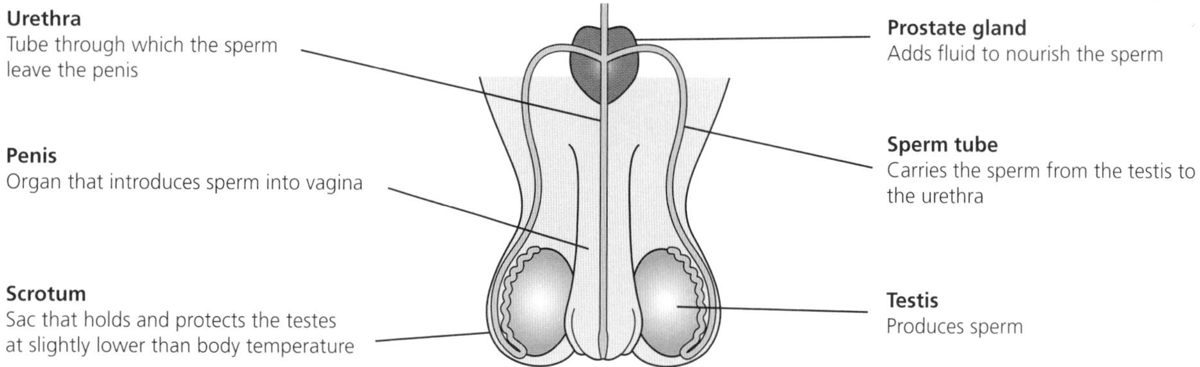
Urethra
Tube through which the sperm leave the penis
Penis
Organ that introduces sperm into vagina
Scrotum
Sac that holds and protects the testes at slightly lower than body temperature
Prostate gland
Adds fluid to nourish the sperm
Sperm tube
Carries the sperm from the testis to the urethra
Testis
Produces sperm

Figure 11.1 The male reproductive system

Sperm are cells highly adapted for their function. They have a flagellum (tail) that allows the sperm to swim to meet the egg. Sperm (and egg cells) are also adapted to their function in being haploid.

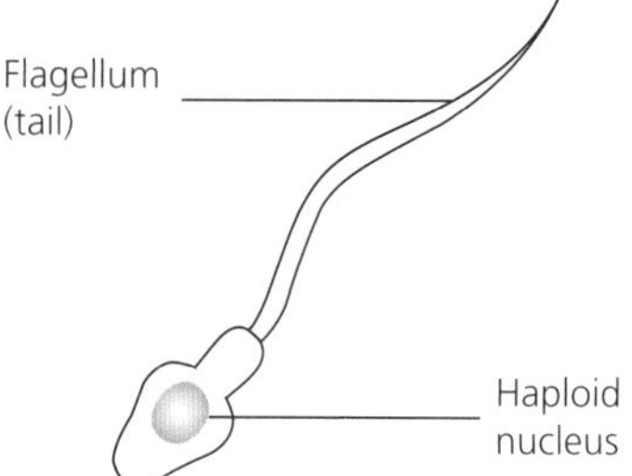
Flagellum (tail)
Haploid nucleus

Figure 11.2 A sperm cell

Sperm – haploid male gametes formed by meiosis.

H Sperm also contain many **mitochondria** for energy production.

The female reproductive system

The female reproductive system is the part of the body that makes and releases eggs **(ova)** and where the foetus will develop if pregnancy results.

Figure 11.3 shows the female reproductive system and describes the role of each part of the system.

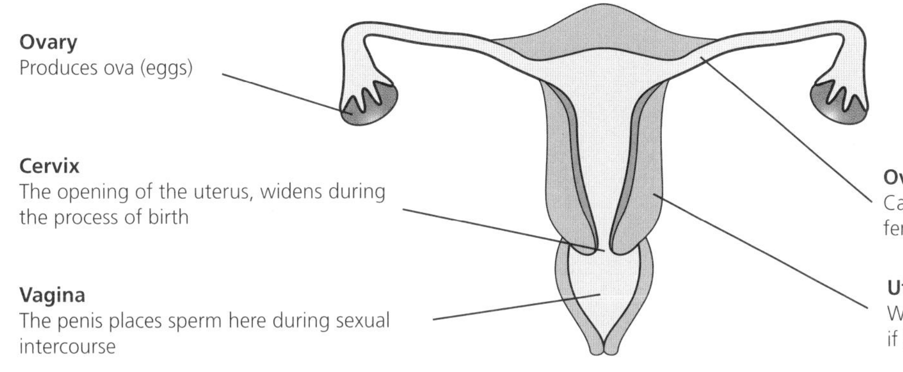

Ovary
Produces ova (eggs)
Cervix
The opening of the uterus, widens during the process of birth
Vagina
The penis places sperm here during sexual intercourse
Oviduct
Carries the ova (eggs) to the uterus, fertilisation takes place here
Uterus
Will nourish the developing foetus if pregnancy results

Figure 11.3 The female reproductive system

Ova (singular ovum) – haploid female gametes formed by meiosis.

Zygote – the first (diploid) cell of the new individual, following fertilisation.

Fertilisation and pregnancy

If a sperm and an ovum meet and **fuse** (join) in an **oviduct**, fertilisation will result. **Fertilisation** involves the **haploid** nuclei of the sperm and ovum fusing and restoring the **diploid** (normal chromosome number) condition. The fertilised egg is the first cell (**zygote**) of the new individual.

After fertilisation, the following sequence of events occur:

- The zygote divides by **mitosis** and grows into a **ball of cells**, referred to as an **embryo**, that develops further as it travels down the oviduct into the uterus.
- In the uterus, the embryo sinks into the thick uterine lining and becomes attached (in a process called **implantation**).
- At the point where the embryo begins to develop in the uterus lining, the **placenta** and **umbilical cord** form.
- A protective membrane, the **amnion**, develops around the embryo. It contains a fluid, the **amniotic fluid**, within which the growing embryo develops. This fluid cushions the delicate developing embryo, which increasingly **differentiates into tissues and organs**. The embryo is referred to as a **foetus** after a few weeks, when it begins to become more recognisable as a baby.
- During pregnancy, useful materials, including **oxygen** and **dissolved nutrients** (e.g. amino acids and glucose), pass from the mother to the foetus through the placenta and umbilical cord. Waste excretory materials including **carbon dioxide** and **urea** pass from the foetus back to the mother.

Figure 11.4 shows the main structures involved in pregnancy.

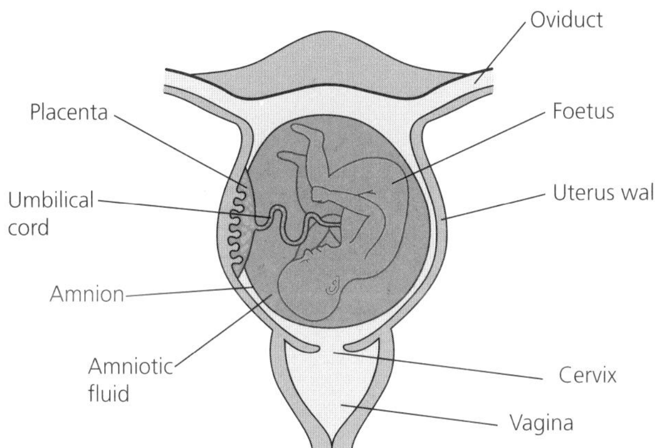

Oviduct
Placenta
Foetus
Umbilical cord
Uterus wall
Amnion
Amniotic fluid
Cervix
Vagina

Figure 11.4 The main structures involved in pregnancy

Exam tip

All gametes are **haploid** – when they combine in fertilisation the **zygote** is **diploid**.

Implantation – the attachment of the embryo (ball of cells) to the uterus lining following fertilisation.

Placenta – the structure that links the uterus wall to the foetus via the umbilical cord. It is here that the exchange of materials takes place between the mother and the foetus.

Exam tip

The placenta is highly adapted for **diffusion**, as it has a **very large surface area** at the point of contact with the uterine lining.

The **surface area** between the uterine wall and the placenta is further increased by small villi (extensions) in the placenta that extend into the uterus wall.

Now test yourself

TESTED

1 State the function of the prostate gland in males.
2 Name the part of the female reproductive system in which fertilisation occurs.
3 What is the function of the amniotic fluid during pregnancy?

You should be aware of the roles of mitosis and meiosis in reproduction and the development of the human embryo.

Example

Explain the roles of meiosis and mitosis in fertilisation and the development of the human embryo.

Answer

- Meiosis produces haploid gametes/gametes with half the number of chromosomes of other cells; it ensures that the diploid number is restored at fertilisation.
- Mitosis maintains chromosome number/diploid number during the growth of the embryo.

Sex hormones and secondary sexual characteristics

Testosterone (produced by the testes in males) and oestrogen (produced by the ovaries in females) are sex hormones that produce secondary sexual characteristics.

- **Testosterone** – produced by **testes** in **males**. Secondary sexual characteristics include:
  - sexual organs enlarge
  - body and pubic hair grows
  - voice deepens and body becomes more muscular
  - sexual awareness and drive increase.
- **Oestrogen** – produced by **ovaries** in **females**. Secondary sexual characteristics include:
  - sexual organs and breasts enlarge
  - pubic hair grows
  - pelvis and hips widen
  - menstruation begins
  - sexual awareness and drive increase.

The menstrual cycle

The process of menstruation (having periods) starts in girls at puberty and continues until the end of a woman's reproductive life. The function of the menstrual cycle is the monthly renewal of the delicate blood-rich lining of the uterus, so that it will provide a suitable environment for the embryo should fertilisation occur.

- The **menstrual cycle** lasts (approximately) 28 days (Figure 11.5).
- The ovum is released (**ovulation**) on day 14 (approximately) – by this time the uterine lining has built up in preparation for pregnancy.
- Sex can result in pregnancy if it occurs in a short window on either side of ovulation.
- **Menstruation** is the breakdown and removal of the blood-rich uterine lining at the end of each cycle.

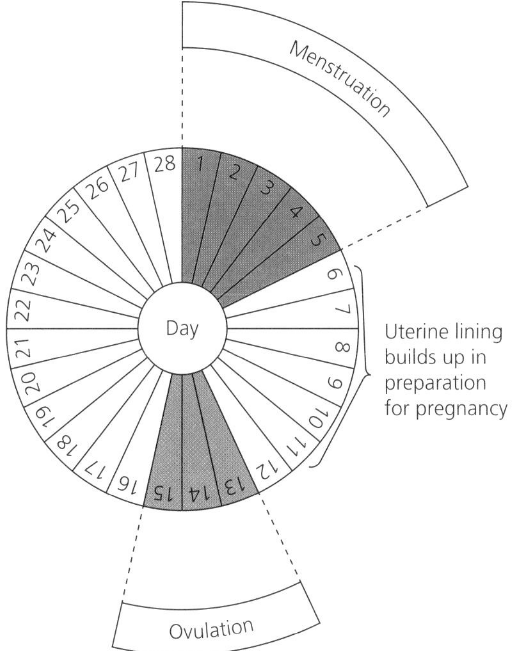

Figure 11.5 The menstrual cycle

The menstrual cycle is controlled by **hormones** including:

- **oestrogen** – this stimulates **ovulation** and starts the **build-up and repair** of the **uterine wall** after menstruation
- **progesterone** – this continues the build-up of the uterine lining after ovulation.

Exam tips

- **Menstruation** occurs when the levels of **both oestrogen and progesterone fall** at the end of the cycle.
- The menstrual cycle stops if pregnancy occurs and only restarts after the baby is born.

Exam tip

If sex takes place a day or so before ovulation, pregnancy can still occur as there will be sperm in the oviduct. Similarly, if it takes place a day or so after ovulation, the egg will still be travelling down the oviduct, so fertilisation can still occur.

Oestrogen – the female sex hormone produced by the ovaries. The hormone that both causes the repair and build-up of the uterus lining following menstruation and stimulates ovulation. It also causes the development of secondary sexual characteristics.

Progesterone – the female hormone that maintains the build-up of the uterus lining and prepares the uterus for pregnancy.

Now test yourself

TESTED

4 Name the male hormone that leads to the development of secondary sexual characteristics.
5 Name the hormone that stimulates ovulation during the menstrual cycle.

Fertility problems and their treatment (infertility)

Problems – there are many causes, including **females** unable to produce eggs or blockages preventing movement of eggs down the oviducts, and **males** having low sperm counts or impotence.

Treatment can involve:

1. giving females **fertility drugs** to increase egg production
2. collecting eggs from ovaries and adding to sperm in a test tube (***in vitro* fertilisation**, or **IVF**)
3. placing **embryos** back into the uterus.

***In vitro* fertilisation** – fertilisation outside the body.

Fertility treatments can raise **ethical issues**, as IVF treatment can be used to screen for abnormalities or for particular characteristics, e.g. selecting the sex of the embryo.

Exam tips

- When replacing the embryos back into the uterus, it is important to strike a balance between increasing the chances of success and avoiding the potential for multiple births. For this reason, only two embryos are often placed back into the uterus following IVF.
- Before replacing the embryos back into the uterus, the mother has to be given **hormones** to ensure that the uterine lining is at a stage of development where **implantation** can occur.

Contraception – preventing pregnancy

REVISED

Pregnancy can be prevented by **contraception**. However, contraception can raise ethical issues for some people.

Table 11.1 summarises the three main types of contraception: **mechanical**, **chemical** and **surgical**.

Table 11.1 Methods of contraception

Type	Example	Method	Advantages	Disadvantages
Mechanical (physical)	Male condom	Acts as a barrier to prevent the sperm entering the woman	Easily obtained and also protects against sexually transmitted infections (STIs) such as chlamydia, gonorrhea and HIV (leading to AIDS) Some STIs can lead to infertility if untreated, e.g. chlamydia	Unreliable if not used properly
	Female condom	Acts as a barrier to prevent the sperm passing up the female reproductive system	Protects against STIs (see above)	Unreliable if not used properly
Chemical	Contraceptive pill	Taken regularly by the woman and prevents the ovaries from releasing ova by changing hormone concentrations	Very reliable	Can cause some side effects, such as weight gain and mood swings, and may increase the risk of blood clots The woman needs to remember to take the pill daily for around 21 consecutive days in each cycle

Type	Example	Method	Advantages	Disadvantages
Chemical (continued)	Implants	Implants are small tubes about 4 cm long that are inserted just under the skin in the arm and release hormones slowly over a long period of time to prevent the development and release of an egg	Very reliable Can work for up to 3 years	Do not protect against STIs Can prevent menstruation taking place
Surgical	Vasectomy (male sterilisation)	Cutting of sperm tubes, preventing sperm from entering the penis	Virtually 100% reliable	Very difficult or impossible to reverse Does not protect against STIs
	Female sterilisation	Cutting of oviducts, preventing ova from moving through the oviduct and being fertilised	Virtually 100% reliable	Very difficult or impossible to reverse Does not protect against STIs

Now test yourself

TESTED

6 Give **one** disadvantage in using condoms as a method of contraception.
7 Explain how male sterilisation prevents pregnancy.

Exam practice

1 (a) Describe the function of the testes. [1]
(b) Describe the passage of sperm from the testes to leaving the male body via the penis. [3]
(c) Give **two** ways in which sperm cells are adapted for their function. [2]
2 Describe the main stages between fertilisation and implantation. [3]
3 (a) What is meant by the term 'ovulation'? [2]
(b) Outline the functions of oestrogen and progesterone. [3]
H 4 (a) Suggest why females need to be given hormones when having IVF treatment. [2]
(b) Suggest **two** reasons why IVF is such an expensive procedure. [2]
(c) Explain the term '*in vitro* fertilisation'. [1]
5 Table 11.2 shows the number of males of different ages having vasectomy operations in one hospital over a year-long period in 2011.

Table 11.2

Age range	Number of patients having a vasectomy
20–29	4
30–39	41
40–49	121
50–59	68
60+	10

(a) Describe the trend shown by the data in the table. [2]
(b) Suggest **one** reason for the large difference between the data for the 20–29 and 40–49 age groups. [1]
(c) Explain how a vasectomy prevents pregnancy. [2]

Answers online

ONLINE

12 Variation and selection

Living organisms that belong to the same species (type) resemble each other, but usually differ in a number of ways – these differences are called **variation**.

Variation

REVISED

Variation can be:

- **genetic** – due to differences in DNA caused by:
  - variation as a result of **sexual reproduction**
  - **mutations** – random changes in the number of chromosomes or structure of a gene.
- **environmental** – due to the environment or lifestyle
- due to a combination of both – for example you have genes for a particular height, but your actual height reached depends on your health and diet.

Variation can be **continuous** or **discontinuous**, as shown in Table 12.1.

Continuous variation – the type of variation characterised by a gradual change in a characteristic across a population.

Discontinuous variation – the type of variation in which all the individuals can be clearly divided into two or more groups and there are no intermediate states.

Table 12.1 Continuous and discontinuous variation

Variation	Description	Examples
Continuous	Gradual change in a feature with no clearly distinct groups – no clear boundaries	Height/length
Discontinuous	Individuals can be placed into distinct groups easily, with no overlap	Tongue rolling/hand dominance

Exam tips

- Genetic and environmental are the *causes* of variation. Continuous and discontinuous are the *types* of variation.
- In exam questions, continuous variation is often represented by a histogram and discontinuous variation by a bar chart.

Discontinuous variation is usually **genetic** – for example, eye colour and blood group. **Continuous** variation is often both **genetic** and **environmental**.

Natural selection

REVISED

All living organisms are **adapted** for living in their normal environment – but some are better adapted than others due to **variation in phenotypes** and are better able to survive. This is called **natural selection**.

Natural selection – the process in which the better adapted individuals survive (at the expense of the less well adapted ones) and pass on their genes to their offspring.

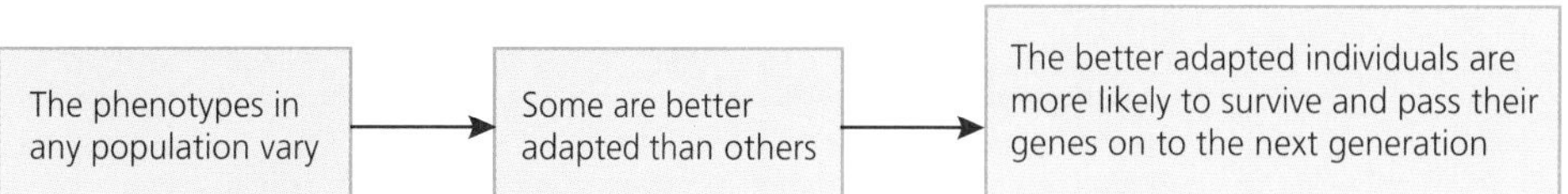
The phenotypes in any population vary
Some are better adapted than others
The better adapted individuals are more likely to survive and pass their genes on to the next generation

Figure 12.1 Natural selection

Exam tips

- Natural selection is very important when there is **competition for resources** – this is because being better, or less well, adapted can make a difference. The individuals that are better adapted may succeed, while those individuals that are less well adapted may fail, e.g. in finding food.
- Natural selection has three key elements:
  - **differences between phenotypes** (e.g. some grey squirrels can run faster than others and escape from predators)
  - **differential survival** (e.g. the fastest squirrels survive and the slower ones get caught)
  - **differential reproductive success** (e.g. the fastest squirrels are able to pass their genes on to the next generation).

One of the best examples of natural selection is **antibiotic resistance in bacteria**. The resistant phenotypes are not killed by **antibiotics** and so survive, but the non-resistant bacteria are killed by antibiotics. The resistant bacteria are then able to survive and pass their (resistant) genes on to future generations.

Antibiotic resistance – an antibiotic-resistant bacterium cannot be killed by at least one type of antibiotic.

Antibiotic – a chemical produced by fungi that kills bacteria.

Exam tips

- Note that the example of antibiotic resistance in bacteria has **different phenotypes** (resistant and non-resistant), **differential survival** (only the antibiotic-resistant bacteria survive) and **differential reproductive success** (only the antibiotic-resistant bacteria pass their genes on to the next generation).
- The use of antibiotics does not *cause* the bacteria to become resistant (some are already resistant due to mutations) – the use of antibiotics creates the conditions in which resistant bacteria are better adapted than non-resistant ones.

Now test yourself

TESTED

1 Give the **two** main causes of variation in populations.
2 State the **three** key features of natural selection.

Provided with any example and suitable data, you need to be able to describe the process of natural selection.

Example

In a typical pasture there may be a few plants that have alleles (forms of a gene) for resistance to high levels of copper in the soil. In these conditions the normal grasses grow better than the copper-resistant variety. However, in areas where the soil is contaminated with copper, the copper-resistant variety may make up over 90% of the plants present. Explain this observation.

Answer

- In copper-contaminated areas the presence of resistant alleles/genes is an advantage.
- Copper-resistant plants are more likely to survive/are fitter/better adapted.
- Copper-resistant plants are more likely to have offspring/pass genes on to next generation.
- The percentage of copper-resistant genes increases over time in the population.

Charles Darwin was the scientist who first explained the idea of natural selection.

H The link between natural selection and evolution

Charles Darwin used his theory of natural selection to explain the process of **evolution**.

- Natural selection can explain how species have **changed gradually** over **time** in a process called evolution.
- This happens because certain features in the species are favoured.
- Eventually the species may be very different from how it started out.
- Evolution is a continuing process – natural selection is always happening and all species change very gradually over a long time period.
- Evolution can also result in the development of **new species**.

Evolution – a continuing process of natural selection that leads to gradual changes in organisms over time, which may lead to the formation of a new species.

There are a number of reasons why not everyone accepts the theory of evolution. These include the fact that:

- it contradicts some **religious beliefs**
- the **very long timescales** involved mean that it is very difficult to see evolution actually happening.

Now test yourself

TESTED

H 3 Define the term 'evolution'.

Extinction

Sometimes entire species may not be well enough adapted to survive in a changing world and can no longer survive – they may become **extinct**, e.g. mammoths and dinosaurs are species that have been extinct for some time.

Extinction – a species is extinct if there are no living members of that species left.

Many organisms are **endangered** (at risk of extinction) due to climate change, hunting by humans, habitat destruction and many other reasons. Currently both mountain gorillas and many species of large cat are examples of endangered species.

Selective breeding

REVISED

For centuries, people have controlled selection in crops and domestic animals by deliberately selecting particular characteristics that are of use to us. This is the process of **selective breeding** (artificial selection).

Selective breeding – the selection and subsequent breeding of organisms chosen by humans for their desirable properties.

Traits selected include increased crop yield or quality, appearance, hardiness, disease resistance and longer shelf life.

The selective breeding of wheat shows the key features of this process. Wheat is a cereal that has been bred over many years to produce:

- a **shorter stalk length** (which is less likely to suffer wind damage and is easier to harvest because of the uniform size)
- a **larger head of grain** (higher yield).

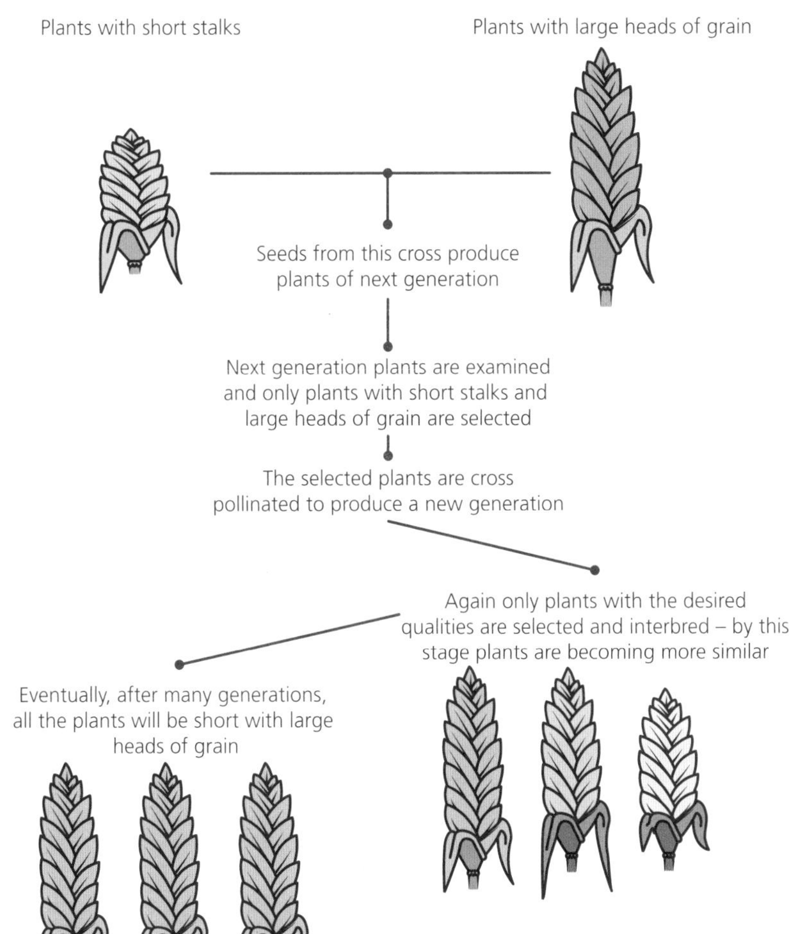

Plants with short stalks
Plants with large heads of grain
Seeds from this cross produce plants of next generation
Next generation plants are examined and only plants with short stalks and large heads of grain are selected
The selected plants are cross pollinated to produce a new generation
Again only plants with the desired qualities are selected and interbred – by this stage plants are becoming more similar
Eventually, after many generations, all the plants will be short with large heads of grain

Figure 12.2 Selective breeding in wheat

Exam tips

- Selective breeding is **not natural selection** – it is not 'nature' doing the selection, it is humans.
- Selective breeding normally takes **many generations** (reproductive cycles) and a long time to reach the stage at which all the animals or plants have the desired characteristics.
- **Dogs** have been selectively bred to produce the many breeds that exist today – each with its own distinctive characteristics.
- **Inbreeding** can also lead to genetic weakness, e.g. some breeds of dog are short lived or have problems with their bones.

Exam practice

1 (a) Explain what is meant by the term 'discontinuous variation'. [1]
(b) Give one example of discontinuous variation. [1]

2 Over time many species of predator and their prey have become more agile and able to run faster. Use your understanding of natural selection to explain why many species of predator have become faster over time. [3]

3 The peppered moth exists in two forms: light coloured and black. In non-polluted areas the light form is well camouflaged on the bark of trees whereas the black form is easily spotted and eaten by birds. In these areas the light forms are more common. In industrial areas where the trees are heavily polluted with soot, the black forms are more common. Explain why. [3]

4 Traditional selective breeding techniques involve breeding selected animals or plants together over a very long period of time until the entire population has all the characteristics desired. Artificial insemination in cattle – the placing of sperm from a prize bull into a cow – is also an example of selective breeding. Suggest **one** way in which this method is similar to 'traditional' selective breeding and **one** way in which it is different. [2]

Answers online

ONLINE

13 Health, disease, defence mechanisms and treatments

Microorganisms and communicable diseases

REVISED

A **communicable disease** is a disease that can be passed from one organism (person) to another, i.e. spread among people.

Exam tip

Communicable diseases are also described as **infectious** diseases.

Communicable disease – a communicable disease is one that can be passed from one organism (person) to another.

Bacteria, **viruses** and **fungi** are the causes of most communicable diseases.

Table 13.1 provides information on some communicable diseases.

Table 13.1 Communicable diseases

Microbe	Type	Spread	Control/prevention/treatment
HIV (which leads to AIDS)	Virus	Exchange of body fluids during sex Infected blood	Using a condom will reduce risk of infection, as will drug addicts not sharing needles Currently controlled by drugs
Colds and flu	Virus	Airborne (droplet infection)	Flu vaccination for targeted groups
Human papilloma virus (HPV)	Virus	Sexual contact	HPV vaccination given to 12- to 13-year-old girls to protect against developing cervical cancer
Salmonella food poisoning	Bacterium	From contaminated food	Always cooking food thoroughly; not mixing cooked and uncooked foods can control spread Treatment with antibiotics
Tuberculosis	Bacterium	Airborne (droplet infection)	BCG vaccination If contracted, treated with drugs, including antibiotics
Chlamydia	Bacterium	Sexual contact	Using a condom will reduce risk of infection Treatment with antibiotics
Athlete's foot	Fungus	Contact	Reduce infection risk by avoiding direct contact in areas where spores are likely to be present, e.g. wear 'flip flops' in changing rooms/swimming pools
Potato blight	Fungus	Spores spread in the air from plant to plant, particularly in humid and warm conditions	Crop rotation and spraying plants with fungicide

Exam tip

Potato blight is a plant disease that affects the potato and similar plants – all the other communicable diseases in Table 13.1 are passed among humans.

The body's defence mechanisms

REVISED

These involve both stopping harmful microorganisms gaining entry to the body and destroying them in the blood.

1 The first stage of defence is stopping microorganisms from entering the body (Table 13.2).

Table 13.2 Stopping microorganisms entering the body

Skin	Barrier that stops microorganisms entering the body
Mucous membranes	Thin membranes in the nose and respiratory system that trap and expel microorganisms
Clotting	Closes wounds quickly to form a barrier that stops microorganisms gaining entry (also prevents loss of blood)

2 The role of **white blood cells** is to destroy microorganisms that have entered the body. There are two main ways this happens:

(a) **Lymphocytes** are white blood cells that produce antibodies when microorganisms enter the blood. Protection by antibodies (Figure 13.1) involves the following:

(i) Microorganisms have special 'marker' chemicals on their surface called **antigens**.

(ii) These antigens cause the lymphocytes (white blood cells) to produce **antibodies**.

(iii) The antibodies are **complementary in shape** (like a lock and key) to the antigens.

(iv) The antibodies latch on to the antigens (microorganisms), linking them together.

(v) This immobilises (clumps) the microorganisms and they can then be destroyed.

(vi) After an infection the body produces **memory lymphocytes** that remain in the body for a very long time – these can respond quickly and produce antibodies if the body is infected again by the same microorganism.

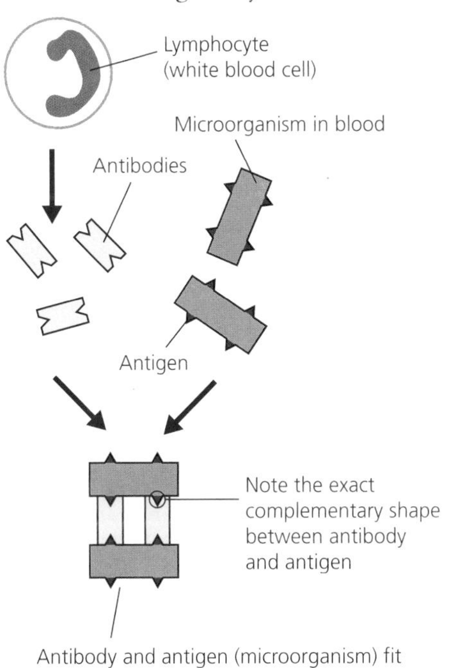
Lymphocyte (white blood cell)
Microorganism in blood
Antibodies
Antigen
Note the exact complementary shape between antibody and antigen
Antibody and antigen (microorganism) fit together and form a 'trapped' clump

Figure 13.1 How antibodies work

Lymphocyte – a type of white blood cell that produces antibodies.

Antigen – a distinctive marker on a microorganism that leads to the body producing specific antibodies.

Antibody – a structure produced by lymphocytes that has a complementary shape (and can attach to) antigens on a particular microorganism.

Memory lymphocyte – a special type of lymphocyte that can remain in the body for many years and produce antibodies quickly when required.

Exam tips

- Clumping the harmful microorganisms (and then destroying them) prevents them from spreading around the body, which leads to reduced symptoms in the patient.
- Because each type of microorganism has different types and **shapes** of antigen, each type of antibody has a unique shape that matches (is complementary to) the antigens. Therefore, there is a different type of antibody for each type of microorganism.

(b) Once the microorganisms are clumped together, they are destroyed by a second type of white blood cell – the **phagocytes**. This process is called **phagocytosis** (Figure 13.2).

Phagocyte – a type of white blood cell that destroys microorganisms by engulfing them and then digesting them (phagocytosis).

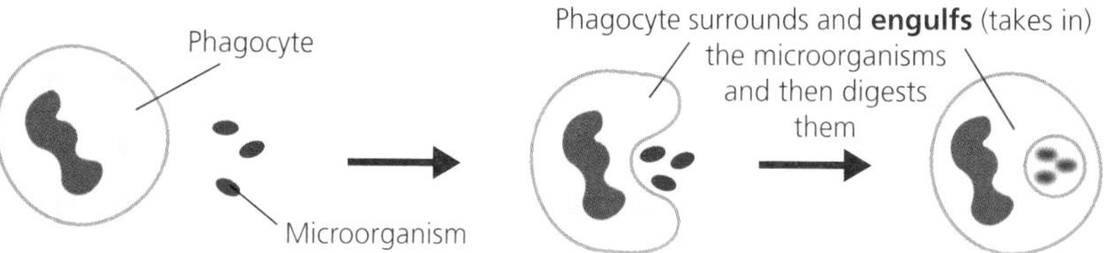

Figure 13.2 Phagocytosis

Now test yourself

TESTED

1 What is meant by the term 'communicable disease'?
2 State the type of microorganism that causes colds and flu.
3 Name the type of white blood cell that produces antibodies.

Primary and secondary responses

Individuals infected by a disease-causing bacterium or virus are often ill for a few days before the antibody numbers are high enough to provide immunity – the primary response.

However, once infected the body is able to produce memory lymphocytes that remain in the body for many years. This means that if infection by the same type of microorganism occurs again, the memory lymphocytes will be able to produce antibodies very quickly to stop the individual catching the same disease again. This is known as the secondary response.

Exam tip

We are often unaware when the secondary response occurs, as we may not show symptoms (catch the disease).

Immunity

Immunity means that antibody levels are high enough (or high enough levels can be produced quickly enough) to combat microorganism infection should it occur. There are two types of immunity: **active immunity** and **passive immunity** (Figure 13.3).

Immunity – freedom from disease.

Active immunity – the type of immunity produced when the body produces antibodies.

Passive immunity – the type of immunity produced by injecting antibodies.

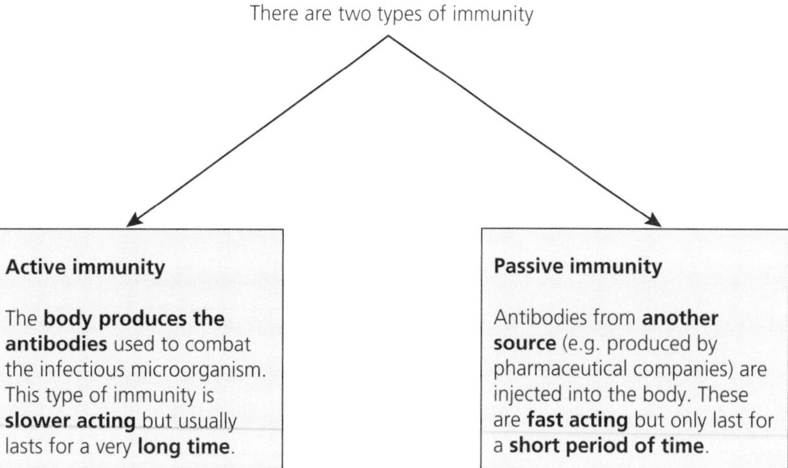

Figure 13.3 Immunity

Exam tip

Passive immunity allows the **rapid** (medical) **treatment** of very serious infections.

Vaccinations

Vaccinations involve the use of **dead** or **modified** disease-causing microorganisms (pathogens) that are injected into the body (Figure 13.4).

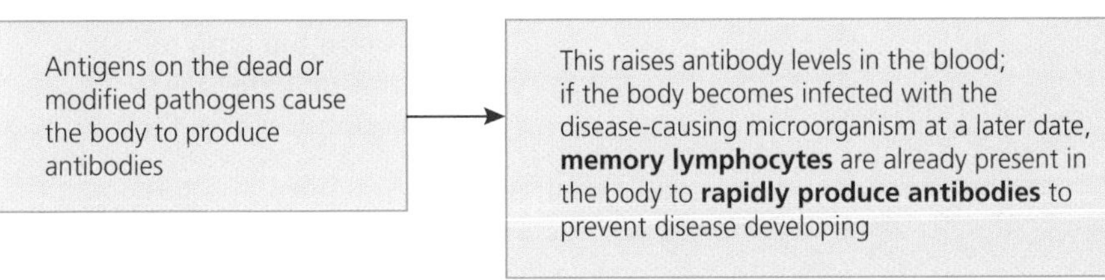

Antigens on the dead or modified pathogens cause the body to produce antibodies
This raises antibody levels in the blood; if the body becomes infected with the disease-causing microorganism at a later date, **memory lymphocytes** are already present in the body to **rapidly produce antibodies** to prevent disease developing

Figure 13.4 Vaccinations

Vaccination – the injection of dead or modified pathogens (disease-causing microorganisms) with the purpose of raising antibody and memory lymphocyte levels in the blood.

Exam tips

- The microorganisms in vaccinations need to be dead or modified in some way, otherwise the vaccination would give you the disease you are trying to avoid.
- The process of antibody action following vaccination is exactly the same as if you had caught the disease – the big difference is you don't get ill first.

H Sometimes we need more than one vaccination to make sure that we remain immune for a reasonable period of time. This is known as a follow-up **booster**. Figure 13.5 shows what happens following a vaccination that involves a booster.

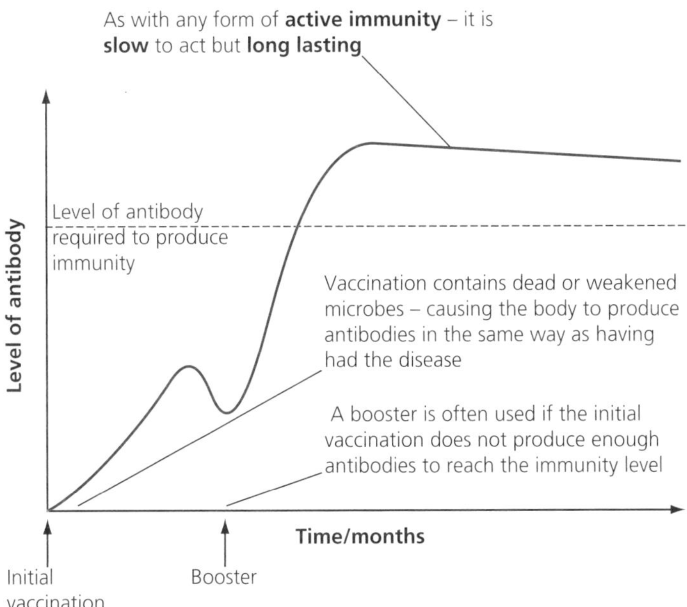

As with any form of **active immunity** – it is **slow** to act but **long lasting**
Level of antibody
Level of antibody required to produce immunity
Vaccination contains dead or weakened microbes – causing the body to produce antibodies in the same way as having had the disease
A booster is often used if the initial vaccination does not produce enough antibodies to reach the immunity level
Time/months
Initial vaccination
Booster

Figure 13.5 Active immunity (by vaccination)

Higher tier candidates need to be able to interpret graphs showing the antibody levels typically produced in active and passive immunity. Examples of these are shown in the graphs in Figures 13.6 and 13.7.

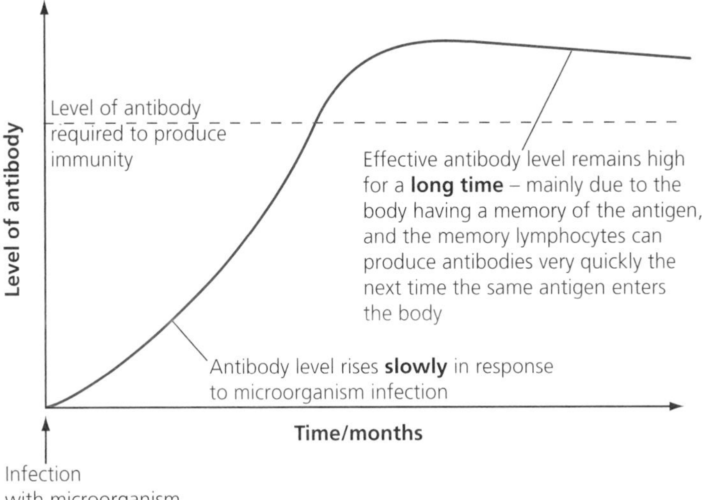

Figure 13.6 Active immunity (acquired by having the disease)

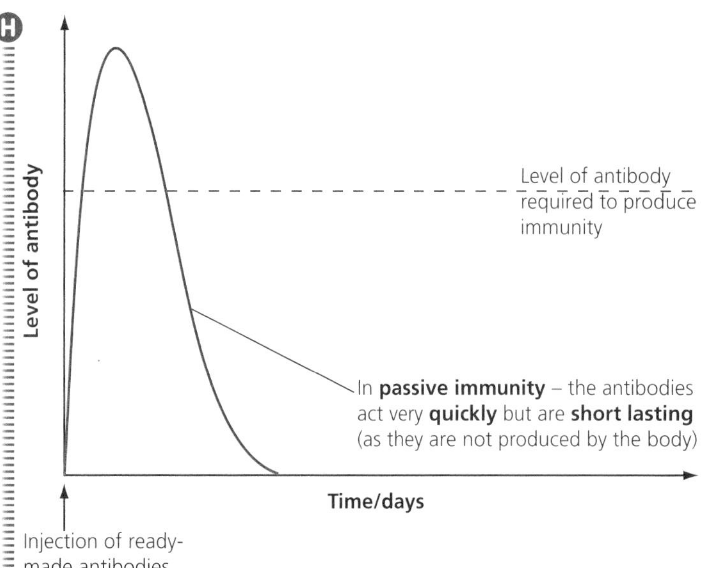
Level of antibody
Level of antibody required to produce immunity
In passive immunity – the antibodies act very quickly but are short lasting (as they are not produced by the body)
Time/days
Injection of ready-made antibodies

Figure 13.7 Passive immunity (by injection of ready-made antibodies)

Now test yourself

4 Define the term 'vaccination'.
5 Name the type of immunity that is fast acting but short lived.

TESTED

Antibiotics

REVISED

Antibiotics, such as penicillin, are chemicals **produced by fungi** that are used against bacterial diseases to **kill bacteria** or **reduce their growth**.

Antibiotic – a chemical produced by fungi that kills bacteria.

Antibiotic resistance

Sometimes bacteria can evolve (change) so that antibiotics no longer have an effect:

- Bacteria can **mutate**.
- Their DNA changes and the bacteria develop new properties.
- This can make them **resistant** to antibiotics.
- Antibiotics will not work against these particular bacteria, or cure diseases caused by them.

Exam tip

Antibiotics can kill bacteria, but they have no effect on viruses.

Overuse of antibiotics has been a major factor in the development of bacterial resistance to antibiotics (and the development of '**superbugs**'). **MRSA** is a type of bacterium that is resistant to most antibiotics – an example of a superbug. Antibiotic-resistant bacteria have been a particular problem in hospitals.

Procedures to reduce the incidence of superbugs include:

- **not overusing antibiotics** when not needed (e.g. against viral diseases)
- increased **hygiene** measures in hospitals, e.g. staff and visitors washing hands or using hand gels
- **isolating patients** infected with MRSA or other 'superbug' infections.

Exam tip

Although MRSA (and other superbugs) have been a particular problem in hospitals, they can occur anywhere and we all must do our part in not overusing antibiotics.

Now test yourself

6 Define the term 'antibiotic'.

TESTED

Aseptic techniques

REVISED

When working with bacteria and fungi in the laboratory, it is very important that great care is taken to avoid:

- contamination of the cultures used
- the growth of unwanted, pathogenic microorganisms.

The procedures used to avoid this are referred to as **aseptic techniques** (Figure 13.8).

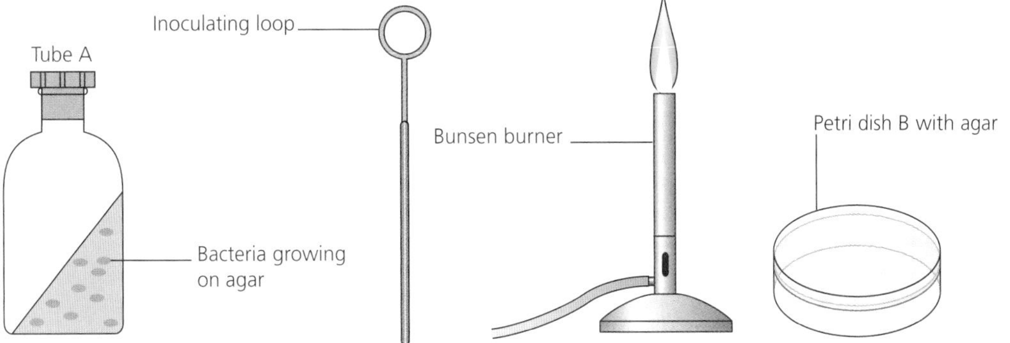

Figure 13.8 Using aseptic techniques

Using aseptic techniques when transferring microorganisms

1. Pass the metal loop through the flame of the Bunsen burner.
2. Allow the metal loop to cool.
3. Remove the lid of the culture bottle (Tube A) and glide the loop over the surface of the agar (without applying any pressure). This is called inoculation.
4. Replace the lid of the culture bottle to prevent contamination. When doing this, 'sweep' the neck of the bottle through the flame to destroy any airborne microorganisms.
5. Spread the microbes over the surface of the agar in the Petri dish (B) by gently gliding the metal loop over the nutrient agar surface (this is called plating). It is important to hold the Petri dish lid at an angle rather than completely removing it, as this will reduce the chance of unwanted microbes from the air entering the dish.
6. The metal loop can then be heated again to a high temperature to ensure that any microorganisms remaining on the loop are destroyed.
7. The Petri dish should be taped (three or four times) and then incubated in an oven at 25°C.
8. When carrying out the transfer it is important to work close to a Bunsen burner as this creates an upward current of air that carries microorganisms in the air away from the area where the microorganisms are being transferred, thus avoiding contamination.
9. When the investigation is complete, it is important to clean all work surfaces and hands and safely dispose of bacterial cultures by following your teacher's instructions. Autoclaving (heating at high temperatures and pressures) will sterilise glass Petri dishes and culture bottles.

Exam tips

- Instead of using a metal loop, it is possible to use sterile disposable plastic loops that do not require heating.
- All the apparatus used, e.g. agar plates, should be sterilised in advance, or disposable sterile plates should be used.

Non-communicable diseases

REVISED

Non-communicable diseases are diseases that are *not* passed from person to person – they are not infectious diseases.

Non-communicable diseases are usually a consequence of inheriting a **combination of genes** that predispose us to developing some conditions, such as cancer, or are due to **lifestyle**, or a combination of both.

Exam tip

Remember that non-communicable diseases can be due to the genes we carry (inherited) or due to our lifestyle, or a combination of both.

Lifestyle factors

Some lifestyle factors that can contribute to non-communicable diseases are listed in Table 13.3.

Table 13.3 The effect of some lifestyle factors on health

Lifestyle factor	Effect
Poor diet	Too much sugar and fat can lead to obesity
Lack of exercise	Reduced exercise can result in less energy being used than taken in, which can lead to obesity
Overexposure to the Sun	Ultraviolet radiation (UV) can cause mutations leading to skin cancer

Lifestyle factors also include the **misuse of drugs**:

- **Alcohol** – drinking too much (especially binge-drinking) can harm health. **Binge drinking** is drinking too much alcohol on any one occasion.
  Drinking too much alcohol can:
  - damage the **liver**
  - affect **foetal development (foetal alcohol syndrome)** during pregnancy.
- **Tobacco smoke** – smoking can seriously damage health, as summarised in Table 13.4.

Table 13.4 Harmful effects of tobacco smoke

Substance in cigarette smoke	Harmful effect(s)
Tar	Causes bronchitis (narrowing of the bronchi and bronchioles in lungs), emphysema (damage to the alveoli, reducing the surface area for gas exchange) and lung cancer (abnormal cell division)
Nicotine	Is addictive and affects the heart rate
Carbon monoxide	Combines with red blood cells to reduce the oxygen-carrying capacity of the blood

Circulatory diseases

Heart disease is a cardiovascular disease affecting the blood vessels of the heart (Figure 13.9).

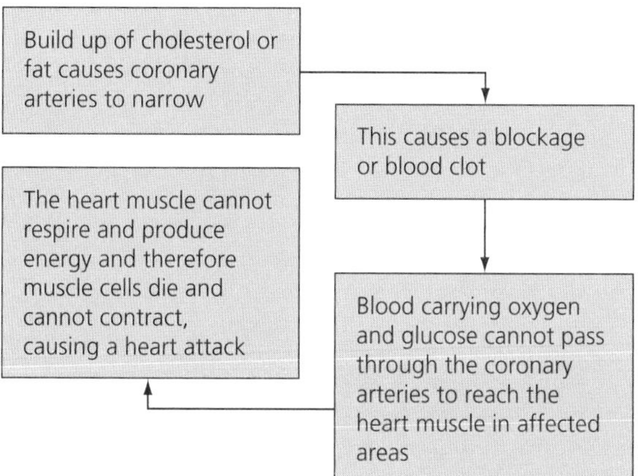

Figure 13.9 How heart attacks happen

Strokes are also circulatory diseases, but they affect the **brain**. They are also caused by blood vessels becoming blocked, resulting in the death of brain cells and reduced brain function.

Lifestyle factors and circulatory disease

The lifestyle factors that increase the risk of heart attacks and strokes are shown in Figure 13.10.

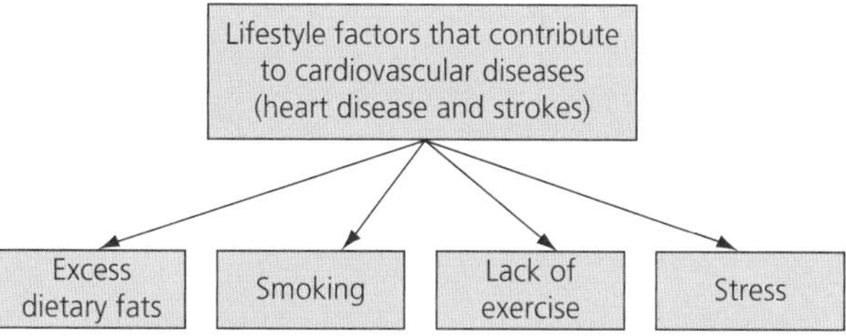

Figure 13.10 Factors that increase the risk of cardiovascular diseases

Treating cardiovascular diseases

- **Angioplasty** and **stents** – angioplasty is a medical technique involving the use of balloon-like structures to hold open diseased arteries so that **stents** (small mesh-like structures) can be inserted into the blood vessels to keep them open.
- Drugs such as statins and aspirin can help protect against cardiovascular disease. **Statins** help reduce blood cholesterol and **aspirin** helps 'thin' the blood and makes it less 'sticky'.

Some of the diseases covered in this section are closely linked. Consequently, many people who are affected by one condition, such as obesity, often suffer from one or more other conditions. For example, people who are obese are more likely to suffer from cardiovascular disease and type 2 diabetes.

Exam tip

The **coronary arteries** are the blood vessels that bring blood to the heart.

Now test yourself

TESTED

7 Describe how too much strong sunlight can harm health.
8 Name **three** medical conditions caused by the tar in cigarette smoke.
9 Give **four** lifestyle factors that can contribute to cardiovascular disease.

Cancer

REVISED

Cancer is caused by uncontrolled cell division. There are two types of cancer tumour:

- **benign** – these do not spread through the body and often have a distinct boundary (are encapsulated)
- **malignant** – these can spread throughout the body (they are not encapsulated).

Causes and relevant lifestyle choices for some cancers are summarised in Figure 13.11.

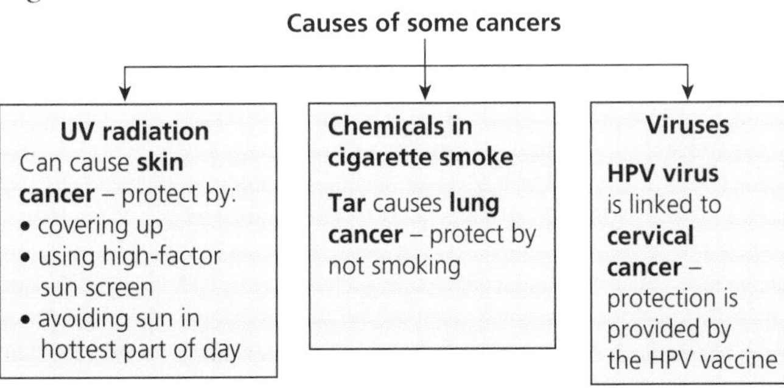

Figure 13.11 Causes of some cancers and lifestyle choices

Cancer – a range of diseases caused by uncontrolled cell division.

Benign tumour – a tumour that is encapsulated and does not spread to other parts of the body.

Malignant tumour – a tumour that is not surrounded by a capsule and is capable of spreading around the body.

The economic costs to the NHS of treating disease

Both communicable and non-communicable disease treatment have high costs to society, including the National Health Service (NHS):

- Many diseases involve **long-term** treatment, e.g. heart disease, cancer.
- They may involve **long stays in hospital**.
- **Expensive drugs** and **medicines** may be required.
- Highly trained and **specialist staff** are required.

Illness also affects families, as people who care for those who are ill may need time off work as well. Additionally, individuals who are ill may not be as productive in the workplace.

Exam practice

1 (a) Which of the following diseases could be cured by taking an antibiotic?
flu **cold** **tuberculosis** **AIDS** **HPV**
Explain your answer. [2]
(b) Mary went to the doctor because she was suffering from a sore throat caused by a bacterial infection. Explain why the doctor gave her an antibiotic rather than a vaccination. [2]

2 (a) Explain how antibodies help protect against disease. [3]
(b) Name the type of immunity produced when the body produces its own antibodies. [1]
(c) Describe the process of phagocytosis. [2]

3 (a) Give **two** lifestyle factors that can contribute to cardiovascular disease. [2]
(b) Explain how stents can be used to treat cardiovascular disease. [2]

4 John would like to give up smoking but finds it hard to stop.
(a) Name the chemical in cigarette smoke that makes it hard to stop. [1]
(b) Explain how the carbon monoxide in cigarette smoke can lead to a smoker having a shortage of energy. [3]

5 (a) Name the two types of cancer tumour. [1]
(b) Give **one** difference between these. [1]

Answers online

ONLINE

14 Atomic structure

Atoms

The structure of atoms

REVISED

- An **atom** is the simplest particle of an **element** that can exist on its own in a stable environment.
- Atoms are made up of three subatomic particles — protons, neutrons and electrons.
- Protons and neutrons are found in the **nucleus** (the centre of the atom) and electrons are found in **shells** orbiting the nucleus.
- The mass of an atom is largely in its nucleus because the mass of an electron is very small compared with a proton or a neutron.
- Atoms are **electrically neutral** (they have no charge). This is because they have equal numbers of protons and electrons. Protons have a positive charge and electrons have a negative charge.
- Neutrons do not have a charge.

Table 14.1 gives the relative masses and relative charges of the three subatomic particles.

Table 14.1 Subatomic particles

Subatomic particle	Relative mass	Relative charge	Location
Proton	1	+1	Nucleus
Neutron	1	0	Nucleus
Electron	$\frac{1}{1840}$	−1	Shells

Exam tip

An incomplete form of Table 14.1 is often given in questions. The most common mistakes are giving 0 as the relative mass of an electron and writing the relative charge of a proton as + (instead of +1) and of an electron as − (instead of −1). Learn this table and be able to identify the names of the subatomic particles from their relative charges and relative masses.

Atomic number and mass number

REVISED

- The **atomic number** and **mass number** of an atom are usually written before the symbol of the element. The mass number is at the top and the atomic number is at the bottom — for example, ${}^{12}_{6}C$, ${}^{39}_{19}K$, ${}^{35}_{17}Cl$, ${}^{31}_{15}P$, ${}^{27}_{13}Al$.
- The atomic number is used to order the elements in the periodic table.
- The electron mass is so small that it does not affect the mass number.

The **atomic number** is the number of protons in (the nucleus of) an atom.

The **mass number** is the total number of protons and neutrons in (the nucleus of) an atom.

Determining numbers of subatomic particles

REVISED

The numbers of subatomic particles **in an atom** can be determined from the atomic number and the mass number:

- number of protons = atomic number
- number of neutrons = mass number − atomic number
- number of electrons = number of protons (in an atom)

Example

Determine the number of each subatomic particle in each of the following atoms:

(a) $^{12}_{6}C$

(b) $^{39}_{19}K$

(c) $^{35}_{17}Cl$

Answer

(a) atomic number = 6 (**6 protons**)
6 electrons because it is an atom
mass number − atomic number = 12 − 6 = 6 (**6 neutrons**)

(b) atomic number = 19 (**19 protons**)
19 electrons because it is an atom
mass number − atomic number = 39 − 19 = 20 (**20 neutrons**)

(c) atomic number = 17 (**17 protons**)
17 electrons because it is an atom
mass number − atomic number = 35 − 17 = 18 (**18 neutrons**)

Exam tip

Remember that atoms have the *same number* of protons and electrons. They do not have to have the same number of protons and neutrons. ($^{12}_{6}C$ does, but this is not always the case.)

You can be asked to identify a particular isotope from the numbers of subatomic particles. For example, if asked to identify a particle that has 9 protons, 9 electrons and 10 neutrons:

- 9 protons means that the atomic number is 9, so it is a particle of fluorine.
- 9 protons and 9 electrons means that it is an atom, so it is an atom of fluorine.
- 9 protons and 10 neutrons means that its mass number is 9 + 10 = 19.
- The atom is $^{19}_{9}F$.

Now test yourself

TESTED

1 Which subatomic particle has a relative charge of −1?
2 What is meant by the term 'mass number'?
3 State the number of protons, electrons and neutrons in an aluminium atom, $^{27}_{13}Al$.

Isotopes and relative atomic mass

Isotopes

REVISED

The relative atomic mass of an element is an average mass of all of the **isotopes** of that element.

- There are two types of chlorine atom with different mass numbers — $^{35}_{17}Cl$ and $^{37}_{17}Cl$.
- There are three types of hydrogen atom with different mass numbers — $^{1}_{1}H$, $^{2}_{1}H$ and $^{3}_{1}H$.
- There are two types of boron atom with different mass numbers — $^{10}_{5}B$ and $^{11}_{5}B$.

Isotopes are atoms that have the same number of protons (so they are atoms of the same element), but they have a different number of neutrons (so they have a different mass number).

Isotopes of the same element have the same chemical properties because they have the same number of electrons in the outer shell.

Calculating relative atomic mass

REVISED

The relative atomic mass (often represented by A_r or RAM) can be calculated from the mass numbers and relative abundances of the isotopes of an element:

$$\text{relative atomic mass} = \frac{\Sigma\,(\text{mass number} \times \text{abundance})}{\Sigma\,(\text{abundance})}$$

where Σ means the sum (added together) for all isotopes.

- The mass of a chlorine atom in a sample of chlorine is an average of the masses of chlorine-35 and chlorine-37 atoms. Only one-quarter of chlorine atoms are chlorine-37, with three-quarters of them chlorine-35, so the average mass of a chlorine atom works out at 35.5.
- If you had 100 chlorine atoms, 75 would have a mass of 35 and 25 would have a mass of 37.
- The relative atomic mass of a chlorine atom is calculated like this:

$$\text{relative atomic mass} = \frac{\Sigma(\text{mass number} \times \text{abundance})}{\Sigma(\text{abundance})}$$

$$\text{relative atomic mass} = \frac{(35 \times 75) + (37 \times 25)}{100} = \frac{3550}{100} = 35.5$$

- The relative atomic mass of chlorine is the average mass of a chlorine atom based on the mass numbers and the relative abundances of the isotopes. The word 'relative' is used because the mass of all atoms is measured relative to the mass of an atom of carbon-12.

Exam tip

Being asked to define the term 'isotope' is a common question. Remember that there are three main parts to the definition — 'Isotopes are *atoms* with the *same atomic number* (or the same number of protons) but with *different mass numbers* (or different numbers of neutrons)'.

Example 1

Boron has only two isotopes, ^{10}B and ^{11}B. ^{11}B has a percentage abundance of 80%. Calculate the relative atomic mass of boron.

Answer

$$\text{relative atomic mass} = \frac{\Sigma(\text{mass number} \times \text{abundance})}{\Sigma(\text{abundance})}$$

As there are only two isotopes ^{10}B must have a relative abundance of 20%:

$$\text{relative atomic mass} = \frac{(11 \times 80) + (10 \times 20)}{100} = \frac{1080}{100} = 10.8$$

Example 2

Lithium has two isotopes, ^{6}Li and ^{7}Li. ^{7}Li has a percentage abundance of 92.5%. Calculate the relative atomic mass of lithium.

Answer

$$\text{relative atomic mass} = \frac{\Sigma(\text{mass number} \times \text{abundance})}{\Sigma(\text{abundance})}$$

As there are only two isotopes, ^{6}Li must have a relative abundance of 7.5%.

$$\text{relative atomic mass} = \frac{(6 \times 7.5) + (7 \times 92.5)}{92.5 + 7.5} = \frac{692.5}{100} = 6.925$$

H **Example 3**

The table below shows the relative abundances of the three different isotopes of magnesium. Calculate the relative atomic mass of magnesium.

Isotope	Relative abundance
^{24}Mg	15.8
^{25}Mg	2.0
^{26}Mg	2.2

Answer

For each isotope multiply the mass by the relative abundance and add these together. Finally divide this number by the sum of the relative abundances.

$$\text{relative atomic mass} = \frac{(24 \times 15.8) + (25 \times 2.0) + (26 \times 2.2)}{15.8 + 2.0 + 2.2} = \frac{379.2 + 50 + 57.2}{20} = \frac{486.4}{20} = 24.32$$

Now test yourself

TESTED

4 What name is given to atoms of the same element with different numbers of neutrons?
5 What is the mass number of an atom that has 18 protons, 18 electrons and 22 neutrons? Identify the element.
H 6 Calculate the relative atomic mass of silicon to one decimal place if the isotopes have the following percentage abundances: ^{28}Si (92.2%), ^{29}Si (4.7%) and ^{30}Si (3.1%).

Exam tip

Often you may be asked to quote the answer to a specific number of decimal places. The answer in Example 3 to one decimal place is 24.3. Abundances are often given as percentage abundances, so Σ(abundance) = 100, but in Example 3 the sum of the abundances is 20.

Atomic theory

Electronic configuration

REVISED

The arrangement of electrons in the shells around the nucleus can be represented either by a diagram or in written format. You will need to remember the following rules as to where the electrons can be:

1 The shells are at increasing distances from the central nucleus. The shell closest to the nucleus is called the first shell — then there are the second, third and fourth shells.
2 The first shell can hold a maximum of 2 electrons.
3 The second and third shells can hold a maximum of 8 electrons.
4 The first shell must fill first, before an electron can be put into the second shell; and the second shell must fill before an electron can be put into the third shell, and so on.
5 Electrons pair up in the shells but only when no other space is available. For example, four electrons in shell 2 would not pair up; but six electrons would have two pairs with the other two electrons unpaired. The two electrons in the first shell must be paired as it can only hold a maximum of two.

In many diagrams '×' is used to represent an electron.

Example 1

Lithium has atomic number 3, so an atom of lithium has 3 electrons.

The first shell takes 2 electrons (paired) and 1 electron goes into the second shell.

The written electronic configuration of lithium is 2,1.

The electronic configuration of lithium is shown here:

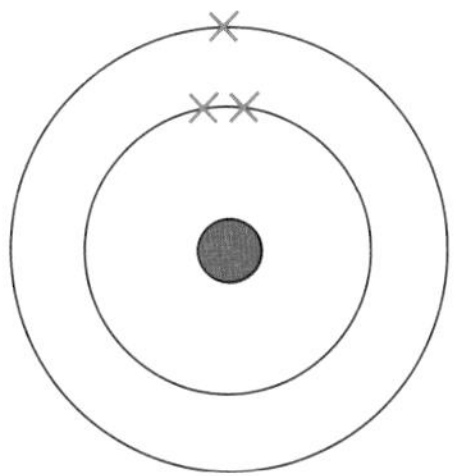

Example 2

Sulfur has atomic number 16, so an atom of sulfur has 16 electrons. The first shell takes 2 electrons, the second shell takes 8 electrons (four pairs) and the third shell takes 6 electrons (two pairs and two singles).

The written electronic configuration of sulfur is 2,8,6.

The drawn electronic configuration of sulfur is shown here:

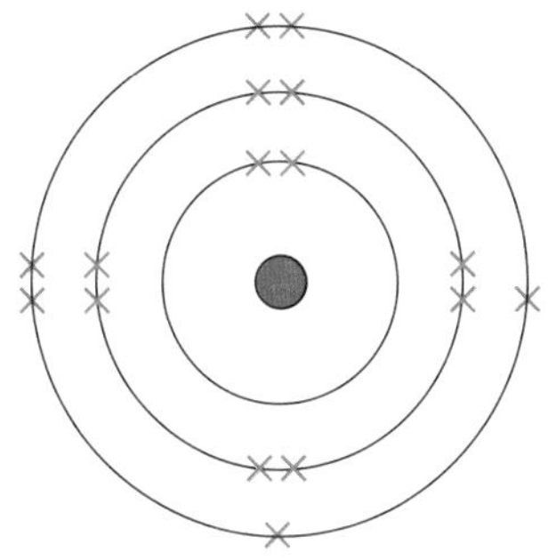

The written and drawn electronic configurations for atoms of elements with atomic number 1 to 20 are shown in Figure 14.1.

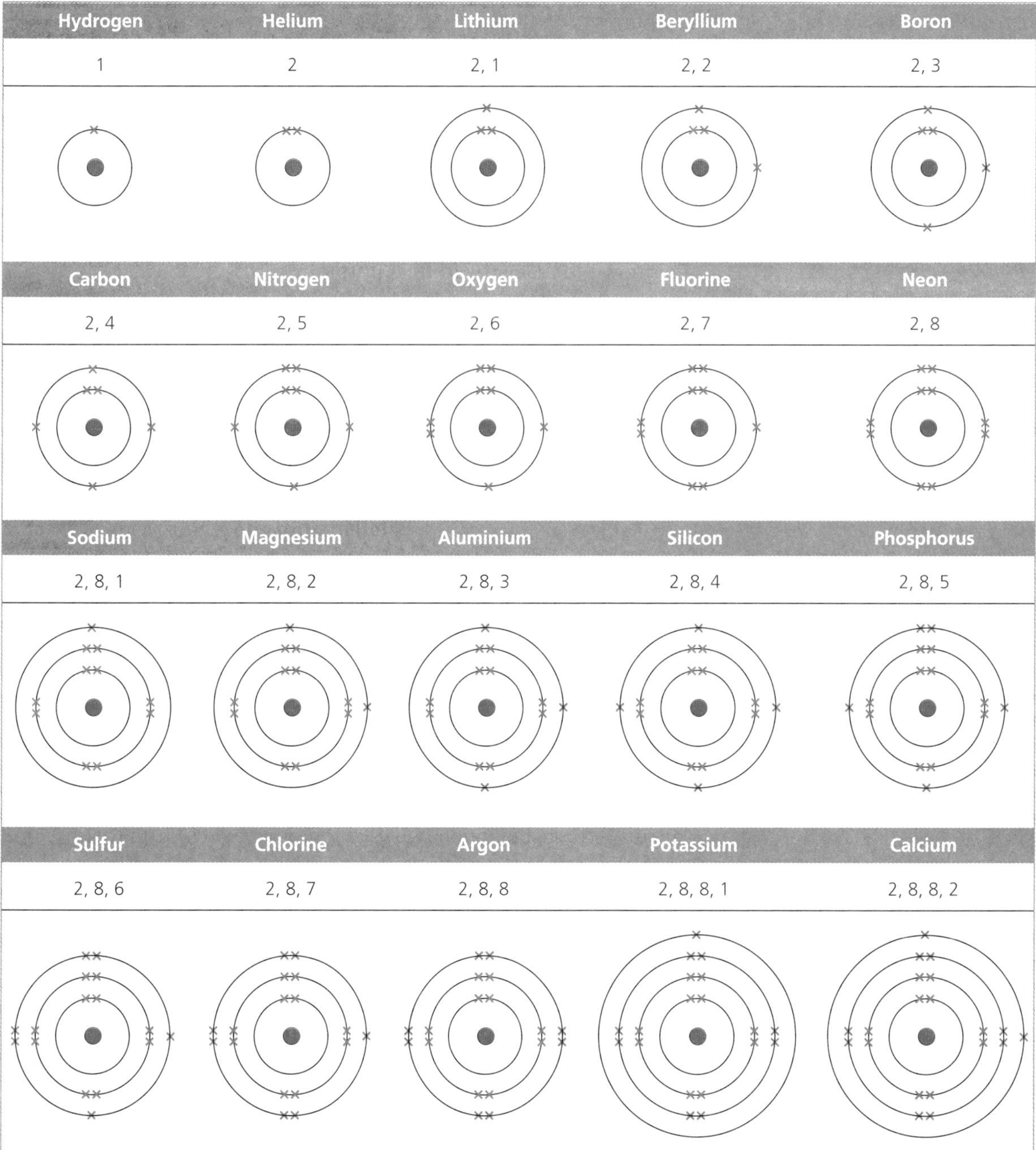

Figure 14.1 Electronic configurations of atoms of elements 1–20

Exam tip

When a question asks you to draw a diagram of an atom, you must write the correct number of protons and neutrons in the nucleus and the correct number and arrangement of electrons in the shells. These questions can be worth up to 4 marks.

If you are asked to show an electronic configuration, read the question twice to check if it has asked for the diagram form or the written form. If the question says 'draw', then make sure you do! If the question does not specify which, then the written form is acceptable.

Electronic configuration from the periodic table

REVISED

The electronic configuration of an atom can be determined from the periodic table by using the group number and the period number:

- The group number gives the number of electrons in the outer shell (except group 0 elements, which have a full outer shell of electrons).
- The period number gives the number of shells in use.

Example 1

Potassium is in period 4 — that means four shells in use:

–,–,–,–

The element is in group 1 — that means one electron in the outer shell:

–,–,–,1

The three inner shells must be full (see rule 4 on page 88).

So the electronic configuration of a potassium atom must be **2,8,8,1**.

Example 2

Nitrogen is in period 2 — that means two shells in use:

–,–

The element is in group 5 — that means five electrons in the outer shell:

–,5

The inner shell must be full.

So the electronic configuration of a nitrogen atom must be **2,5**.

Noble gases

The atoms of noble gases all have full outer shells:

- helium, He: 2
- neon, Ne: 2,8
- argon, Ar: 2,8,8

A full outer shell is stable and this makes all the noble gases unreactive.

Now test yourself

TESTED

7 Write the electron configuration of an oxygen atom.
8 An element is in group 2 and period 3. Identify the element and write the electronic configuration of an atom of the element.
9 An atom of element X has the electronic configuration 2,8,5. Identify element X.

Compounds and ions

Compounds

REVISED

- A **compound** is a substance formed when two or more elements are chemically combined. The elements must be different for it to be classified as a compound.
- Some common examples of compounds are:

A **compound** is a substance formed when two or more *different* elements are chemically combined.

- ○ Water, H_2O, contains the elements hydrogen and oxygen chemically combined
- ○ Sodium chloride, NaCl, contains the elements sodium and chlorine chemically combined
- ○ Carbon dioxide, CO_2, contains the elements carbon and oxygen chemically combined.

- A compound is different from a mixture where the components of the mixture are not chemically combined and so may be separated easily using separation techniques (page 154).
- If the atoms in a substance are all the same, even if they are chemically combined, the substance is an element, for example O_2 contains two oxygen atoms chemically combined but it is an element as it only contains one element chemically combined.

Exam tip

All of the diatomic elements (H_2, N_2, O_2, F_2, Cl_2, Br_2 and I_2) have atoms chemically combined, but they are elements, not compounds. This is a common mistake.

Ions

REVISED

An ion is a charged particle:

- H A positive ion is called a **cation**.
- A negative ion is called an **anion**.

A **cation** is a positive ion.

An **anion** is a negative ion.

A simple ion is formed when an atom loses or gains electrons to achieve a full outer shell. This is the electronic configuration of the closest noble gas. A full outer shell of electrons is stable.

- The overall charge on an ion depends on the number of positive protons and the number of negative electrons.
- Simple positive ions have the same name as the atom from which they are formed — for example, a hydrogen ion is H^+, a sodium ion is Na^+ and a magnesium ion is Mg^{2+}.
- Simple negative ions change their ending to '**-ide**' — for example, chlorine forms chloride ions (Cl^-) and oxygen forms oxide ions (O^{2-}).

There are two main types of examination question on ions:

- those that ask you to determine the number of each subatomic particle in an ion
- those that ask you to to work out the formula of the ion, including its charge, from the number of subatomic particles.

Example 1

Determine the numbers of protons and electrons, and the electronic configuration of an oxide ion, O^{2-}.

Answer

- The atomic number of oxygen is 8, so there are **8 protons**.
- An atom of oxygen would therefore have 8 electrons, but the ion has a 2– charge so it has two extra electrons, making **10 electrons**.
- The electronic configuration of the oxide ion is **2,8**.

Example 2

Write the formula, including the charge, of the simple ion that has 13 protons and has the electronic configuration 2,8.

Answer

- 13 protons means atomic number 13, so it is an aluminium ion (**Al**).
- The electronic configuration 2,8 means that there are **10 electrons**.
- 13 positive protons and 10 negative electrons make +13 – 10 = +3, so the ion has charge of **3+**.
- It is an aluminium ion, with formula Al^{3+}.

Exam tip

This type of question is often given in table form. All the data may be given and the identity of the particles asked for, or some data may be missing and you are asked to fill in the missing information.

Letters are used to represent atoms and ions but you will be reminded that these are not the symbols of elements.

Remember that the electronic configuration does not give you the identity of the atom or ion — it must be the number of protons or the atomic number.

Worked example

(a) A, B and C are atoms or ions of different elements. Fill in the missing information in the table. [3 marks]

Particle	Atomic number	Mass number	Number of protons	Number of neutrons	Number of electrons	Electronic configuration
A	12	24				2,8
B			18	22	18	
C	7			7		2,8

(b) Identify A, B and C and include the charge on any ions. [3 marks]

Answer

(a)

Particle	Atomic number	Mass number	Number of protons	Number of neutrons	Number of electrons	Electronic configuration
A	12	24	12	12	10	2,8
B	18	40	18	22	18	2,8,8
C	7	14	7	7	10	2,8

[1 mark for each correct row]

(b) For A, atomic number = 12 so it is a magnesium particle. Number of protons = 12.
Number of electrons = 10 (2+8 from the electronic configuration). So it is a magnesium ion, Mg^{2+}. [1]
For B, atomic number = 18 so it is an argon particle. Number of protons = 18.
Number of electrons = 18. So it is an argon atom, Ar. [1]
For C, atomic number = 7 so it is a nitrogen particle. Number of protons = 7.
Number of electrons = 10 (2+8 from the electronic configuration). So it is a nitride ion, N^{3-}. [1]

Exam tip

- The atomic number and the number of protons must be the same.
- The mass number is the total of the number of protons added to the number of neutrons.
- The number of electrons must be the same as the total of the electronic configuration.

Now test yourself

TESTED

10 State the number of protons and electrons present in a sulfide ion.
11 State the charge on an ion that has 23 protons and 20 electrons.
12 What is the electronic configuration of a N^{3-} ion?

Exam practice

1 What is meant by the term 'atomic number'? [1 mark]

2 State the relative mass and relative charge of:
(a) a proton
(b) an electron
(c) a neutron. [3 marks]

3 ^{35}Cl and ^{37}Cl are isotopes of chlorine. What is meant by the term 'isotope'? [2 marks]

4 Explain why atoms are electrically neutral. [1 mark]

5 Write electronic configurations for these atoms:
(a) P
(b) Li
(c) O
(d) K
(e) Ar
(f) He
(g) Al
(h) Na [8 marks]

6 Write electronic configurations for these ions:
(a) Na^+
(b) F^-
(c) Al^{3+}
(d) O^{2-}
(e) K^+ [5 marks]

7 Consider the following ions.
Cl^- Al^{3+} Li^+ O^{2-} F^- H^-
(a) Which ions have an electronic configuration that is the same as a Ne atom? [1 mark]
(b) Which ion contains 13 protons? [1 mark]
(c) Which ion has 18 electrons? [1 mark]
(d) What is the name of the H^- ion? [1 mark]

8 An ion has 15 protons, 18 electrons and 16 neutrons.
(a) State the charge on the ion. [1 mark]
(b) Name the element from which the ion is formed. [1 mark]
(c) Name the ion. [1 mark]

9 Draw a labelled diagram of a ^{40}Ca atom showing the number and position of all subatomic particles. [4 marks]

H 10 Iron has four isotopes, which are shown below with their percentage abundances:
^{54}Fe (5.8%) ^{56}Fe (91.8%) ^{57}Fe (2.2%) ^{58}Fe (0.2%)
Calculate the relative atomic mass of iron to 1 decimal place. [3 marks]

Answers online

ONLINE

15 Bonding, structures and nanoparticles

Bonding

Types of bonding

REVISED

The way in which molecules and structures are held together is called **bonding**. There are three main types of bonding — **ionic**, **covalent** and **metallic**.

- Ionic bonding occurs in compounds composed of a metal and a non-metal — such as sodium chloride and magnesium oxide. Ionic compounds contain ions.
- Covalent bonding occurs between non-metal atoms — this can be in compounds, such as hydrogen chloride and water, or in elements, such as chlorine (Cl_2) and carbon (graphite).
- Metallic bonding occurs in metals.
- All bonding involves electrons, and these are usually represented by dots (•) or crosses (×) in bonding diagrams — these are called **dot-and-cross diagrams**. Remember that all electrons are the same — using dots and crosses shows which atoms the electrons come from.

Formation of ionic compounds

REVISED

Ionic compounds are compounds that contain a metal — they are said to have ionic bonding. Examples are sodium chloride (NaCl), magnesium oxide (MgO), calcium chloride ($CaCl_2$), potassium oxide (K_2O) and lithium fluoride (LiF).

- Ionic compounds are made up of ions, which are **charged particles**.
- H An ionic compound contains **positive ions** (cations) and **negative ions** (anions).
- Simple ions are those formed when an atom of an element gains or loses electrons — for example, Na^+, O^{2-}, Al^{3+} and Br^-.
- Molecular ions are charged particles that contain more than one atom. The molecular ions required are given on the back of the *Data Leaflet* — they include the ammonium ion, NH_4^+, sulfate ion, SO_4^{2-}, hydroxide ion, OH^- and nitrate ion, NO_3^-.
- The hydroxide ion is the only molecular ion you need to know that ends in -ide. All other ions that end in -ide are simple negatively charged ions.

Exam tip

The ammonium ion (NH_4^+) is a positive ion and ammonium compounds such as ammonium chloride and ammonium sulfate are also ionic compounds, even though they do not contain a metal. All acids such as hydrochloric acid are covalent compounds, but when dissolved in water they produce ions.

Formation of ions from atoms

REVISED

When an ionic compound forms from the atoms of its elements, a **transfer of electrons** occurs. Metal atoms lose electrons and give them to non-metal atoms, which accept electrons. Each will lose or gain enough electrons to give it a full outer shell and make it more stable:

- When a metal atom loses electrons, it becomes a positively charged ion.
- When a non-metal atom gains electrons, it becomes a negatively charged ion.

You can be asked to show the formation of an ionic compound from the atoms of the elements from which it is composed. These compounds will contain either a group 1 or group 2 metal (Li, Na, K, Mg or Ca) with a group 6 or group 7 non-metal (O, S, F or Cl). Beryllium will not be used because its chemistry is unusual.

There are four possible combinations:

- a group 1 metal with group 7 non-metal — for example LiF, LiCl, NaF, NaCl, KF, KCl
- a group 2 metal with group 7 non-metal — for example MgF_2, $MgCl_2$, CaF_2, $CaCl_2$
- a group 1 metal with group 6 non-metal — for example Li_2O, Li_2S, Na_2O, Na_2S, K_2O, K_2S
- a group 2 metal with group 6 non-metal — for example MgO, MgS, CaO, CaS

If other examples appear in a question, it is simply a matter of changing the initial electronic configuration — the way in which each type of compound is formed is the same.

An **ionic bond** is an attraction between oppositely charged ions.

Example 1

Sodium chloride

- A sodium atom has an electronic configuration of 2,8,1.
- When it reacts with a chlorine atom (electronic configuration 2,8,7), the one outer electron of the sodium atom is transferred to the outer shell of the chlorine atom.
- Each sodium ion now has only 10 electrons (2,8), but it has 11 protons in its nucleus, so it has charge of 1+.
- The sodium atom is written as 'Na'; the sodium ion formed is written as 'Na^+'.
- Simple positive ions have the same name as the atom — so Na^+ is called a **sodium ion**.
- Each chloride ion now has 18 electrons (2,8,8) but it has 17 protons in its nucleus so it has charge of 1–.
- The chlorine atom is written as 'Cl'; the chloride ion formed is written as 'Cl^-'.
- Simple negative ions change the end of their name to -ide, so it is called a **chloride ion**.
- The sodium ions and chloride ions are attracted to each other and form an **ionic compound**.
- The **ionic bond** is the attraction between oppositely charged ions in an ionic compound.
- The dot-and-cross diagram below summarises the process of ion formation in sodium chloride.

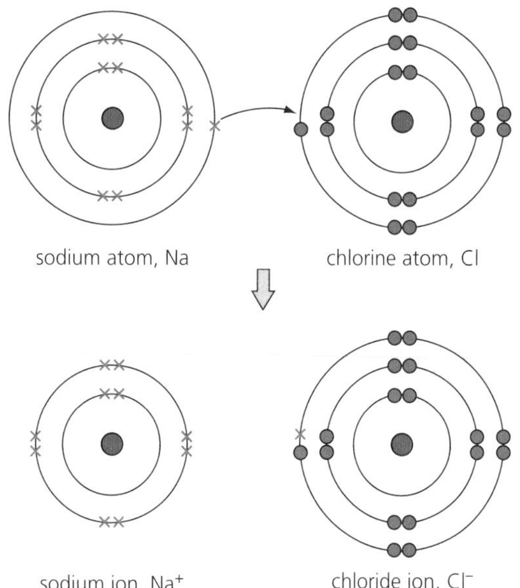

Exam tip

When showing the formation of an ionic compound, always use dots and crosses to make it clear where the transferred electrons are coming from and where they are going to. In this example you can see that a sodium electron ends up in a chloride ion. Make sure the charges on the ions are shown and that the numbers of each ion present match up.

Exam tip

- When an ion has a single positive charge, it is correct to write this as '+'. Do not write '1+' or '+1'. Similarly, when an ion has a single negative charge, write this as '–', not '1–' or '–1'.
- For ions with a higher charge, always write these in the same way as those on the back of the *Data Leaflet* — for example 2+ (not +2) and 3– (not –3).

The compound formed is called **sodium chloride** (because it contains sodium ions and chloride ions). There is a lot of information here that helps in working out the **formula** of the compound:

- Each sodium atom loses one electron and each chlorine atom gains one electron, so only one atom of each is required and the formula of sodium chloride is **NaCl**.
- Because a sodium ion is Na^+ and a chloride ion is Cl^-, only one ion of each is required, so the compound has no overall charge.

Exam tip

Remember that ionic compounds do not have a charge — but ions do have a charge.

Exam tip

Questions often ask about the electronic configurations of atoms and of ions. The electronic configuration of the Na^+ ion is 2,8 (see above), but this is the same as the electronic configuration of a Ne atom. The difference is the number of protons in the nucleus. The number of electrons can help to identify an unknown atom, but be careful if it is an ion because it will have lost or gained electrons. You can work out how many it has lost or gained from its charge.

Example 2

Calcium chloride

- Two chlorine atoms are required for each calcium atom because each calcium atom loses two electrons and each chlorine atom gains one electron:

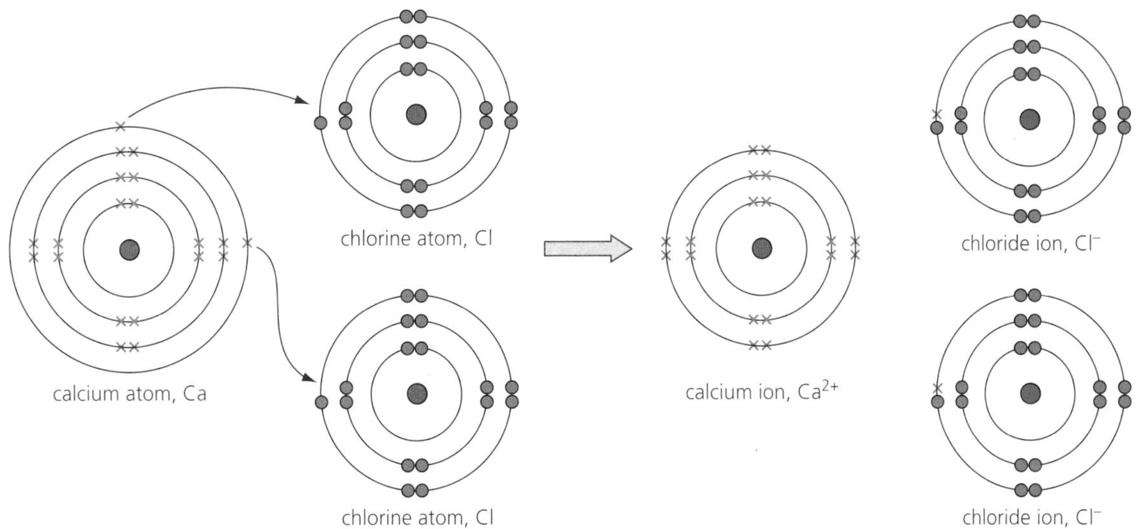

- A calcium ion is Ca^{2+} because it has only 18 electrons but 20 protons (atomic number = 20).
- Chloride ions are Cl^-, as described earlier.
- The formula of calcium chloride is $\mathbf{CaCl_2}$.

Example 3

Magnesium oxide

- Only one magnesium atom is required for each oxygen atom because each magnesium atom loses two electrons and each oxygen atom gains two electrons:

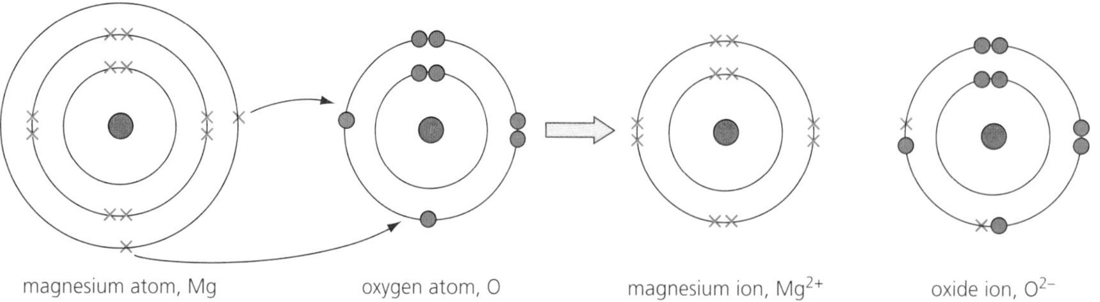

- A magnesium ion is Mg^{2+} because it has only 10 electrons (2,8) but 12 protons (atomic number = 12).
- An oxide ion is O^{2-} because it has 10 electrons (2,8) but only eight protons (atomic number = 8).
- The formula of magnesium oxide is **MgO**.

Example 4

Potassium sulfide

- Two potassium atoms are needed for each sulfur atom because each potassium atom loses one electron and each sulfur atom gains two electrons:

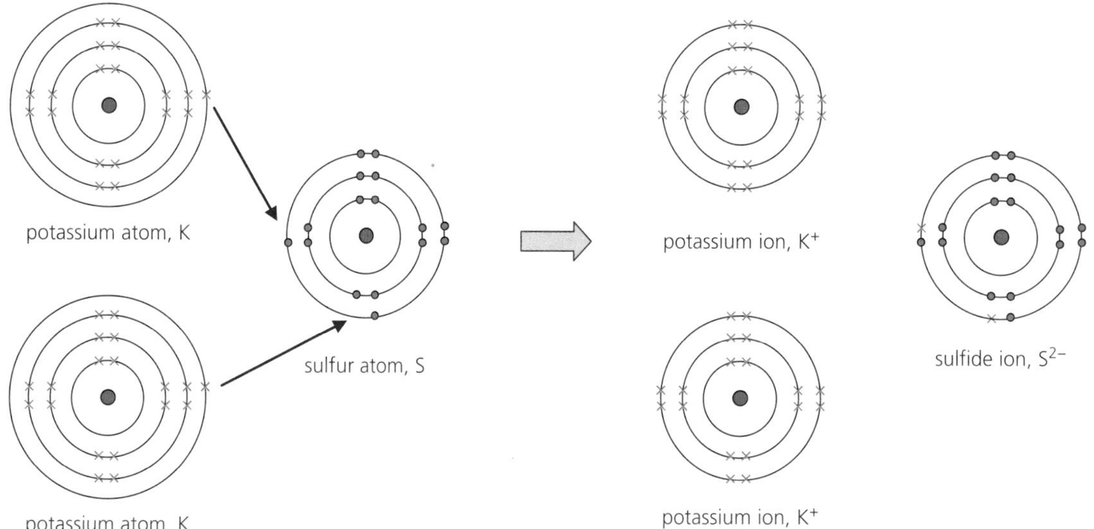

- A potassium ion is K^+ because it has only 18 electrons (2,8,8) but 19 protons (atomic number = 19).
- A sulfide ion is S^{2-} because it has 18 electrons (2,8,8) but only 16 protons (atomic number = 16).
- The formula of potassium sulfide is $\mathbf{K_2S}$.

Exam tip

The charge on a simple ion can be checked by looking at the group number of the element:

Group 1 elements form ions with a + charge.

Group 2 elements form 2+ ions.

Group 3 elements form 3+ ions.

Group 4 elements do not generally form ions.

Group 5 elements form ions with a 3– charge.

Group 6 elements form ions with a 2– charge.

Group 7 elements form ions with a – charge.

Group 0 elements do not form ions because they have full outer shells.

Exam tip

If you are asked to show all the subatomic particles in an atom or ion, label the nucleus and write how many protons and neutrons are present. Then add the shells showing the electronic configuration.

Now test yourself

TESTED

1. What are the names of the two ions in NaCl?
2. Write the formula of the compound formed between Li^+ and S^{2-} ions.
3. Write the electronic configuration of the atoms and ions formed when calcium atoms react with fluorine atoms. Write the charges on the ions.

Ionic bonding and structure

REVISED

- An ionic bond is the attraction between oppositely charged ions.
- Ionic compounds have a structure described as an ionic lattice (Figure 15.1).

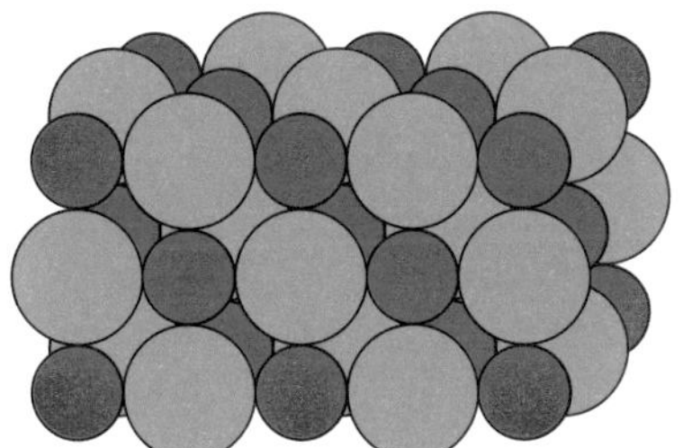

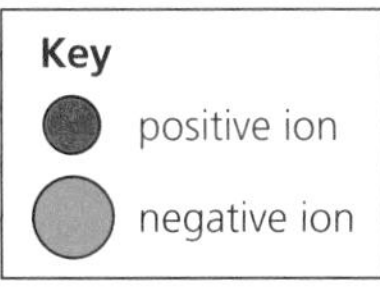

Figure 15.1 An ionic lattice structure

- The ionic lattice is a three-dimensional network of positive ions and negative ions held together by the ionic bonds (electrostatic forces) between the oppositely charged ions.

Properties of ionic compounds

REVISED

Ionic bonds are strong and it is these bonds that give ionic compounds their properties.

- In their normal, solid state, ionic compounds cannot conduct electricity because the charged particles (the ions) cannot move and carry the charge.

- When an ionic compound is melted (it is described as **molten**) the ions can move and carry charge, so molten ionic compounds conduct electricity.
- Most ionic compounds are soluble in water.
- When an ionic compound dissolves in water (it is described as **aqueous**) the ions can move and carry charge, so aqueous ionic compounds conduct electricity.
- Ionic compounds have high melting points and boiling points and are all solids at room temperature and pressure because substantial energy is required to break the strong ionic bonds between the ions.
- Ionic compounds are brittle because, when struck, the layers of ions move and an ion is then beside an ion of similar charge. The similar charge ions repel each other, which shatters the crystal structure.

Now test yourself

TESTED

4 State the structure and bonding present in potassium chloride.
5 What is meant by an ionic bond?
6 Explain why a molten ionic compound can conduct electricity.

Exam tip

Learn the properties of ionic compounds carefully. A question can ask you to identify the type of an unknown compound just from its properties. If it has a high melting point and does not conduct electricity when solid, but does when molten or dissolved in water, then it is an ionic compound.

Covalent bonding

REVISED

- A **covalent bond** is formed between non-metal atoms. It typically occurs in substances that only contain non-metallic elements.
- Two or more atoms covalently bonded together is a **molecule**.
- A single **covalent bond** is formed by a shared pair of electrons.
- There are several non-metallic elements that are **diatomic**, which means that there are two atoms covalently bonded in a molecule. The element should be written with a small '$_2$' after the symbol — for example, H_2.
- The diatomic elements are H_2, N_2, O_2, F_2, Cl_2, Br_2, I_2 (and At_2). So, the symbol for chlorine is written 'Cl', but the element chlorine should be written 'Cl_2'.
- Hydrogen molecules contain a single covalent bond between the hydrogen atoms, involving a shared pair of electrons. One electron comes from the H atom on the left and the other from the H atom on the right (Figure 15.2).
- The covalent bond in hydrogen molecules can be shown as 'H–H', where the dash represents the single covalent bond.
- If two pairs of electrons are shared between two non-metal atoms, this is a double **covalent bond**. This is represented by a double dash, for example O=O.
- A triple covalent bond (three shared pairs of electrons) is shown as a triple dash between the atoms, for example N≡N.
- Non-metal atoms share electrons to complete their outer shells.
  Note: it is important to understand that shared electrons count as outer shell electrons for *both* atoms.

A **covalent bond** is formed by a shared pair of electrons.

Diatomic means that there are two atoms covalently bonded in a molecule.

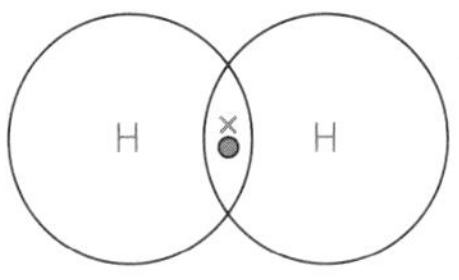

Figure 15.2 Hydrogen, a diatomic element

Figure 15.3 shows some dot-and-cross diagrams for molecules with only single covalent bonds.

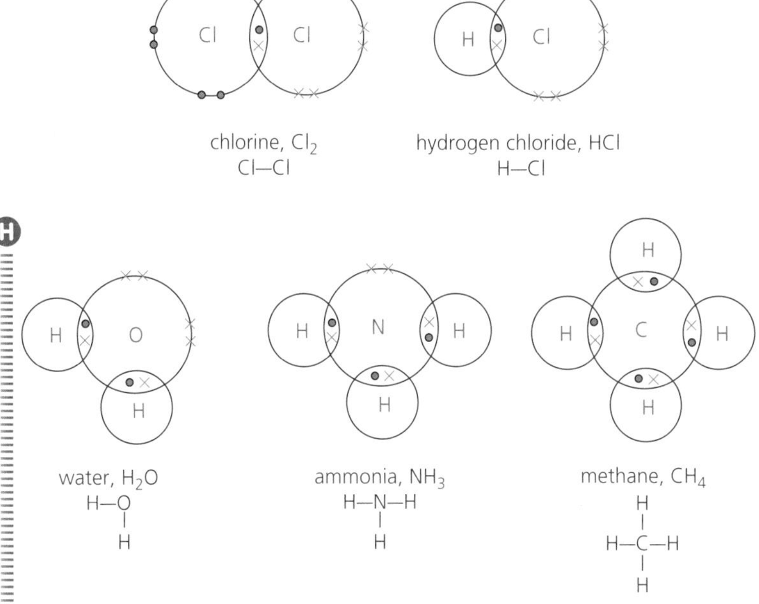

Figure 15.3 Substances with single covalent bonds

A pair of electrons that is part of a covalent bond is called a **bonding pair** of electrons (Figure 15.4). A pair of electrons that is not involved in a bond is called a **lone pair** of electrons. You need to be able to label lone pairs of electrons.

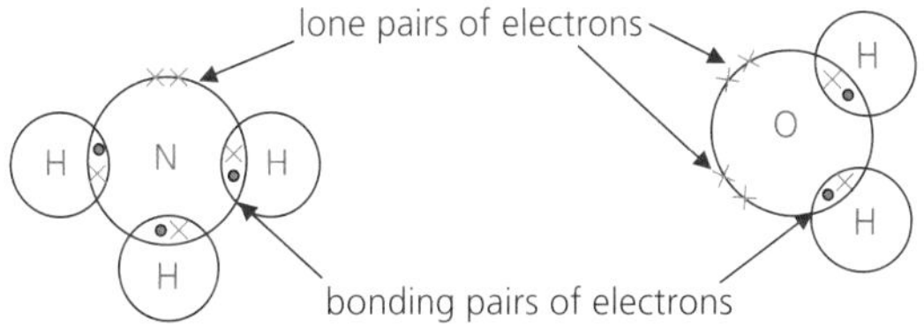

Figure 15.4 Lone pairs and bonding pairs of electrons

The dot-and-cross diagrams in Figure 15.5 show molecules with multiple covalent bonds.

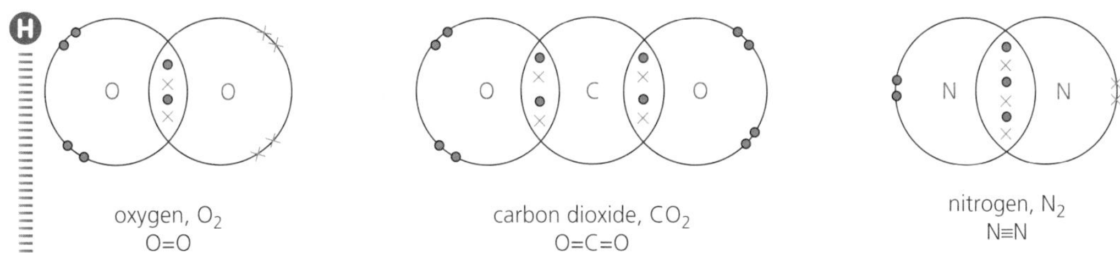

Figure 15.5 Substances with multiple covalent bonds

- All of the molecules shown are called molecular covalent molecules and their structure is described as **molecular** or **simple**.
- Chlorine and oxygen are examples of molecular covalent elements — they are **diatomic**.
- Water and methane are examples of molecular covalent compounds.

Exam tip

Diagrams of covalent bonding need to show only the outer shell electrons. Do not show the inner shells of electrons — you would lose marks if these were incorrect. Make sure you pair up the electrons in the lone pairs and also draw a proper dot-and-cross diagram. Any shared pairs of electrons at GCSE are shown as ×●.

Properties of molecular covalent substances

REVISED

Covalent bonds are strong and require substantial energy to break. However, molecular covalent substances have weak forces of attraction between their molecules. These weak forces are called van der Waals' forces of attraction. The physical properties of molecular covalent substances depend on the weak van der Waals' forces of attraction. The physical properties of molecular covalent substances are as follows:

- They have low melting and boiling points (and so are gases, liquids or low-melting-point solids at room temperature) because little energy is required to break the weak van der Waals' forces of attraction between the molecules.
- They do not conduct electricity because they have no charged particles (electrons or ions) to move and carry charge.
- They are mostly insoluble in water or have a low solubility in water. (There are exceptions to this — for example, hydrogen chloride and ammonia are soluble in water because they react with it.)
- Carbon dioxide and iodine are typical examples of molecular covalent substances.

Now test yourself

TESTED

7 What is a covalent bond?
8 How many lone pairs are present in a molecule of (a) water, (b) methane, (c) ammonia?
9 What name is given to the weak forces of attraction between simple covalent molecules?

H Metallic bonding

REVISED

The bonding in a metal is called **metallic bonding** and is the attraction between delocalised electrons and the positive ions in a regular lattice (Figure 15.6). The ions are arranged in layers, as shown, with the delocalised electrons moving between the layers.

Metallic bonding is the attraction between delocalised electrons and the positive ions in a regular lattice.

Properties of metals

The structure and bonding of metals are used to explain their properties:

- Metals (mostly) have high melting points and are normally solids (except mercury) because substantial energy is needed to break the strong metallic bonds between the positive ions and the delocalised electrons.
- Metals conduct electricity because the delocalised electrons can move and carry charge.
- When metals are molten they will still conduct electricity because the electrons are still delocalised and can move and carry charge.
- Metals are **malleable** (can be hammered into shape) and **ductile** (can be drawn out into wires) because the layers can slide over each other without disrupting the bonding.
- An **alloy** is a mixture of two or more elements, at least one of which is a metal. An alloy will have metallic properties but the properties are often 'better' than the original metal(s) from which it was formed — for example, stainless steel is an alloy of iron with chromium and it does not rust easily, unlike iron.

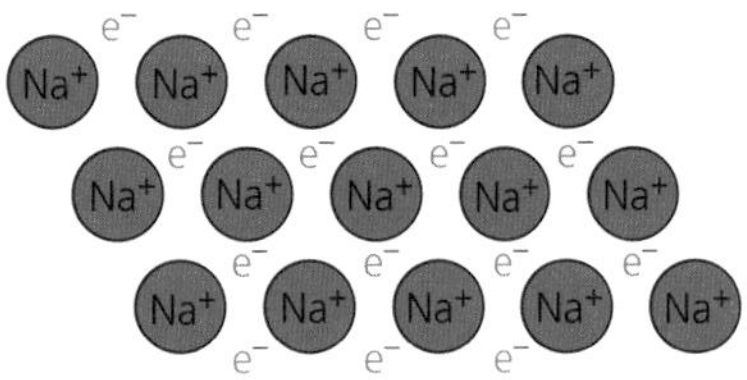

Figure 15.6 Metallic bonding

Now test yourself

TESTED

10 Explain why metals like copper are ductile.
11 What is an alloy?

Giant covalent structures

REVISED

Some covalently bonded substances have **giant structures** (Figure 15.7).

- Carbon (diamond) and carbon (graphite) have giant covalent structures.
- Graphene is a single sheet of carbon atoms from graphite.
- Diamond, graphite and graphene are forms of the element carbon.
- Carbon atoms can form four covalent bonds as they have four unpaired electrons in their outer shell.
- Different forms of the same element in the same physical state are called **allotropes**.
- Diamond, graphite and graphene are allotropes of carbon.

An **allotrope** is a different form of the same element in the same physical state.

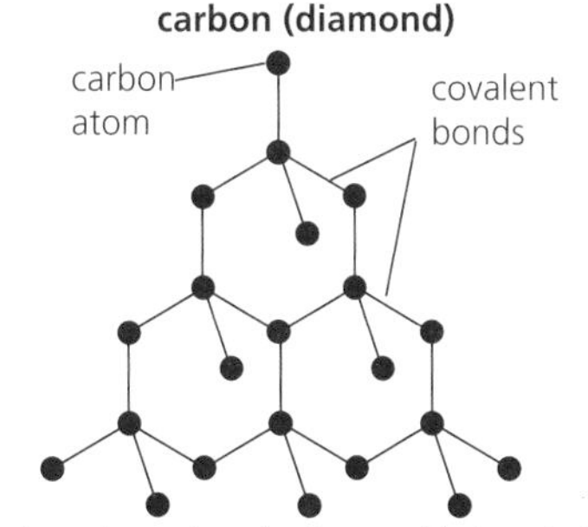

Each carbon atom in diamond is bonded to four other carbon atoms by strong covalent bonds. The basic arrangement of bonds is called **tetrahedral**.

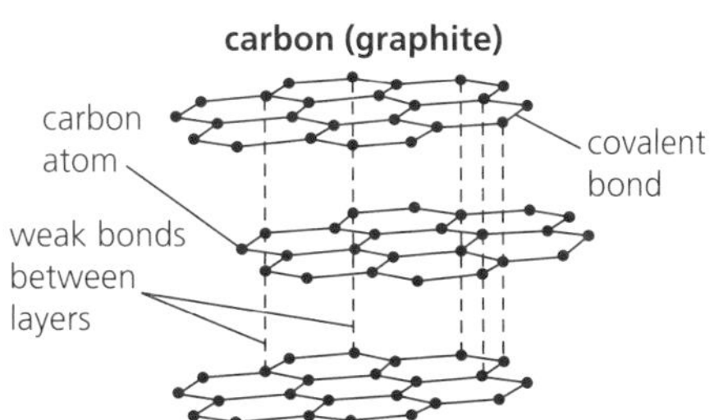

Each carbon atom in graphite is bonded to three other carbon atoms by strong covalent bonds. The atoms are bonded in layers with weak bonds between the layers.The weak bonds between the layers are caused by **delocalised** electrons.

Figure 15.7 The giant covalent structures of the allotropes diamond and graphite

Properties of giant covalent substances

Covalent bonds are strong and substantial energy is needed to break them. The large number of strong covalent bonds in giant covalent substances can be used to explain their properties:

- Both diamond and graphite have a very high melting point (and boiling point) and are solids at room temperature and pressure because substantial energy is needed to break the strong covalent bonds.
- Graphite conducts electricity because it has delocalised electrons in its structure — these can move and carry charge. The structure of graphite is mostly disrupted when it is molten and it can no longer conduct electricity as effectively because there are fewer delocalised electrons to move and carry charge.
- Diamond does not conduct electricity because it does not have delocalised electrons.
- Graphite can act a lubricant because the layers can slide over each other. This also explains its use in pencils — layers slide off when the pencil is moved against paper.
- Both diamond and graphite are insoluble in water.
- Diamond is the hardest naturally occurring substance on Earth. This is caused by the strong covalent bonds throughout the rigid, tetrahedral, three-dimensional, giant covalent structure.

- Because it is so hard, diamond is used in cutting tools (like drill tips and saws) and can drill through rock and cut through metals.

Graphene

Graphene is a single-atom-thick layer of graphite (Figure 15.8).

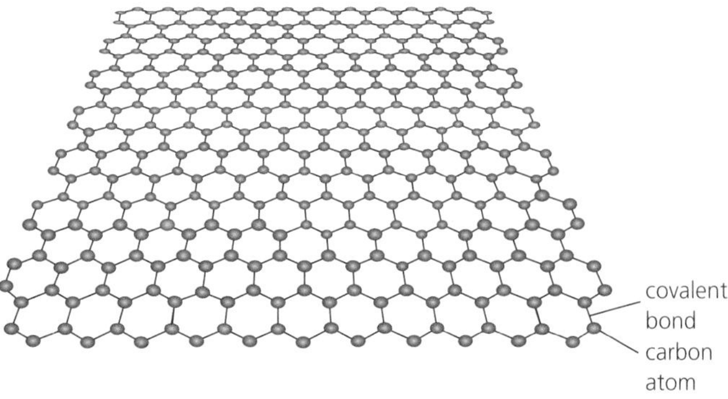

Figure 15.8 Graphene

The physical properties of graphene are as follows:

- It has a low density because it is only made up of a single layer of carbon atoms.
- It is 100 times stronger than steel due to the rigid structure of hexagons and strong covalent bonds.
- It is flexible and can stretch by up to 25% of its original length because the flat layer of carbon atoms can bend without breaking the covalent bonds.
- It is an excellent conductor of heat (better than any metal).
- It is an excellent conductor of electricity because each carbon atom is covalently bonded to three others, meaning that each carbon atom has an unbonded electron that is delocalised throughout the structure. These delocalised electrons can move and carry charge.
- Graphene is transparent because it is so thin. It transmits around 98% light through it, compared with a single sheet of glass, which allows transmisson of around 85%.

Uses of graphene

Graphene is an excellent conductor of electricity and has a low density, so is being used to replace traditional carbon in batteries, making them much lighter and smaller and also faster to recharge. Much less graphene is needed compared with carbon.

Graphene is also being used in solar cells because it is transparent, strong, conducts electricity and also makes the solar cells lighter.

Now test yourself

TESTED

12 Name three allotropes of carbon.
13 Explain why graphite conducts electricity.
14 State two properties of graphene.

Summary of bonding and structures

REVISED

It is important to understand the structure of a substance as well as its bonding.

- Bonding and structure are *not* the same. A substance can be covalently bonded but it may have a molecular structure or a giant structure.

- Structure and bonding are used to explain many of the properties of substances.
- Table 15.1 shows the four main types of structure, with information on their physical properties.

Table 15.1

Type of structure	Metals	Ionic compounds	Molecular (simple) covalent	Giant covalent
Common examples	Magnesium Sodium	Sodium chloride Magnesium oxide	Carbon dioxide Iodine	Diamond Graphite Graphene
Bonding	Metallic	Ionic	Covalent within the molecules and van der Waals' forces of attraction between the molecules	Covalent (graphite has weak bonds between the layers in its structure)
Electrical conductivity	Conducts electricity when solid and molten	Does not conduct electricity when solid but does conduct electricity when molten or dissolved in water	Does not conduct electricity	Diamond does not conduct electricity but graphite and graphene do conduct electricity
Melting point	Mostly high	High	Low	High
Solubility in water	Insoluble in water (some metals will react with water)	Mostly soluble in water	Mostly insoluble in water (some dissolve and some react with water)	Insoluble in water

Exam tip

Get familiar with the properties of ionic compounds, molecular covalent substances, giant covalent substances and metals — particularly to do with melting point (and boiling point) and whether or not they conduct electricity. This information can be vital in identifying the type of an unknown substance.

Worked example 1

Explain why metals such as magnesium are malleable. [2 marks]

Answer

Layers of positive ions can slide over each other [1] without disrupting the bonding [1].

H Worked example 2

The table below gives information on four substances — A, B, C and D.

Substance	Melting point/°C	Electrical conductivity of solid substance	Electrical conductivity of molten substance
A	2800	Poor	Good
B	−7	Poor	Poor
C	660	Good	Good
D	3500	Good	Poor

Which substance is likely to be:

(a) a metal [1 mark]

(b) an ionic compound [1 mark]

(c) graphite [1 mark]

(d) a molecular covalent substance? [1 mark]

Answer

(a) C [1]

(b) A [1]

(c) D [1]

(d) B [1]

Exam tip

Classification of structures is a common question, so it is vital to remember the properties of all four types of substance. If you cannot recall them, look again at pages 100–105.

Nanoparticles

- A nanometre (nm) is a unit of length equal to 10^{-9} m.
- 1 million nanometres make up a millimetre (1 mm = 1 000 000 nm) or 1000 million nm make up 1 metre (10^9 nm = 1 m).
- Nanoscience, or nanotechnology, uses particles with sizes in the range 1–100 nm, which consist of only a few hundred atoms. For comparison, a human hair is 80 000 nm wide.

Uses and risks of nanoparticles

REVISED

Nanoparticles have been used in modern suncreams. Suncreams work by absorbing some of the ultraviolet radiation from the Sun, which can cause skin cancer. With the depletion of the ozone layer over the last few decades, the number of skin cancer cases has been increasing. Older suncreams used bulk zinc oxide and titanium oxide, which left a white residue on the skin. The most recent sun protection creams use zinc oxide and titanium oxide **nanoparticles**.

The benefits of the nanoparticle suncreams are that they:

- rub on as a clear film, which creates a clear surface on the skin
- provide better skin coverage

- give more effective protection against the ultraviolet radiation from the Sun
- do not break down on exposure to the Sun, which is a problem with bulk particles.

These properties are due to the larger surface area-to-volume ratio of nanoparticles.

Risks of nanoparticles

- Breathing in nanoparticles may cause inflammation of the lungs, and the nanoparticles may be absorbed into the bloodstream.
- Nanoparticles in the body could cause cell damage.
- Nanoparticles from suncreams wash off and enter the water environment. They could also have harmful effects on the organisms that live in water or organisms that consume the water.

Most of the risk is centred around the unknown. More research is needed because nanoparticles have only been used in recent years.

Now test yourself

TESTED

15 State one use of nanoparticles.
16 State one risk from the use of nanoparticles.

Exam practice

1 What is an alloy? [2 marks]
2 State two reasons why graphene is used in rechargeable batteries in place of carbon. [2 marks]
3 What is an ionic bond? [2 marks]
4 State the type of bonding found in the following:
(a) potassium fluoride [1 mark]
(b) carbon dioxide [1 mark]
(c) iron [1 mark]
5 The diagram below shows the bonding in ammonia.

ammonia, NH_3

(a) On the diagram label a lone pair of electrons. [1 mark]
(b) State the bonding present in a molecule of ammonia. [1 mark]
(c) Name the forces of attraction that are present between ammonia molecules. [1 mark]
(d) Write the chemical formula for ammonia. [1 mark]
6 Explain how a single covalent bond is formed. [2 marks]
7 Draw a dot-and-cross diagram to show the bonding in the following molecules. Only outer-shell electrons need to be shown:
(a) Cl_2 [1 mark]
H (b) H_2O [1 mark]
(c) CO_2 [1 mark]
8 Chlorine boils at –34°C. Explain why chlorine has a low boiling point. [3 marks]
9 Explain the bonding and structure in diamond. [3 marks]
10 Explain why graphite conducts electricity. [2 marks]
H 11 Describe the structure and bonding in a metal like sodium. [4 marks]
12 From the following list of substances:

ammonia	sodium	graphene	potassium iodide	sodium chloride
diamond	lithium chloride	iron	hydrogen	carbon dioxide

(a) Which substances are metals? [1 mark]
(b) Which substances conduct electricity at room temperature? [1 mark]
(c) Which substances have molecular covalent structures? [1 mark]
13 Sodium chloride has a melting point of 797°C. It will not conduct electricity in the solid form, but will when molten.
(a) Explain why sodium chloride solution conducts electricity, but solid sodium chloride does not. [3 marks]
(b) State the type of structure shown by sodium chloride. [1 mark]
(c) Explain why sodium chloride has a high melting point. [2 marks]

Answers online

ONLINE

16 Symbols, formulae and equations

Valencies

Valency and group number

REVISED

To be able to work out the formula of a compound, you need to use valencies. 'Valency' is the combining power of an element or an ion. It can be worked out from the group number of the elements, as shown in Table 16.1.

Table 16.1

Group	Valency
1	1
2	2
3	3
4	4
5	3
6	2
7	1
0	None

Note that elements in group 0 do not form compounds, so they have no valency. Hydrogen always has a valency of 1.

The valencies of molecular ions

REVISED

The way in which atoms become ions was covered earlier. Molecular ions have valencies as shown in Table 16.2. These are the same as the charge on the molecular ion.

Table 16.2

Name	Formula	Valency
Ammonium	NH_4^+	1
Butanoate	$C_3H_7COO^-$	1
Carbonate	CO_3^{2-}	2
Dichromate	$Cr_2O_7^{2-}$	2
Ethanoate	CH_3COO^-	1
Hydrogencarbonate	HCO_3^-	1
Hydroxide	OH^-	1
Methanoate	$HCOO^-$	1
Propanoate	$C_2H_5COO^-$	1
Nitrate	NO_3^-	1
Sulfate	SO_4^{2-}	2
Sulfite	SO_3^{2-}	2

Exam tip

For unfamiliar ions, the valency is the same as the charge. For example, the thiosulfate ion, $S_2O_3^{2-}$, has a valency of 2; the Cu^+ ion has a valency of 1.

The valency of a transition metal (elements in the middle block of the periodic table (page 123) is given in roman numerals after the name of the element.

The valencies of transition elements

REVISED

The valencies of transition elements are given in Table 16.3.

Table 16.3

Transition metal	Valency of metal
Iron(II) compounds	2
Iron(III) compounds	3
Copper(II) compounds	2
Cobalt(II) compounds	2
Silver(I) compounds	1

Exam tip

Transition metal ions without a roman numeral in their name usually have a valency of 2. The zinc in 'zinc chloride' has a valency of 2. This is important because some exam questions do not give the roman numeral. The only exception to this in GCSE is silver compounds, in which the silver has a valency of 1. Often silver compounds are written without the (I). Silver can have other valencies but (I) is the only valency used in GCSE chemistry.

Formulae

How to work out the formula of a compound

REVISED

The formula of a compound can be determined using valencies by the following method.

1	Compound name	potassium	chloride
2	Convert to symbols	K	Cl
3	Look up the valencies	1 (in group 1)	1 (in group 7)
4	Write the valencies above the symbols	$\overset{1}{K}$	$\overset{1}{Cl}$
5	Cancel down if necessary	(no need here)	
6	Cross over	$\overset{1}{K}$ ⤨ $\overset{1}{Cl}$	
7	Write with crossed-over numbers (Use brackets here if a molecular ion is multiplied by 2 or more.)	K_1Cl_1	
8	Ignore the '1's (so K_1 = K)	KCl	

Exam tip

The cancel down step is the most often forgotten step.

Example 1

Use the method shown above to work out the formulae of these compounds:

(a) magnesium chloride
(b) calcium chloride
(c) iron(III) chloride
(d) calcium hydroxide

Answer

(a) Magnesium chloride	Mg	Cl
Valencies	2	1
No cancel down	2	1
Cross-over	Mg_1Cl_2	
Formula	$MgCl_2$	
(b) Calcium oxide	Ca	O
Valencies	2	2
Cancel down	1	1
Cross-over	Ca_1O_1	
Formula	CaO	
(c) Iron(III) chloride	Fe	Cl
Valencies	3	1
No cancel down	3	1
Cross-over	Fe_1Cl_3	
Formula	$FeCl_3$	
(d) Calcium hydroxide	Ca	OH
Valencies	2	1
No cancel down	2	1
Cross-over	$Ca_1(OH)_2$	
Formula	$Ca(OH)_2$	

Exam tip

For ionic compounds like those in Example 1, the formulae of the compounds can be worked out using valency or the charges on the ions. The compounds do not have a charge so the total positive charge must cancel out the total negative charge. For (a), magnesium chloride, the magnesium ion is Mg^{2+} and the chloride ion is Cl^- so $2Cl^-$ are needed to cancel out the 2+ charge, so the formula is $MgCl_2$. For (b), calcium oxide, the calcium ion is Ca^{2+} and the oxide ion is O^{2-} so one of each is needed and calcium oxide is CaO.

Covalent compounds are compounds containing only non-metals. So we cannot use ions to determine the formulae of covalent compounds, but the rules of valency still apply.

Exam tip

The formulae of some compounds do not need working out. For example, carbon dioxide is CO_2 — the 'di' indicates two oxygen atoms in the compound. Sulfur trioxide is SO_3. As you work through the examples, you will get quicker at working out formulae and it will become like a new language to you — the language of chemistry!

Example 2

Use the valency method to work out the formulae of the following compounds:

(a) Hydrogen chloride
(b) Methane
(c) Ammonia
(d) Water

Answer

(a) Hydrogen chloride	H	Cl
Valencies	1	1
No cancel down	1	1
Cross-over	H_1Cl_1	
Formula	HCl	
(b) Methane (C and H)	C	H
Valencies	4	1
No cancel down	4	1
Cross-over	C_1H_4	
Formula	CH_4	
(c) Ammonia (N and H)	N	H
Valencies	3	1
No cancel down	3	1
Cross-over	N_1H_3	
Formula	NH_3	
(d) Water (H and O)	H	O
Valencies	1	2
No cancel down	1	2
Cross-over	H_2O_1	
Formula	H_2O	

Exam tip

Some formulae have to be learned — such as ammonia (NH_3), water (H_2O), organic compounds like methane (CH_4) and acids such as sulfuric acid (H_2SO_4), hydrochloric acid (HCl) and nitric acid (HNO_3).

Remember that ammonia is NH_3, but ammonium is NH_4^+.

Now test yourself

TESTED

1 Write the formula of chromium(III) nitrate.
2 What is the name of $Fe_2(SO_4)_3$?
3 The iodate ion is IO_3^-. Write the formula for magnesium iodate.

Atoms in formulae

REVISED

Questions can be asked about the total number of atoms in a given formula or the total number of a specific type of atom.

Example 1

How many atoms are there in one molecule of sulfuric acid, H_2SO_4?

Answer

There are two hydrogen atoms, one sulfur atom and four oxygen atoms — so there are seven atoms in one molecule.

Exam tip

A question could be asked about an unusual molecule — but just add up all the atoms needed.

Example 2

How many hydrogen atoms are there in one molecule of $CH_3CH_2CH_2OH$?

Answer

The answer is simply the total of all the hydrogen atoms in one molecule — so the answer is eight atoms.

Equations

Word equations and balanced symbol equations

REVISED

- Word equations are a long way of representing reactions.
- Balanced symbol equations are a more convenient method of doing this.
- Equations have to be balanced because no atoms are lost or gained during a chemical reaction — the atoms are just reorganised. There should always be the same number of each type of atom when an equation is balanced.

Exam tip

You may be given a word equation and asked to write a balanced symbol equation for the reaction. Write the formulae of all the reactants and products, and then balance the equation.

Example 1

Write the balanced symbol equation for the reaction of sodium hydroxide with hydrochloric acid.

Answer

1 Write the word equation for the reaction:

sodium hydroxide(aq) + hydrochloric acid(aq) → sodium chloride(aq) + water(l)

2 Write the formulae of the substances that react together and the substances that are produced:
Reactants: sodium hydroxide, NaOH(aq), and hydrochloric acid, HCl(aq)
Products: sodium chloride, NaCl(aq), and water, $H_2O(l)$

3 Substitute the formulae into the word equation to give you the initial symbol equation:

$NaOH(aq) + HCl(aq) \rightarrow NaCl(aq) + H_2O(l)$

→

Exam tip

Sometimes state symbols are given to show the physical state of each reactant and each product. The symbols below are used to indicate this:

- (aq) means aqueous (dissolved in water)
- (l) means liquid
- (s) means solid
- (g) means gas

They should be written in brackets immediately after the chemical, with no space.

4 It is important to balance all symbol equations so that there are the same numbers of each type of atom on both sides of the equation arrow. In the above equation:

Atoms involved	Reactant side (left-hand side)	Product side (right-hand side)
Na	1	1
O	1	1
H	2	2
Cl	1	1

You can see that there is the same number of each type of atom on both sides of the equation. This symbol equation is already balanced. So the balanced symbol equation is:

$NaOH(aq) + HCl(aq) \rightarrow NaCl(aq) + H_2O(l)$

Note: The state symbols are not always necessary and you should only include them when asked to do so in a question. This equation can be written:

$NaOH + HCl \rightarrow NaCl + H_2O$

Exam tip

Be careful to use an arrow and not an equal sign (=) in your equation. This is a chemical equation, not a mathematical one!

Example 2

Write the balanced symbol equation for the reaction of magnesium with hydrochloric acid.

Answer

1 Write the word equation for the reaction:

magnesium + hydrochloric acid → magnesium chloride + hydrogen

2 Write the formulae of the reactants and products:
Reactants: magnesium, Mg, and hydrochloric acid, HCl
Products: magnesium chloride, $MgCl_2$, and hydrogen, H_2

3 Write the initial symbol equation:

$Mg + HCl \rightarrow MgCl_2 + H_2$

Atoms involved	Reactant side (left-hand side)	Product side (right-hand side)
Mg	1	1
H	1	2
Cl	1	2

It can be seen clearly that the above symbol equation is not balanced, but it is important to remember that none of the formulae can be changed to balance an equation.

To balance an equation, balancing numbers are written in front of specific formulae and the whole formula then becomes multiplied by this number. For example, '2HCl' balances the above equation because it gives us two H atoms and two Cl atoms on the left-hand side. So the balanced symbol equation is:

$Mg + 2HCl \rightarrow MgCl_2 + H_2$

Example 3

Write the balanced symbol equation for the reaction of calcium hydroxide with nitric acid.

Answer

1 Write the word equation for the reaction:

calcium hydroxide + nitric acid → calcium nitrate + water

2 Use the formulae in the initial symbol equation:

$Ca(OH)_2 + HNO_3 \rightarrow Ca(NO_3)_2 + H_2O$

3 For balancing purposes, the nitrate, NO_3, stays intact (i.e. does not break up) in the reaction and so can be considered as a unit. (If a molecular ion breaks up then its atoms must be considered separately.)

Atoms involved	Reactant side (left-hand side)	Product side (right-hand side)
Ca	1	1
O (ignoring the 'O' in nitrate	2	1
H	3	2
NO_3	1	2

This needs balancing:

- We need two 'NO_3' on the left-hand side, so '$2HNO_3$' is needed on the left.
- Now there are four Hs on the left-hand side, but only two on the right-hand side, so '$2H_2O$' is needed on the right.
- Oxygen is now balanced.

So the balanced symbol equation is:

$Ca(OH)_2 + 2HNO_3 \rightarrow Ca(NO_3)_2 + 2H_2O$

Exam tip

Watch hydroxides — many errors occur with calcium hydroxide and copper(II) hydroxide, where the brackets are left out. Calcium hydroxide is $Ca(OH)_2$ and **not** $CaOH_2$.

Example 4

Write the balanced symbol equation for the reaction of calcium carbonate with hydrochloric acid.

Answer

1 Write the word equation for the reaction:

calcium carbonate + hydrochloric acid → calcium chloride + carbon dioxide + water

2 Use the formulae in the initial symbol equation:

$CaCO_3 + HCl \rightarrow CaCl_2 + CO_2 + H_2O$

Atoms involved	Reactant side (left-hand side)	Product side (right-hand side)
Ca	1	1
C	1	1
O	3	3
H	1	2
Cl	1	2

→

It can be seen that there are two H atoms and two Cl atoms on the right-hand side, while the left-hand side has only one of each. So, '2HCl' is needed to balance.

So the balanced symbol equation is:

$$CaCO_3 + 2HCl \rightarrow CaCl_2 + CO_2 + H_2O$$

Example 5

Write the balanced symbol equation for the reaction of sodium with water.

Answer

1 Write the word equation for the reaction:

sodium + water → sodium hydroxide + hydrogen

2 Use the formulae in the initial symbol equation:

$$Na + H_2O \rightarrow NaOH + H_2$$

Atoms involved	Reactant side (left-hand side)	Product side (right-hand side)
Na	1	1
H	2	3
O	1	1

There are three H on the right-hand side and only two H on the left, so we need '2NaOH' to give four H on right-hand side. Then '$2H_2O$' on the left-hand side to make four H on both sides. Finally, '2Na' on the left-hand side to balance the two Na now present on the right-hand side. So the balanced symbol equation is:

$$2Na + 2H_2O \rightarrow 2NaOH + H_2$$

Example 6

Write the balanced symbol equation for the following reaction of ethene with oxygen, forming carbon dioxide and water (ethene is C_2H_4).

Answer

1 Write the word equation for the reaction:

ethene + oxygen → carbon dioxide + water

2 Use the formulae in the initial symbol equation:

$$C_2H_4 + O_2 \rightarrow CO_2 + H_2O$$

Atoms involved	Reactant side (left-hand side)	Product side (right-hand side)
C	2	1
H	4	2
O	2	3

There are two C on the left-hand side and one C on the right, so we need '$2CO_2$'. There are four H on the left-hand side and two H on the right, so we need '$2H_2O$' on the right-hand side. Finally we need '$3O_2$' on the left-hand side to balance oxygen. So the balanced symbol equation is:

$$C_2H_4 + 3O_2 \rightarrow 2CO_2 + 2H_2O$$

Exam tip

Balanced symbol equations can be worth 2 or 3 marks. If balancing numbers are needed, there are 3 marks; if balancing numbers are not needed, the equation is worth 2 marks. State symbols are worth an extra mark if they are asked for in the question.

Remember that diatomic elements, such as oxygen, should always be written as 'O_2' in a balanced symbol equation.

If a substance works out to have '½' as a balancing number, you can multiply all the balancing numbers by 2 to get whole numbers — but you do not have to.

Exam tip

Writing the incorrect formula for sodium sulfate or sodium carbonate is a common error. Sodium sulfate is Na_2SO_4 and sodium carbonate is Na_2CO_3. Sometimes the small 2 after the sodium gets left out, which loses a mark for the formula and the balancing in the equation.

Now test yourself

TESTED

4 Write a balanced symbol equation for the reaction of copper with oxygen.
5 Write a balanced symbol equation for the reaction of magnesium hydroxide with ammonium chloride, forming ammonia (NH_3), magnesium chloride and water.
6 Balance the following equation:

$$NH_3 + O_2 \rightarrow NO + H_2O$$

H Ionic equations

REVISED

Some chemical reactions, which involve ionic compounds, are actually reactions between some of the ions involved in the mixture. The balanced symbol equation can be rewritten as an ionic equation, leaving out the ions that do not take part in the reaction.

For each ionic substance in the reaction, write the ions present below it and how many of each ion is present. If an ion appears on both sides of the equation (in the same state), it should not be included in the ionic equation.

(Sometimes state symbols have to be included to show exactly which ions have changed and which have not.)

Example 1

Write the ionic equation for the reaction between hydrochloric acid and sodium hydroxide.

Answer

The balanced symbol equation is shown below, along with the particles (ions and any molecules) present below the equation:

HCl	+	$NaOH$	→	$NaCl$	+	H_2O
H^+		Na^+		Na^+		
Cl^-		OH^-		Cl^-		H_2O

The Na^+ and Cl^- ions are on both sides of the equation, in the same numbers, and so are not part of the ionic reaction. The ionic equation is:

$$H^+ + OH^- \rightarrow H_2O$$

Ions that do not take part in a reaction are called spectator ions.

Exam tip

This equation contains water as a molecule because ions can combine to form molecules. Be aware that ionic equations may *only* contain ions but some ions can combine to form molecules, or molecules can split to form ions.

H

Example 2

Write the ionic equation for the reaction between copper(II) sulfate solution and sodium hydroxide solution.

Answer

The balanced symbol equation is shown below, along with the particles (ions and any molecules) present below the equation:

$CuSO_4(aq)$	+	$2NaOH(aq)$	→	$Cu(OH)_2(s)$	+	$Na_2SO_4(aq)$
$Cu^{2+}(aq)$		$2Na^+(aq)$		$Cu^{2+}(s)$		$2Na^+(aq)$
$SO_4^{2-}(aq)$		$2OH^-(aq)$		$2OH^-(s)$		$SO_4^{2-}(aq)$

$2Na^+(aq)$ and $SO_4^{2-}(aq)$ are on both sides of the equation, so are not part of the ionic equation.

The ionic equation is:

$$Cu^{2+}(aq) + 2OH^-(aq) \longrightarrow Cu(OH)_2(s)$$

Example 3

Write the ionic equation for the reaction between potassium chloride solution and silver nitrate solution.

Answer

The balanced symbol equation is shown below, along with the particles (ions and any molecules) present below the equation:

$KCl(aq)$	+	$AgNO_3(aq)$	→	$AgCl(s)$	+	$KNO_3(aq)$
$K^+(aq)$		$Ag^+(aq)$		$Ag^+(s)$		$K^+(aq)$
$Cl^-(aq)$		$NO_3^-(aq)$		$Cl^-(s)$		$NO_3^-(aq)$

$K^+(aq)$ and $NO_3^-(aq)$ are on both sides of the equation, so are not part of the ionic equation.

The ionic equation is:

$$Cl^-(aq) + Ag^+(aq) \rightarrow AgCl(s)$$

Example 4

Write the ionic equation for the reaction between zinc and copper(II) sulfate solution to form zinc sulfate and copper.

Answer

The balanced symbol for the reaction is shown below, along with the particles (ions and molecules) present below the equation:

$Zn(s)$	+	$CuSO_4(aq)$	→	$ZnSO_4(aq)$	+	$Cu(s)$
$Zn(s)$		$Cu^{2+}(aq)$		$Zn^{2+}(aq)$		$Cu(s)$
		$SO_4^{2-}(aq)$		$SO_4^{2-}(aq)$		

$SO_4^{2-}(aq)$ is on both sides of the equation, so is not part of the ionic equation.

The ionic equation is:

$$Zn(s) + Cu^{2+}(aq) \rightarrow Zn^{2+}(aq) + Cu(s)$$

Exam tip

State symbols are most often not required but they may help you to work out which are the spectator ions in a reaction.

Half-equations

REVISED

A half-equation involves electrons. In some chemical reactions (called redox reactions in Unit 2), electrons are transferred. You will meet these reactions during the GCSE course. Some examples are given here.

Example 1

Magnesium reacts with copper(II) sulfate solution to form magnesium sulfate and copper.

The balanced symbol for the reaction is shown below, along with the particles (ions and molecules) present below the equation:

$Mg(s)$	+	$CuSO_4(aq)$	→	$MgSO_4(aq)$	+	$Cu(s)$
$Mg(s)$		$Cu^{2+}(aq)$		$Mg^{2+}(aq)$		$Cu(s)$
		$SO_4^{2-}(aq)$		$SO_4^{2-}(aq)$		

$SO_4^{2-}(aq)$ is on both sides of the equation, so is not part of the ionic equation.

The ionic equation is:

$$Mg(s) + Cu^{2+}(aq) \rightarrow Mg^{2+}(aq) + Cu(s)$$

In this example magnesium atoms transfer electrons to the copper(II) ions. The half-equations are:

$$Mg \rightarrow Mg^{2+} + 2e^-$$

$$Cu^{2+} + 2e^- \rightarrow Cu$$

The magnesium loses electrons and the copper(II) ion gains electrons.

Example 2

In the reaction of magnesium with hydrochloric acid, magnesium chloride and hydrogen are formed.

The balanced symbol for the reaction is shown below, along with the particles (ions and molecules) present below the equation:

$Mg(s)$	+	$2HCl(aq)$	→	$MgCl_2(aq)$	+	$H_2(g)$
$Mg(s)$		$2H^+(aq)$		$Mg^{2+}(aq)$		$H_2(g)$
		$2Cl^-(aq)$		$2Cl^-(aq)$		

$2Cl^-(aq)$ are on both sides of the equation, so do not form part of the ionic equation.

The ionic equation is:

$$Mg(s) + 2H^+(aq) \rightarrow Mg^{2+}(aq) + H_2(g)$$

In this example magnesium atoms transfer electrons to the hydrogen ions. The half-equations are:

$$Mg \rightarrow Mg^{2+} + 2e^-$$

$$2H^+ + 2e^- \rightarrow H_2$$

Now test yourself

TESTED

7 Write the ionic equation for the reaction between barium chloride solution and sodium sulfate solution to form solid barium sulfate and sodium chloride solution. Include state symbols.
8 Write the ionic equation for the reaction between magnesium and copper(II) chloride, forming magnesium chloride and copper.
9 Write the ionic equation and the half-equations for the reaction of zinc with copper(II) oxide forming zinc oxide and copper.

Exam practice

1 Write the formulae of these simple compounds:
 (a) sodium fluoride
 (b) magnesium oxide
 (c) potassium oxide
 (d) barium chloride [4 marks]
2 Write the formulae of these transition metal compounds:
 (a) copper(II) chloride
 (b) zinc oxide
 (c) copper(II) sulfate
 (d) iron(III) hydroxide [4 marks]
3 Write formulae for:
 (a) sodium carbonate
 (b) sodium hydrogencarbonate [2 marks]
4 Name the following compounds:
 (a) CO_2
 (b) KNO_3
 (c) $CuCO_3$
 (d) HF
 (e) $MgSO_4$ [5 marks]
5 Write the formulae of these compounds:
 (a) ammonium sulfate
 (b) sulfur dioxide
 (c) calcium hydrogencarbonate
 (d) aluminium sulfate [4 marks]
6 Sodium thiosulfate has the formula $Na_2S_2O_3$. Write the formula, including the charge, for the thiosulfate ion. [1 mark]
7 Name the following ions:
 (a) OH^-
 (b) O^{2-}
 (c) Cl^-
 (d) Al^{3+}
 (e) SO_4^{2-} [5 marks]
8 Write a balanced symbol equation for these reactions:
 (a) potassium hydroxide + sulfuric acid → potassium sulfate + water [3 marks]
 (b) calcium + oxygen → calcium oxide [3 marks]
 (c) aluminium + chlorine → aluminium chloride [3 marks]
9 Write a balanced symbol equation to represent the thermal decomposition of copper(II) carbonate into copper(II) oxide and carbon dioxide. [2 marks]
10 Ethane (C_2H_6) reacts with oxygen according to the word equation:

 ethane + oxygen → carbon dioxide + water

 Write a balanced symbol equation for this reaction. [3 marks]
11 Barium chloride solution reacts with potassium sulfate solution to produce solid barium sulfate, and potassium chloride remains in solution.
 (a) Write a balanced symbol equation for the reaction. Include state symbols. [4 marks]
 (b) Write an ionic equation for the reaction. [2 marks]

12 Write a balanced symbol equation for these reactions:

(a) calcium hydroxide + hydrochloric acid → calcium chloride + water [3 marks]

(b) aluminium oxide + sulfuric acid → aluminium sulfate + water [3 marks]

(c) zinc + hydrochloric acid → zinc chloride + hydrogen [3 marks]

13 Write a balanced symbol equation for the reaction between nitrogen and hydrogen to produce ammonia. [3 marks]

H 14 Write an ionic equation for the reaction between zinc ions and hydroxide ions to produce zinc hydroxide. [3 marks]

15 (a) Convert the following balanced symbol equations into ionic equations by removing any spectator ions.

(i) $Mg + CuSO_4 \rightarrow MgSO_4 + Cu$ [2 marks]

(ii) $Zn + 2HCl \rightarrow ZnCl_2 + H_2$ [3 marks]

(iii) $CaCO_3 + 2HCl \rightarrow CaCl_2 + CO_2 + H_2O$ [3 marks]

(b) For (i) and (ii), write half-equations for the reactions. [12 marks]

Answers online

ONLINE

17 The periodic table

Organisation of the periodic table

The periodic table lists all known elements. You need to know definitions of an **element**, **compound**, **atom** and **molecule**.

History of development of the periodic table

REVISED

In 1869 Dmitri Mendeleev arranged the elements in order of atomic mass (or atomic weight as it was called in his time) and recognised repeating patterns. However, he left gaps for undiscovered elements and switched the mass order of the elements (for example, iodine and tellurium) to fit the patterns in the table. Using his periodic table, Mendeleev was able to predict the properties of undiscovered elements. (The noble gases had still not been discovered.)

An **element** is a substance that consists of only one type of atom and cannot be broken down into anything simpler by chemical means.

A **compound** is a substance that consists of two or more elements chemically combined.

An **atom** is the simplest particle of an element that can exist on its own in a stable environment.

A **molecule** is a particle that consists of two or more atoms chemically bonded together.

Features of the modern periodic table compared with Mendeleev's

- The modern periodic table is arranged in order of atomic number/ Mendeleev's table was arranged in order of atomic mass (or atomic weight).
- There are more elements in the modern periodic table (and no gaps)/ there were fewer elements in Mendeleev's table (with some gaps).
- There is a block of transition metals in the modern periodic table/there was no block of transition metals in Mendeleev's table.
- The noble gases are in the modern periodic table/the noble gases were not in Mendeleev's table.
- There are actinides and lanthanides in the modern periodic table/there were no actinides or lanthanides in Mendeleev's table.

Exam tip

You may be asked for features of Mendeleev's table that are different from the modern periodic table. Your answers must relate to Mendeleev's table and not the modern one. You will not gain marks for answers about the modern periodic table. Also, if you provide an answer common to both the modern periodic table and Mendeleev's table, you will not gain marks.

Now test yourself

TESTED

1 State two features of the modern periodic table that are different from Mendeleev's table of elements.
2 What is meant by the term element?
3 What is an atom?

Worked example

Describe the work of Mendeleev in the development of the modern periodic table. [3 marks]

Answer

- He arranged elements in order of atomic mass (or atomic weight).[1]
- He left gaps for undiscovered elements. [1]
- He changed the order of certain elements to suit properties. [1]

Exam tip

The most common mistake in this question is to state that Mendeleev arranged the elements in order of atomic number. Remember that Mendeleev also changed the order of some of the elements, such as iodine and tellurium, to suit their properties better.

Groups and periods

REVISED

The first period only contains two elements — hydrogen and helium (Figure 17.1).

A **period** is a horizontal row in the periodic table.

A **group** is a vertical column in the periodic table.

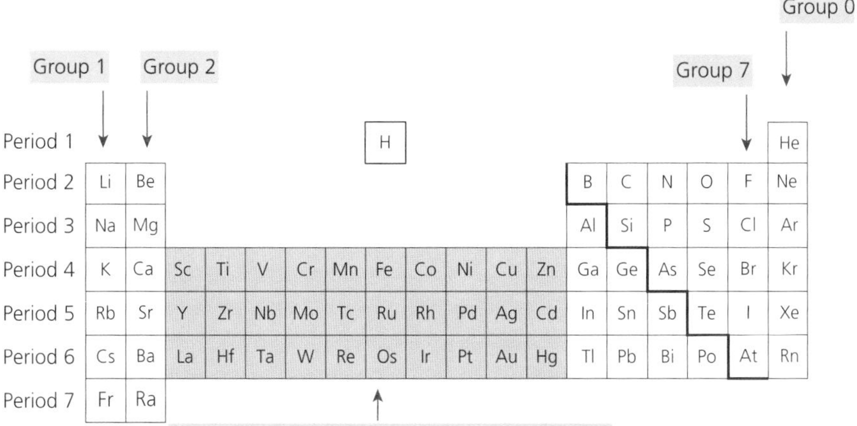

Figure 17.1 The periodic table

Exam tip

Sometimes an incomplete form of the periodic table is given containing only a few of the elements and in the questions that follow you are asked to use *only* the elements given. Many students include other elements in their answers that cannot be accepted.

Exam tip

A common question is to name or write the symbol of an element in a certain period and a certain group of the periodic table. Remember to count period 1. For example, chlorine is the element in period 3 and group 7. A common mistake is to write bromine — having forgotten to count period 1.

Metals and non-metals

REVISED

- The elements to the left of the thick black line in Figure 17.1 are **metals**.
- The elements to the right of the thick black line in Figure 17.1 are **non-metals**.
- Metals conduct heat and electricity. They are ductile (can be drawn out into wires) and malleable (can be hammered into shape). Metals also generally have high melting points and are sonorous (ring when struck).
- Non-metals generally do not conduct heat or electricity (graphite and graphene are exceptions). They are not ductile or malleable, but instead are often brittle solids and generally have low melting points (diamond, graphite and graphene are exceptions). Non-metals are non-sonorous.

Exam tip

The thick black line in Figure 17.1 divides the metals (left of the line) from the non-metals (to the right). In the exam it is useful to write 'metals' on the left of the periodic table and 'non-metals' on the right.

Names of groups

REVISED

- **Group 1 (I)** elements are called the **alkali metals**. Group 1 is a group of reactive metals.
- **Group 2 (II)** elements are called the **alkaline earth metals**. Group 2 is a group of reactive metals, but they are less reactive than the elements in group 1.
- **Group 7 (VII)** elements are called the **halogens**. Group 7 is a group of reactive non-metals.
- **Group 0** (sometimes called **Group 8**) elements are called the **noble gases**. Group 0 is a group of unreactive non-metals.

Elements with similar properties appear in the same group.

The chemical properties of elements in the same group are similar because atoms of elements in the same group have the same number of electrons in their outer shell, so they react similarly.

Exam tip

The names of group 1 and group 2 are often confused. Common errors include calling group 1 the *alkaline* metals and group 2 the *alkali* earth metals.

Exam tip

Be careful with *same* and *similar*. Same means identical. Similar means largely the same, but there will be some variation, such as in the properties of the elements in a group.

Now test yourself

TESTED

4 What name is given to group 1 of the periodic table?
5 What name is given to group 2 of the periodic table?
6 What name is given to group 7 of the periodic table?

Solids, liquids and gases

REVISED

Of all the known elements at room temperature and pressure:

- 11 are gases
- two are liquids
- the rest are solids.

The 11 gases are hydrogen, nitrogen, oxygen, fluorine, chlorine and the noble gases (helium, neon, argon, krypton, xenon and radon). The two liquids are bromine (a **non-metal**) and mercury (a **metal**).

Questions are often asked about changes of state of the elements, and also of some compounds. You must know how to define each change of state, and how to use melting and boiling point data to determine the state of elements and their compounds at specific temperatures.

Diatomicity

REVISED

Some elements exist as **diatomic** molecules.

When writing the formulae of these elements in a balanced symbol equation, they should be written with a little '$_2$' after the symbol to indicate two atoms joined as a molecule — for example, H_2 and O_2.

There are seven diatomic elements to remember: hydrogen, H_2; nitrogen, N_2; oxygen, O_2; fluorine, F_2; chlorine, Cl_2; bromine, Br_2; and iodine, I_2.

Exam tip

To remember the diatomic elements use: **H**ow **N**o **O**ne **F**orgot **Cl**aire's **Br**illiant **I**dea.

Diatomic means that there are two atoms covalently bonded in a molecule.

Exam tip

If asked for the symbol of an element or the symbol for an atom of the element, it is correct to write Cl or H, rather than Cl_2 or H_2.

Groups of the periodic table

Group 1 (I)

REVISED

Group 1 (or sometimes group I) is a group of reactive metals on the left-hand side of the periodic table. The elements in this group are also called the alkali metals. The following list gives some properties and features of the elements in this group.

- All are very reactive metals.
- All are soft, easily cut metals, exposing a shiny surface.
- The shiny surface rapidly tarnishes (goes dull) in air.
- All have low density (the first three, Li, Na and K, are less dense than water, so they float on water).
- All have relatively low melting points for metals.
- All are stored under oil to prevent reaction with oxygen and moisture in the air.
- All react vigorously with water and burn in air.
- All the atoms have one electron in their outer shell.
- All form simple ions with a charge of +, for example Na^+, K^+
- All have a valency of 1.
- All conduct electricity.

Worked example

State the electronic configurations of sodium and potassium, and use these to explain what atoms of the two elements have in common. [3 marks]

Answer

Sodium 2,8,1 [1]; potassium 2,8,8,1 [1]

The atoms have one electron in their outer shell. [1]

(It is important to state the electronic configurations to gain all the marks in this question.)

Trends going down group 1

- The reactivity increases.
- The melting point decreases.

Reactivity of group 1 elements

H

- All group 1 elements have similar chemical properties because their atoms have one electron in their outer shell.
- When the atoms react, they lose this outer shell electron to form an ion. The ion has a full outer shell of electrons and is stable. The ion has a single positive charge.
- A half-equation can be written for the formation of the ion from the atom.
- An example using sodium is:

Na	→	Na^+	+	e^-
sodium atom		sodium ion		outer shell electron lost

- The half-equation for potassium is: $K \rightarrow K^+ + e^-$
- Half-equations involve electrons. They are called half-equations because they show only half of the reaction. They do not show where the lost electron is gained.

Explaining the trend in reactivity

- When an atom of a group 1 element reacts, it loses its outer shell electron.
- The elements become more reactive down the group and this is because the outer shell electron is lost more readily.
- The reason for this is the increasing distance of the electron from the attractive power of the positive nucleus. The further the electron is from the nucleus, the more readily it is lost.
- Also, there are more electrons between the outer electron and the nucleus and these inner electrons shield the outer electrons from the attractive power of the nucleus.

Figure 17.2 shows the electronic configurations of lithium and sodium, with the distance from the nucleus labelled. Note that sodium has more inner electrons shielding the outer electron from the nucleus.

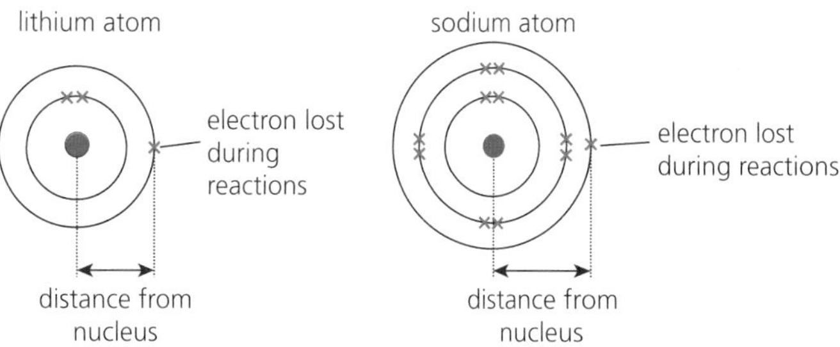

Figure 17.2 Electronic configurations of lithium and sodium

Now test yourself

TESTED

7 Which of lithium, sodium and potassium is the most reactive? Explain why.
8 Name two diatomic elements.
H 9 Write a half-equation for the formation of a lithium ion from a lithium atom.

Reactions of group 1 elements with water

The elements of group 1 react with water to form a metal hydroxide in solution and hydrogen gas. The general equation is:

group 1 metal + water → group 1 metal hydroxide + hydrogen

For example:

sodium + water → sodium hydroxide + hydrogen

Rubidium and caesium are highly reactive and their reactions with water are explosive. These group 1 elements are not generally used in laboratories, but you need to know the products of the reactions, be able to write balanced symbol equations for the reactions and understand why they are not used because of safety. Francium is highly radioactive and very rare and so its chemistry is generally not examined.

Managing risk when using group 1 elements

- When reacting the elements in group 1 with water, a small piece of lithium, sodium or potassium is used to ensure that the reaction is controlled.
- Residual oil from the storage vessel is removed from the surface of the metal using filter paper or towels. Due to the heat of the reaction, the oil can burn and release a choking smoke during the reaction.

- The small piece of the group 1 element is added to a large trough of water placed behind a safety screen. Gloves and safety goggles should be worn because the group 1 elements form a corrosive substance on contact with water on the skin or in the eye. The metal element should not be handled directly. Tongs can be used to add it to the water.

Reactions of group 1 elements with water

Table 17.1 summarises the reactions of group 1 elements with water.

Table 17.1

Metal	Observations for the reaction with water	Balanced symbol equation
Li	Floats on the surface Moves about the surface Fizzes/gas given off Eventually disappears Heat released Colourless solution formed	$2Li + 2H_2O \rightarrow 2LiOH + H_2$
Na	Floats on surface Forms a silvery ball/melts Moves about the surface Fizzes/gas given off Eventually disappears Heat released Colourless solution formed	$2Na + 2H_2O \rightarrow 2NaOH + H_2$
K	Floats on the surface Moves about the surface Burns with a lilac flame Fizzes/gas given off Eventually disappears Small explosion/crackle Heat released Colourless solution formed	$2K + 2H_2O \rightarrow 2KOH + H_2$

Almost all group 1 compounds are white solids at room temperature and pressure and they dissolve in water to form colourless solutions. (A few group 1 compounds are coloured solids if they also contain a transition metal such as potassium dichromate, $K_2Cr_2O_7$, which is orange.) The solutions formed in the reactions above are colourless because they are solutions of group 1 compounds — lithium hydroxide, sodium hydroxide and potassium hydroxide.

Exam tip

The trend in reactivity in group 1 can be used to predict how fast and vigorous reactions will be. You may also be given information on the reactions and asked to determine a reactivity series for elements in group 1. Make sure you can name the products for the reactions of group 1 elements and also write balanced symbol equations for the reactions.

Now test yourself

TESTED

10 Name the products of the reaction of rubidium with water.
11 What colour of flame is observed when potassium reacts with water?
12 Write a balanced symbol equation for the reaction of caesium with water.

Group 7 (VII)

REVISED

Group 7 (or sometimes group VII) is a group of reactive non-metals on the right-hand side of the periodic table. The elements in this group are also called the halogens. The following list gives some properties and features of the elements in this group.

- All are reactive non-metals.
- All are coloured.
- All exist as diatomic molecules.
- All react with group 1 elements to form solid, white, ionic compounds.
- All atoms of group 7 elements have seven electrons in their outer shell.
- All form simple halide ions with a charge of −, for example Cl^-, Br^-.
- All have a valency of 1.
- They do not conduct electricity.
- All the halogens are toxic (page 144).

Trends going down group 7

- Colour intensity of the elements darkens and melting point increases.
- The melting point increases down the group and this explains why there is a clear trend from gas to liquid to solid.

Physical properties of group 7 elements

Table 17.2 lists the physical properties of group 7 elements.

Table 17.2

Group 7 element	Colour and state at room temperature and pressure
Fluorine, F_2	Yellow gas
Chlorine, Cl_2	Yellow-green gas
Bromine, Br_2	Red-brown liquid
Iodine, I_2	Grey-black solid

Exam tip

Learn the colours and states of the halogens carefully because mistakes are common.

- Iodine gas is purple and the grey-black solid iodine sublimes to form the purple gas when heated. Sublimation is the change of state from solid to gas on heating.
- The test for chlorine gas uses damp universal indicator paper, which changes to red and then bleaches white in the presence of chlorine.
- H Based on trends, it would be expected that astatine is a black solid. It is hard to compare the toxicity of the halogens because they change from gases to liquids to solid, but in general the toxicity decreases (fluorine is highly toxic even at very low concentrations; iodine can be used as a sterilising agent for wound dressings because it is toxic to bacteria, but not to humans in low concentrations).
- Reactivity decreases down the group.
- A halogen will displace those below it in the group.

Now test yourself

TESTED

13 Name the most reactive element in group 7.
14 What is the colour and state of bromine at room temperature and pressure?
15 State the test for chlorine gas.

Reactivity of group 7 elements

All group 7 elements have similar chemical properties because their atoms have seven electrons in their outer shell. When the atoms react, they gain one electron to form an ion — the ion has a full outer shell of electrons and is stable. The ion has a single negative charge.

A half-equation can be written for the formation of the ion from the atom, or from a diatomic molecule of the halogen. The negative ions formed are called halide ions (fluoride, chloride, bromide and iodide).

An example using chlorine is:

Cl	+	e^-	→	Cl^-
chlorine atom		outer shell electron gained		chloride ion

The half-equation for iodine is:

$I + e^- \rightarrow I^-$

Half-equations can also be written for the formation of ions from a molecule of the diatomic halogen — for example:

$Cl_2 + 2e^- \rightarrow 2Cl^-$

Note that two electrons are required — one for each of the atoms in Cl_2.

Remember that half-equations involve electrons. They are called 'half-equations' because they show only half of the reaction. They do not show from where the electron has been lost.

Note: In the reaction between sodium and chlorine, each sodium atom loses one electron and each chlorine atom gains one electron. This can be represented by two half-equations:

$Na \rightarrow Na^+ + e^-$

$Cl + e^- \rightarrow Cl^-$

Exam tip

When a half-equation for the formation of a halide ion is asked for, make sure you check carefully if it is from a halogen atom or a halogen molecule.

Now test yourself

TESTED

16 Explain how a fluorine atom becomes a fluoride ion.
17 Write a half-equation for the formation of a chloride ion from a chlorine atom.
18 Write a half-equation for the formation of bromide ions from a bromine molecule.

Explaining the trend in reactivity

- When an atom of a group 7 element reacts, it gains one outer shell electron, which completes its outer shell.
- The elements become less reactive down the group because the atom gains the electron less readily.
- The reason for this is the distance of the outer shell from the attractive power of the positive nucleus. If the electron is gained into a shell close to the nucleus then it is gained very readily.

- Also, for the smaller atoms there are fewer electrons between the outer shell and the nucleus, and these inner electrons shield the incoming electron from the attractive power of the nucleus.
- Figure 17.3 shows the electronic configurations of fluorine and chlorine, with the distance from the nucleus labelled. Note that fluorine has fewer inner electrons shielding the incoming electron from the nucleus.

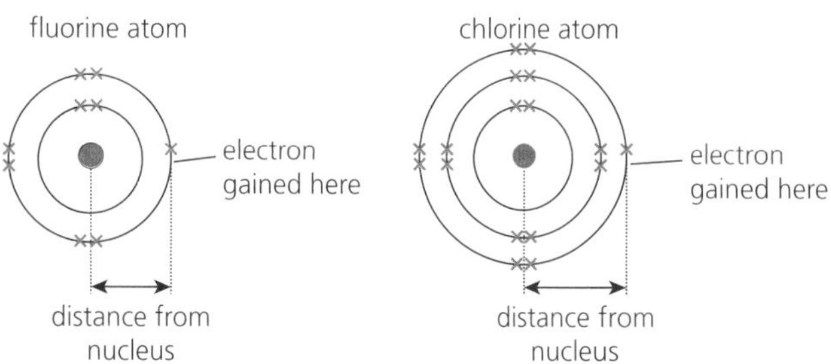

Figure 17.3 Electronic configurations of fluorine and chlorine

Displacement reactions of halogens with halide ions

Halogens can displace other halide ions (of less reactive halogens) from solution.

Chlorine displacing bromide ions from solution

Observations: The yellow-green gas dissolves in the solution and the solution changes from colourless to orange.

Example: chlorine + potassium bromide → potassium chloride + bromine

Balanced symbol equation: $Cl_2 + 2KBr \rightarrow 2KCl + Br_2$

Ionic equation: $Cl_2 + 2Br^- \rightarrow 2Cl^- + Br_2$

> **Exam tip**
>
> You could be asked to write half-equations for the reactions occurring here. For this reaction the half-equations are:
>
> $Cl_2 + 2e^- \rightarrow 2Cl^-$
>
> $2Br^- \rightarrow Br_2 + 2e^-$

Chlorine displacing iodide ions from solution

Observations: The yellow-green gas dissolves in the solution and the solution changes from colourless to brown.

Note: Iodine in an aqueous solution of iodide ions is brown.

Example: chlorine + sodium iodide → sodium chloride + iodine

Balanced symbol equation: $Cl_2 + 2NaI \rightarrow 2NaCl + I_2$

Ionic equation: $Cl_2 + 2I^- \rightarrow 2Cl^- + I_2$

Bromine displacing iodide ions from solution

Observations: Bromine in solution is orange and a solution of iodine in iodide ions is brown, so the solution darkens, but it is only a slight change. (However, iodine dissolves in hexane to form a purple solution, so shaking the solution with hexane will change the hexane to a purple colour.)

Example: bromine + magnesium iodide → magnesium bromide + iodine

Balanced symbol equation: $Br_2 + MgI_2 \rightarrow MgBr_2 + I_2$

Ionic equation: $Br_2 + 2I^- \rightarrow 2Br^- + I_2$

It should also be noted that iodine shows no reaction with solutions containing chloride and bromide ions, and that bromine shows no reaction with solutions containing chloride ions.

- The metal halide used can be any soluble metal chloride, bromide or iodide.
- Fluorine is never used in experiments in school laboratories because it is exceptionally toxic, corrosive and oxidising, and it is also difficult to isolate.
- Chlorine water (chlorine dissolved in water) and bromine water are often used in these experiments because they are safer than the toxic elements and easier to handle. The observations do not include the yellow-green colour of chlorine in these reactions because chlorine water is virtually colourless. Bromine water is orange.
- A displacement reaction occurs when a more reactive element becomes an ion and causes a less reactive element to change from an ion to an atom (for the halogens, the atoms formed combine to form a molecule).

Exam tip

The trend in reactivity in group 7 can be used to predict which displacement reactions will occur. You may also be given information about the reactions and be asked to determine a reactivity series for elements in group 7.

Make sure that you can name the products for the displacement reactions of group 7 elements, and also write balanced symbol equations and ionic equations for the reactions.

Now test yourself

TESTED

19 Write a balanced symbol equation for the reaction that occurs when chlorine gas reacts with potassium iodide solution.
20 What colour change in the solution is observed when chlorine reacts with sodium bromide solution?
21 Write an ionic equation for the reaction between bromine water and potassium iodide solution.

Applying trends to unfamiliar elements

REVISED

An unfamiliar element is usually one low down in the periodic table. Examples of elements commonly used are rubidium, caesium and astatine. Answering questions about these means that you just apply the trends you have learned from those covered above.

Exam tip

The difference in the trend in reactivity of the group 1 elements compared with the group 7 elements is often confused. Questions are set that ask for the most reactive element in group 7. Fluorine is the correct answer. Many students incorrectly state astatine or iodine.

Group 0

REVISED

Group 0 (or sometimes group 8) is a group of unreactive (or non-reactive) non-metals at the extreme right-hand side of the periodic table. The elements in this group are also called the noble gases. The following list gives some properties and features of the elements in this group.

- All are unreactive non-metals.
- All are colourless gases at room temperature.
- All atoms of group 0 elements have eight electrons in their outer shell.
- Group 0 elements do not have a valency because they have a full outer shell of electrons. This makes them unreactive and stable.
- The boiling points of the noble gases increase down the group.

Transition metals

REVISED

The transition metals (or transition elements) are a block of elements between group 2 and 3 in the periodic table. The following are some features of transitions metals:

- Transition metals can form ions with different charges such as iron(II), Fe^{2+} and iron(III), Fe^{3+}.
- Transition metal compounds are often coloured, whereas compounds of group 1 elements are mostly white. For example, copper(II) oxide is black, copper(II) carbonate is green, hydrated copper(II) sulfate is blue and copper(II) salts are usually blue in solution.

Now test yourself

TESTED

22 What colour is copper(II) nitrate solution?
23 What colour is copper(II) oxide?

Exam practice

1 What are the common names given to:
 (a) group 1
 (b) group 2
 (c) group 7
 (d) group 0? [4 marks]

H 2 Write a half-equation to represent the formation of a potassium ion from a potassium atom. [2 marks]

3 Which element is in period 3 and group 5? [1 mark]

4 State what would be observed when potassium reacts with water. [4 marks]

5 Give the colours and states at room temperature and pressure of the following elements:
 (a) fluorine
 (b) chlorine
 (c) bromine
 (d) iodine
 (e) neon [5 marks]

6 Write the formula of the following simple ions:
 (a) rubidium
 (b) iodide
 (c) potassium
 (d) fluoride [4 marks]

7 The atoms of an element have the electronic configuration 2,8,6. In which group of the periodic table would this element be found? [1 mark]

8 Using only the list of elements below, answer the questions that follow:

sodium	magnesium	oxygen	calcium	fluorine
iron	carbon	nitrogen	sulfur	bromine

 (a) Which four elements are diatomic? [4 marks]
 (b) Name one non-metal which is solid at room temperature and pressure. [1 mark]
 (c) Which element would form an ion with a simple charge of +? [1 mark]
 (d) Which element is a transition metal? [1 mark]
 (e) Which three elements are gases? [3 marks]

9 A piece of freshly cut sodium is shiny but tarnishes in air. What does 'tarnish' mean? [1 mark]

10 Which is the most reactive halogen? [1 mark]

11 State the colours of the following solid substances.
 (a) potassium chloride
 (b) iodine

→

(c) copper(II) oxide
(d) hydrated copper(II) sulfate
(e) copper(II) carbonate [5 marks]

H 12 Write a half-equation for the formation of chloride ions from a chlorine molecule. [3 marks]

13 State three safety precautions that are required when carrying out the reaction of sodium with water. [3 marks]

14 Sodium reacts with water and during the reaction a small piece of sodium floats on the surface of the water.
(a) Write a balanced equation for the reaction of sodium with water. [3 marks]
(b) Explain why sodium floats on the water. [1 mark]

15 State two differences between the periodic table developed by Mendeleev and the modern periodic table. [2 marks]

Answers online

ONLINE

18 Quantitative chemistry I

Formula mass

Relative atomic mass

REVISED

- The **relative atomic mass** (**A_r** or **RAM**) of an atom is the mass of the atom measured relative to the mass of an atom of carbon-12 (page 87).
- The mass of an atom of carbon-12 is taken to be exactly 12 and the mass of all other atoms are compared with this.
- It is important to remember that the A_r (or RAM) of an element is a weighted mean (or average) of the mass numbers of all the isotopes of the element, taking abundance of isotopes into account. This is why chlorine has a relative atomic mass (A_r or RAM) of 35.5 (page 87).
- **Relative formula mass** (**M_r** or **RFM**) is the total of the values of the A_r (or RAM) of all the atoms present in the formula of a substance. M_r can be used in place of A_r.
- **Relative molecular mass** (**RMM**) is used for molecular covalent substances that exist as molecules, such as H_2O, CO_2, O_2 and N_2, but M_r can be used in place of RFM and RMM.
- A_r will represent relative atomic mass and M_r will represent relative formula mass throughout this book; RAM, RFM and RMM are equally acceptable and will be seen in papers and mark schemes from the previous specification.
- M_r is the general term used for the relative formula mass of all substances in this book.

The **relative atomic mass** (A_r) of an atom is the mass of the atom compared with that of the carbon-12 isotope, which has a mass of exactly 12, and is a weighted mean of the mass numbers.

The mole

REVISED

- One **mole** of a substance is the standard measurement of amount — 1 mole of any substance contains 6×10^{23} particles.
- The mass of 1 mole of a substance is equal to the M_r of the substance in grams.
- Mole is abbreviated to mol.
- The number of moles of a substance can be determined in various ways, depending on whether it is a solid, a gas or in solution.
- If you are given a certain mass of a substance, you need to divide the mass in grams by the substance's M_r to calculate the number of moles of the substance:

$$\text{moles} = \frac{\text{mass (g)}}{M_r}$$

Relative formula mass (M_r)

REVISED

- The M_r of a substance can be calculated by adding up all the relative atomic masses (A_r) of all the atoms in the formula.
- Relative atomic masses can be found on the periodic table in the *Data Leaflet*.

Example

What is the M_r of:

(a) water, H_2O

(b) carbon dioxide, CO_2

(c) lead(II) nitrate, $Pb(NO_3)_2$

(d) hydrated copper(II) sulfate, $CuSO_4.5H_2O$?

Answer

(a) H_2O contains:

2 H atoms $A_r(H) = 1$

1 O atom $A_r(O) = 16$

M_r of $H_2O = (2 \times 1) + 16 = 18$

(b) CO_2 contains:

1 C atom $A_r(C) = 12$

2 O atoms $A_r(O) = 16$

M_r of $CO_2 = 12 + (2 \times 16) = 44$

(c) $Pb(NO_3)_2$ contains:

1 Pb atom $A_r(Pb) = 207$

2 N atoms $A_r(N) = 14$

6 O atoms $A_r(O) = 16$

M_r of $Pb(NO_3)_2 = 207 + (2 \times 14) + (6 \times 16) = 331$

(d) $CuSO_4.5H_2O$ contains:

1 Cu atom $A_r(Cu) = 64$

1 S atom $A_r(S) = 32$

4 O atoms $A_r(O) = 16$

5 H_2O $M_r(H_2O) = 18$

M_r of $CuSO_4.5H_2O = 64 + 32 + (4 \times 16) + (5 \times 18) = 250$

Exam tip

When you calculate M_r values, remember that all the atoms inside brackets in a formula are multiplied by the small number outside the bracket.

Exam tip

The M_r calculations shown do not require brackets to be mathematically correct but they are there to show how the different A_r values (and M_r of water) are used to calculate the overall M_r.

Exam tip

When calculating the M_r of hydrated substances, remember that the degree of hydration tells us that there are a certain number of water molecules in one 'molecule' of the substance. The total M_r of all these water molecules has to be included in the M_r — for example, hydrated nickel(II) sulfate is $NiSO_4.7H_2O$; its M_r is 59 + 32 + (4 × 16) + (7 × 18) = 281.

Percentage of an element in a compound by mass

REVISED

From the formula of a compound, we can calculate the **percentage by mass** of each of the elements in the compound. Remember that all the percentages of the elements in a compound must add up to 100%.

$$\% \text{ of element X in a compound} = \frac{\text{number of atoms of X in compound} \times A_r \text{ of X}}{M_r \text{ of compound}} \times 100$$

Example 1

Calculate the percentage by mass of carbon in ethane, C_2H_6, and ethene, C_2H_4.

Name	Ethane	Ethene
Formula	C_2H_6	C_2H_4
M_r	30	28
Number of carbon atoms × A_r	2 × 12 = 24	2 × 12 = 24
% of carbon by mass in substance	$\frac{24}{30} \times 100 = 80\%$	$\frac{24}{28} \times 100 = 85.7\%$

Example 2

Calculate the percentage by mass of nitrogen in ammonium nitrate, NH_4NO_3.

Name	Ammonium nitrate
Formula	NH_4NO_3
M_r	80
Number of nitrogen atoms × A_r	2 × 14 = 28
% of nitrogen by mass in substance	$\frac{28}{80} \times 100 = 35\%$

Exam tip

Be careful with ammonium nitrate as there are two nitrogen atoms, but one is in the ammonium ion and the other is in the nitrate ion.

Now test yourself

TESTED

1 Calculate the M_r of ammonium carbonate.
2 What is the M_r of chromium(III) hydroxide?
3 Calculate the percentage of aluminium in aluminium sulfate, $Al_2(SO_4)_3$.

Interchanging M_r and moles

REVISED

The mole expression we used earlier can be written in three ways:

$$\text{moles} = \frac{\text{mass (g)}}{M_r}$$

$$\text{mass (g)} = \text{moles} \times M_r$$

$$M_r = \frac{\text{mass (g)}}{\text{moles}}$$

The mass of 1 mole of a substance is the M_r in grams:

- The M_r of H_2O is 18, so 1 mole of H_2O has a mass of 18 g.
- The M_r of CO_2 is 44, so 1 mole of CO_2 has a mass of 44 g.
- The M_r of $Pb(NO_3)_2$ is 331, so 1 mole of $Pb(NO_3)_2$ has a mass of 331 g.
- The M_r of $CuSO_4.5H_2O$ is 250, so 1 mole of $CuSO_4.5H_2O$ has a mass of 250 g.

This means that we can work through the following calculations:

- 2 moles of H_2O have a mass of 36 g because mass = moles × M_r = 2 × 18 = 36.

- 0.5 moles of CO_2 has a mass of 22 g because mass = moles × M_r = 0.5 × 44 = 22.
- 0.01 moles of $Pb(NO_3)_2$ has a mass of 3.31 g because mass = moles × M_r = 0.01 × 331 = 3.31.
- 0.2 moles of $CuSO_4.5H_2O$ has a mass of 50 g because mass = moles × M_r = 0.2 × 250 = 50.

Exam tip

Watch out for masses given in kilograms (kg) (1 kg = 1000 g) or tonnes (t) (1 tonne = 1 000 000 g). The mass that is divided by the M_r must be in grams (g).

Working out how many moles

- 9 g of H_2O is 0.5 moles because moles $= \frac{\text{mass (g)}}{M_r} = \frac{9}{18} = 0.5$ mol.
- 88 g of CO_2 is 2 moles because moles $= \frac{\text{mass (g)}}{M_r} = \frac{88}{44} = 2$ mol.
- 66.2 g of $Pb(NO_3)_2$ is 0.2 moles because moles $= \frac{\text{mass (g)}}{M_r} = \frac{66.2}{331} = 0.2$ mol.
- 31.75 g of $CuSO_4.5H_2O$ is 0.127 moles because moles $= \frac{\text{mass (g)}}{M_r} = \frac{31.75}{250} = 0.127$ mol.

Working out the M_r

- 21 g of an unknown substance contains 0.25 moles, so its M_r is 84, because $M_r = \frac{\text{mass (g)}}{\text{moles}} = \frac{21}{0.25} = 84$.

The calculation of M_r can often help to identify a substance or an element in the substance.

H Using balanced symbol equations quantitatively

Measuring and calculating quantities

REVISED

'Quantitatively' means measuring and calculating quantities.

No atoms are gained or lost in a balanced symbol equation, so the equation can be read quantitatively. This allows mole calculations to be carried out.

A balanced symbol equation such as the one below for the thermal decomposition of lead(II) nitrate can be read quantitatively:

$$2Pb(NO_3)_2 \rightarrow 2PbO + 4NO_2 + O_2$$

This equation shows that when 2 moles of $Pb(NO_3)_2$ are heated to constant mass, they break down to produce 2 moles of PbO, 4 moles of NO_2 and 1 mole of O_2. Remember that the lead(II) nitrate is heated to constant mass to ensure that all of it decomposes.

If there is a different number of moles of $Pb(NO_3)_2$ to start with, the balancing numbers in the equation still give the ratio of how many moles react and how many moles of products are made.

There are three steps to follow:

Step 1: Using the mass of one of the reactants, which will be given to you, calculate the number of moles of this substance by dividing by its M_r.

Step 2: Using the balancing numbers in the equation, calculate the number of moles of the substance asked about in the question.

Step 3: Convert the number of moles of this substance to its corresponding mass by multiplying by its M_r.

Example 1

Lead(II) nitrate, $Pb(NO_3)_2$, undergoes thermal decomposition according to the equation:

$$2Pb(NO_3)_2 \rightarrow 2PbO + 4NO_2 + O_2$$

3.31 g of $Pb(NO_3)_2$ were heated to constant mass. Calculate the mass of PbO formed.

Answer

Method 1

$$M_r \text{ of } Pb(NO_3)_2 = 207 + (14 \times 2) + (16 \times 6) = 331$$

$$\text{moles} = \frac{\text{mass (g)}}{M_r} =$$

In the balanced symbol equation, 2 moles of $Pb(NO_3)_2$ form 2 moles of PbO. So 0.01 mole of $Pb(NO_3)_2$ forms 0.01 mole of PbO.

$$\text{mass} = \text{moles} \times M_r$$

$$M_r \text{ of PbO} = 207 + 16 = 223$$

$$\text{mass of PbO formed} = \text{moles} \times M_r = 0.01 \times 223 = 2.23\text{ g}$$

H

Method 2

This type of calculation can be set out in a table below the balanced symbol equation:

Balanced equation	$2Pb(NO_3)_2$	$\rightarrow$	2PbO	+	$4NO_2$	+	O_2
Mass	3.31 g		2.23 g***				
M_r	331		223**				
Moles	0.01		0.01*				

Put in the mass you have been given (3.31 g) and calculate the M_r value of that substance (331). Divide the mass by the M_r to calculate the number of moles (0.01). All this is shown in the column below $Pb(NO_3)_2$.

Then calculate the other moles using the balancing numbers. 0.01 mol of $Pb(NO_3)_2$ produces 0.01 mol of PbO (*).

Calculate the M_r of PbO (**) and multiply it by the number of moles to determine the mass of PbO (***).

Exam tip

The type of calculation shown in this example often appears in exams. The most common mistake is to calculate the M_r and then multiply it by the balancing number before calculating the number of moles. Remember that the M_r is for one formula unit. The balancing numbers are for that specific equation. In this example, the error would be to use '662' as the M_r of $Pb(NO_3)_2$ because it has a '2' in front of it in the equation.

Exam tip

Remember, when using the table method to work down to moles, go across using the balancing numbers and work up to the mass.

Example 2

27 kg of aluminium were heated in a stream of oxygen until constant mass was achieved. Calculate the mass, in kg, of aluminium oxide formed.

$$4Al + 3O_2 \rightarrow 2Al_2O_3$$

→

Answer

Method 1

27 kg of aluminium is 27 000 g.

M_r of Al is 27.

$$\text{moles of Al} = \frac{\text{mass (g)}}{M_r} = \frac{27\,000}{27} = 1000\,\text{mol}$$

In the balanced symbol equation, 4 moles of Al form 2 moles of Al_2O_3. So 1000 moles of Al form 500 moles of Al_2O_3.

$$M_r \text{ of } Al_2O_3 = (2 \times 27) + (3 \times 16) = 102$$

$$\text{Mass of } Al_2O_3 \text{ formed} = \text{moles} \times M_r = 500 \times 102 = 51\,000\,\text{g} = 51\,\text{kg}$$

Method 2

The table shows step 1 in the first column, step 2 working out other moles* and step 3 working out M_r ** and mass***.

Balanced equation	4Al	+	$3O_2$	→	$2Al_2O_3$
Mass	27 000 g				51 000 g ***
M_r	27				102**
Moles	1000				500*

Exam tip

Some questions are set using kilograms, so ensure that you know to use mass in grams. Always convert to grams before working out moles by dividing a mass by the M_r. This shows the use of scale — particularly when dealing with industrial processes. There will be 1 mark available for converting from the mass unit given to the mass in grams.

Remember that to convert from kilograms to grams you should multiply by 1000; to convert from tonnes to grams you should multiply by 1 000 000.

Example 3

7.15 g of hydrated sodium carbonate, $Na_2CO_3.10H_2O$, were heated in an evaporating basin. Calculate the mass of anhydrous sodium carbonate that remained on heating to constant mass.

$$Na_2CO_3.10H_2O \rightarrow Na_2CO_3 + 10H_2O$$

Answer

Method 1

$$M_r \text{ of hydrated sodium carbonate} = (2 \times 23) + 12 + (3 \times 16) + 10 \times (2 + 16) = 286$$

$$\text{moles} = \frac{\text{mass (g)}}{M_r} = \frac{7.15}{286} = 0.025 \text{ mol of } Na_2CO_3.10H_2O.$$

In the balanced symbol equation, 1 mol $Na_2CO_3.10H_2O$ forms 1 mol of Na_2CO_3. So 0.025 mol of $Na_2CO_3.10H_2O$ forms 0.025 mol of Na_2CO_3.

$$M_r \text{ of } Na_2CO_3 = (2 \times 23) + 12 + (3 \times 16) = 106$$

$$\text{mass of } Na_2CO_3 \text{ formed} = \text{moles} \times M_r = 0.025 \times 106 = 2.65\,\text{g}$$

Method 2

The table shows step 1 in the first column, step 2 working out other moles* and step 3 working out M_r ** and mass ***.

Balanced equation	$Na_2CO_3.10H_2O$	→	Na_2CO_3	+	$10H_2O$
Mass	7.15 g		2.65 g***		
M_r	286		106**		
Moles	0.025		0.025*		

Exam tip

The equation for this reaction simply removes all the water of crystallisation as the substance is heated to constant mass. Na_2CO_3 is anhydrous sodium carbonate.

Now test yourself

TESTED

4 Calculate the mass of copper(II) oxide obtained from heating 0.31 g of copper(II) carbonate to constant mass.

$CuCO_3 \rightarrow CuO + CO_2$

5 Calculate the mass of aluminium needed to form 1.53 g of aluminium oxide.

$4Al + 3O_2 \rightarrow 2Al_2O_3$

6 Calculate the mass of silver formed when 40.8 g of silver nitrate are heated to constant mass to form silver, oxygen and nitrogen dioxide.

$2AgNO_3 \rightarrow 2Ag + O_2 + 2NO_2$

Limiting reactant

REVISED

When two substances are mixed together to react there may be one substance that is used up before the other one and so some of the other substance will remain.

- The reactant that is used up completely is called the limiting reactant.
- The substance that is left over is said to be the substance in excess.
- The moles of the limiting reactant determine the moles of the products formed.

Example 1

Copper reacts with sulfur to form copper(II) sulfide according to the equation:

$Cu + S \rightarrow CuS$

9.6 g of copper is mixed with 6.4 g of sulfur and the reaction mixture heated.

(a) Calculate the number of moles of copper in 9.6 g.
(b) Calculate the number of moles of sulfur in 6.4 g.
(c) Which reactant is in excess?
(d) Calculate the number of moles of copper(II) sulfide formed.
(e) Calculate the mass of copper(II) sulfide formed.

Answer

Method 1

(a) M_r of Cu = 64, so moles = $\frac{\text{mass (g)}}{M_r} = \frac{9.6}{64} = 0.15\,\text{mol}$

(b) M_r of S = 32, so moles = $\frac{\text{mass (g)}}{M_r} = \frac{6.4}{32} = 0.2\,\text{mol}$

(c) 1 mol of Cu reacts with 1 mol of S in the balanced symbol equation. 0.15 mol of Cu would react with 0.15 mol of S but we have 0.2 mol of S, so S is in excess and Cu is the limiting reactant.

(d) As 0.15 mol of Cu reacts, 0.15 mol of CuS will form (and 0.2 − 0.15 = 0.05 mol of S will be left over).

(e) mass = moles × M_r = 0.15 × 96 = 14.4 g

→

Method 2

This can also be done using an amended table. The reacting row shows how much of the reactants are used up (shown as a minus (–) number because they are lost) and how much of the product is formed (shown as a + number). The remaining moles are what is left over after the reaction.

Balanced equation	Cu	+	S	→	CuS
Mass	9.6 g		6.4 g		
M_r	64		32		96
Moles	0.15		0.2		0
Reacting	–0.15		–0.15		+0.15
Remaining	0		0.05		0.15
Mass	0		0.05 × 32 = 1.6 g		0.15 × 96 = 14.4 g

This also shows the mass of sulfur left over. The reacting line is in the same ratio as the balancing numbers in the equation.

Example 2

100 g of iron(III) oxide were reacted with 50.4 g of carbon monoxide. Calculate the mass of iron formed.

$$Fe_2O_3 + 3CO \rightarrow 2Fe + 3CO_2$$

(a) Calculate the number of moles of iron(III) oxide in 100 g.
(b) Calculate the number of moles of carbon monoxide in 50.4 g.
(c) Which reactant is in excess?
(d) Calculate the number of moles of iron formed.
(e) Calculate the mass of iron formed.

Answer

Method 1

(a) M_r of Fe_2O_3 = 160, so moles = $\frac{\text{mass (g)}}{M_r} = \frac{100}{160}$ = 0.625 mol

(b) M_r of CO = 28, so moles = $\frac{\text{mass (g)}}{M_r} = \frac{50.4}{28}$ = 1.8 mol

(c) 1 mol of Fe_2O_3 reacts with 3 mol of CO in the balanced symbol equation.
0.625 mol of Fe_2O_3 would react with 3 × 0.625 = 1.875 mol of CO, but we have 1.8 mol of CO, so CO is the limiting reactant.
Think of this the other way: 1.8 mol of CO reacts with $\frac{1.8}{3}$ = 0.6 mol of Fe_2O_3, but we have 0.625 mol, so there is more Fe_2O_3 than can react — i.e. Fe_2O_3 is in excess.

(d) As 1.8 mol of CO reacts, $\frac{1.8}{3} \times 2$ = 1.2 mol of Fe will form
(0.625 – 0.6 = 0.025 mol of Fe_2O_3 will be left over).

(e) mass = moles × M_r = 1.2 × 56 = 67.2 g of iron formed

Method 2

This can also be done using an amended table. The reacting row shows how much of the reactants are used up (shown as a minus (–) number because they are lost) and how much of the product is formed (shown as a + number). The remaining moles are what is left over after the reaction.

H

Balanced equation	Fe_2O_3	+	3CO	→	2Fe	+	$3CO_2$
Mass (g)	100		50.4				
M_r	160		28		56		44
Moles	0.625		1.8		0		0
Reacting	−0.6		−1.8		+1.2		+1.8
Remaining	0.025		0		1.2		1.8
Mass	0.025 × 160 = 4 g		0		1.2 × 56 = 67.2 g		1.8 × 44 = 79.2 g

This also shows the mass of Fe_2O_3 left over and the mass of carbon dioxide formed. The reacting line is in the same ratio as the balancing numbers in the equation.

Now test yourself

TESTED

7 1 mol of magnesium reacts with 1 mol of sulfur:

$Mg + S \rightarrow MgS$

If 1 mol of Mg is present and 1.5 mol of sulfur, which reactant is limiting?

8 Calcium reacts with oxygen according to the equation:

$2Ca + O_2 \rightarrow 2CaO$

If 2.0 g of calcium react with 1.0 g of oxygen, which is the reactant in excess?

9 Potassium reacts with sulfur to form potassium sulfide:

$2K + S \rightarrow K_2S$

15.6 g of potassium react with 8 g of sulfur. What is the limiting reactant and what mass of K_2S is formed?

Percentage yield

REVISED

During a chemical reaction, the calculated number of moles or the calculated mass of the product formed is called the **theoretical yield**. It is the number of moles or mass you would expect to be produced if the reaction went to completion. However, many chemical reactions do not give the expected amount of product and the number of moles or the mass you obtain is called the **actual yield**. This is what you obtain experimentally.

- The **percentage yield** is the percentage of the theoretical yield that is achieved in the reaction. It is calculated using the expression:

$$\text{percentage yield} = \frac{\text{actual yield}}{\text{theoretical yield}} \times 100$$

- The actual yield and theoretical yield may be in moles or as a mass, usually in grams, as long as the units are the same.
- The reasons why the percentage yield is not 100% are often asked. The main reasons are:
  - loss by mechanical transfer (from one container to another)
  - loss during a separating technique — for example, filtration or using a separating funnel
  - the reaction is reversible
  - side-reactions occurring
  - the reaction not being complete.

Worked example

A sample of 3.72 g of copper(II) carbonate was heated in a crucible. The copper(II) carbonate decomposed to give copper(II) oxide and carbon dioxide according to the equation below:

$CuCO_3 \rightarrow CuO + CO_2$

(a) Calculate the theoretical yield of copper(II) oxide from this reaction. [3 marks]

(b) 1.8 g of copper(II) oxide were actually obtained. Calculate the percentage yield of this reaction using your answer to part (a). [2 marks]

(c) Suggest one reason why the percentage yield was not 100%. [1 mark]

Answer

(a)

Balanced equation	$CuCO_3$	→	CuO	+	CO_2
Mass (g)	3.72		2.4 [1]		
M_r	124		80		
Moles	0.03 [1]		0.03 [1]		

The theoretical yield is 2.4 g.

(b) Percentage yield = $\frac{\text{actual yield}}{\text{theoretical yield}} \times 100 = \frac{1.8}{2.4} \times 100 = 75\%$ [2]

(c) The reaction was not complete. [1]

Exam practice

1 With what atom are the masses of all atoms compared? [1 mark]

2 Calculate the M_r of these compounds:
(a) H_2SO_4 (b) $Ca(OH)_2$ (c) $Al_2(SO_4)_3$ (d) K_2CO_3 (e) $FeCl_3$ [5 marks]

3 Calculate the mass of calcium oxide, CaO, that would be produced by heating a 5 g sample of calcium carbonate, $CaCO_3$, to constant mass. $CaCO_3 \rightarrow CaO + CO_2$ [3 marks]

4 What mass of magnesium oxide would be produced when 1.2 g of magnesium powder is burned completely in oxygen? $2Mg + O_2 \rightarrow 2MgO$ [3 marks]

5 Iron(III) oxide reacts with sodium according to the equation:

$Fe_2O_3 + 6Na \rightarrow 3Na_2O + 2Fe$

40 g of iron(III) oxide are reacted with 46 g of sodium.
(a) Calculate the number of moles of iron(III) oxide in 40 g. [1 mark]
(b) Calculate the number of moles of sodium in 46 g. [1 mark]
(c) Which reactant is the limiting reactant? [1 mark]
(d) Calculate the number of moles of iron formed. [1 mark]
(e) Calculate the mass of iron formed. [1 mark]

6 Calculate the percentage of water of crystallisation in these hydrated compounds:
(a) copper(II) chloride-2-water [2 marks]
(b) $Na_2CO_3.10H_2O$ [2 marks]

7 A compound has the empirical formula HO. Its M_r is 34. Determine the molecular formula of the compound. [1 mark]

8 A sample of 1.55 g of phosphorus was heated in chlorine to form solid phosphorus(V) chloride, PCl_5. Phosphorus reacts according to the equation: $2P + 5Cl_2 \rightarrow 2PCl_5$
(a) Calculate the theoretical yield of phosphorus(V) chloride. [3 marks]
(b) 8.34 g of phosphorus(V) chloride were obtained. Calculate the percentage yield using your answer to part (a). [2 marks]
(c) Suggest one reason why the percentage yield was not 100%. [1 mark]

Answers online ONLINE

19 Acids, bases and salts

Working with acids, bases and salts

Hazard symbols

REVISED

Hazard symbols (Figure 19.1) are displayed on bottles of chemicals, including acids and alkalis and some salts. They are used rather than words because they are immediately recognisable and there are no language issues with symbols — they are the same worldwide.

- Toxic chemicals represent a serious risk of causing death by poisoning.
- Corrosive chemicals can burn and destroy living tissue.
- Explosive chemicals may explode if heated, exposed to a flame or knocked.
- Flammable chemicals, in contact with air, may catch fire easily.
- The caution hazard symbol indicates that the chemical may cause an allergic skin reaction.

Exam tip

When carrying out a risk assessment always consider the hazards associated with the reactants and products of a chemical reaction.

toxic

corrosive

explosive

flammable

caution

Figure 19.1 Hazard symbols

Risk assessment

REVISED

Any risk to an individual should be minimised by having safety procedures in place to avoid any hazard when using chemicals. For example:

- With corrosive substances, personal protective equipment (safety glasses, gloves and lab coat) should be worn.
- With toxic substances, contact with the substance should be minimised by using personal protective equipment, and they should be used in a fume cupboard if appropriate.
- With flammable substances, there should be no naked flames in the laboratory.
- With substances labelled 'caution', care should be taken in handling to avoid spills and personal protective equipment should be worn.

Indicators

REVISED

- Some **indicators** are solutions but some commonly used ones are also available in paper form for ease of use (e.g. red and blue litmus and universal indicator).
- The paper form of universal indicator is often simply called 'pH paper'.
- Indicator papers are most often used by placing them on a white tile and using a glass rod to put a drop of the solution being tested on the paper.
- Gases can be tested using damp indicator paper held in the gas.

An **indicator** is a chemical that gives a **colour change** in acidic, alkaline and neutral solutions.

Indicator colours

Indicators commonly used are litmus, methyl orange and phenolphthalein. The colours of these in acidic, alkaline and neutral solution are given in Table 19.1.

Table 19.1

Indicator	Colour in acid solution	Colour in neutral solution	Colour in alkaline solution
Red litmus	Red	Red	Blue
Blue litmus	Red	Blue	Blue

To determine whether a substance is acidic or alkaline, both litmus indicators may need to be used:

- If a solution is added to blue litmus and it remains blue, it could be a neutral solution or an alkaline solution.
- Red litmus would need to be used to confirm if the solution was neutral or an alkali — red litmus changes to blue if the solution is an alkali and stays red if it is a neutral solution.

Now test yourself

TESTED

1 What is an indicator?

Concentration of solutions

REVISED

Concentration is measured in **moles per dm³** (mol/dm³) — so the concentration of a solution is the number of moles of a solute dissolved in 1 dm³ of solution. The solvent is usually water.

- 2 mol/dm³ hydrochloric acid is twice as concentrated as 1 mol/dm³ hydrochloric acid.
- 2 mol/dm³ hydrochloric acid has 2 mol of HCl dissolved in 1 dm³ of the solution.

In this section you only need to be aware that these are the units in which concentration is measured.

H Often the terms **dilute** and **concentrated** are used to describe the concentration of solutions of acids and alkalis. A dilute solution has a low concentration of the acid or alkali and a concentrated solution has a high concentration of the acid or alkali.

pH

REVISED

A **pH** (always written lower case 'p' and upper case 'H') value is a numerical value used to indicate how acidic or alkaline a solution is. It is a scale that runs from 0 to 14:

- A pH value of 7 is neutral.
- A pH value less than 7 is acidic.
- a pH value more than 7 is alkaline.

Universal indicator can be used to measure the pH of a solution.

- A few drops of universal indicator are added to the solution being tested and the colour observed. The pH is determined by comparing the colour with a pH colour chart.
- If the paper indicator is used, a few drops of the solution are dropped onto the paper — again, the observed colour is compared with a pH colour chart.

Exam tip

A common question is to describe how to determine the pH of a solution. You need to use universal indicator solution or pH paper, and observe the colour obtained. You determine the pH by comparing the colour obtained with a colour chart.

The pH scale

Table 19.2 gives the expected colours with universal indicator for strong acids, weak acids, neutral solutions, weak alkalis and strong alkalis. Common examples are also given.

Table 19.2

pH	0 1 2	3 4	5 6	7	8 9	10 11	12 13 14
Colour	Red	Orange	Yellow	Green	Green-blue	Blue	Dark blue/ Purple
Strength	Strong acid	Weak acid		Neutral	Weak alkali		Strong alkali
Examples	Hydrochloric acid Sulfuric acid Nitric acid	Ethanoic acid Carbonic acid		Water	Ammonia		Sodium hydroxide Potassium hydroxide
Common solutions	Gastric juice	Vinegar Lemon juice		Salt solution	Blood Seawater		Oven cleaner

A pH meter is an electronic device that gives a numerical value of pH, to at least one decimal place — for example, the pH of 0.5 mol/dm^3 ethanoic acid is 2.52. This would make ethanoic acid a weak acid because the pH is greater than 2 but less than 7.

Validity, reliability and accuracy

REVISED

- **Validity** is part of the overall design of the experiment. The better the design, the more valid it is.
- **Reliability** depends on whether or not the same result could be obtained again if the experiment were to be repeated.
- **Accuracy** is the degree to which a value obtained from a measurement is close to the correct value.
- Red litmus can identify a substance as alkaline, but it stays red for both neutral substances and acids. Blue litmus can identify a substance as acidic, but stays blue with both neutral substances and alkalis. Both red and blue litmus should be used to identify a substance as neutral — using just red litmus to identify a neutral solution would not be a *valid* experiment.
- Universal indicator can identify a substance as acidic, alkaline or neutral, and also whether it is a weak or strong acid or alkali, and give an approximate pH value based on the scale on a colour chart.
- An experiment to determine the pH of a solution using red and blue litmus would not be *valid* because red and blue litmus would not give enough information to determine pH, unless the solution is neutral.
- A pH meter gives the most *accurate* information because it gives a numerical value for pH to one or more decimal places — so a solution with a pH of 3.45 would be orange with universal indicator, but there would be no clear indication of where exactly it would be in the 3–4 pH range. A pH meter gives a more *accurate* value of pH.
- The results of these experiments would be *reliable* because they are able to be reproduced.

Acids and alkalis

Formulae of acids and alkalis

REVISED

Acids

Hydrochloric acid is **H**Cl. Sulfuric acid is **H**$_2$SO$_4$. Ethanoic acid is CH$_3$COO**H**. Carbonic acid is **H**$_2$CO$_3$. Nitric acid is **H**NO$_3$.

- All acids contain hydrogen atoms.
- Some or all of these hydrogen atoms are 'acidic' (shown in bold above).
- The acidic hydrogen atoms can ionise in water to form hydrogen ions.
- H If the acid ionises completely in water, the acid is a **strong acid**.
- If the acid ionises partially in water, the acid is a **weak acid**.

An equation can be written for an acid ionising in water. For example:

$HCl \rightarrow H^+ + Cl^-$

$H_2SO_4 \rightarrow 2H^+ + SO_4^{2-}$

$CH_3COOH \rightleftharpoons H^+ + CH_3COO^-$

Note that strong acids (like HCl and H_2SO_4) use a full arrow as they ionise completely in water whereas a weak acid like CH_3COOH uses a reversible arrow as it only partially ionises in water. H_2SO_4 has two acidic hydrogen ions which both ionise in water.

A **strong acid** ionises completely in water.

A **weak acid** ionises partially in water.

Alkalis

Ammonia is NH_3. Sodium hydroxide is NaOH. Potassium hydroxide is KOH.

- All alkalis contain, or can produce, hydroxide ions in water.
- Some alkalis contain hydroxide ions and, when they dissolve in water, the hydroxide ions are released into the solution.
- Ammonia reacts with water to produce hydroxide ions:

  $NH_3 + H_2O \rightleftharpoons NH_4^+ + OH^-$

- H If an alkali ionises completely in water, the alkali is a **strong alkali**.
- If an alkali ionises partially in water, the alkali is a **weak alkali**.

A **strong alkali** ionises completely in water.

A **weak alkali** ionises partially in water.

Ions in acids and alkalis

REVISED

- All acids dissolve in water, producing hydrogen ions in solution, $H^+(aq)$.
- H The higher the concentration of $H^+(aq)$ ions in an acidic solution, the lower the pH value of that solution. 2 mol/dm^3 hydrochloric acid has a higher $H^+(aq)$ concentration than 1 mol/dm^3 hydrochloric acid, so 2 mol/dm^3 hydrochloric acid would have a lower pH than 1 mol/dm^3 hydrochloric acid.
- All alkalis dissolve in water, producing hydroxide ions in solution, $OH^-(aq)$.
- **Neutralisation** is the reaction between an acid and an alkali, producing a salt and water. It can be represented by the general equation:

  acid + alkali → salt + water

- H The equation for neutralisation can be written as an ionic equation:

  $H^+(aq) + OH^-(aq) \rightarrow H_2O(l)$

Neutralisation is the reaction between an acid and an alkali, producing a salt and water.

Exam tip

The ionic equation for neutralisation is common in questions, and the most common mistake is to leave out the state symbols.

Now test yourself

TESTED

2 What ion is present in alkalis?

H 3 Write the ionic equation for neutralisation including state symbols.

4 Name a strong acid.

Bases and alkalis

REVISED

A **base** is a substance that reacts with an acid, producing a salt and water.

- Common bases are metal oxides and metal hydroxides. Copper(II) oxide (CuO), magnesium oxide (MgO), potassium hydroxide (KOH) and sodium hydroxide (NaOH) are all bases because they are metal oxides and hydroxides.
- An **alkali** is a soluble base.
- The most common alkalis are sodium hydroxide (NaOH), potassium hydroxide (KOH), calcium hydroxide ($Ca(OH)_2$) and ammonia (NH_3).
- Ammonia is not a metal oxide or hydroxide, but a solution of ammonia in water contains hydroxide ions, OH^-, so it is an alkali.

Exam tip

Check the back of the *Data Leaflet* for solubility information. The majority of oxides and hydroxides are insoluble in water, so most are 'just' bases and only a few can be called alkalis. Don't forget that ammonia is an alkali.

Reactions of acids

REVISED

Acids produce **salts** when they react.

- Acids are solutions containing hydrogen ions, H^+, and a negative ion.
- The negative ion (anion) is what combines with a metal ion (or ammonium ion) to form the salt.
- Hydrochloric acid, HCl, contains hydrogen ions, H^+, and **chloride** ions, Cl^-. So when hydrochloric acid reacts, it forms a **chloride**.
- Sulfuric acid, H_2SO_4, contains hydrogen ions, H^+, and **sulfate** ions, SO_4^{2-}. So when sulfuric acid reacts, it forms a **sulfate**.
- Nitric acid, HNO_3, contains hydrogen ions, H^+, and **nitrate** ions, NO_3^-. So when nitric acid reacts, it forms a **nitrate**.
- Ethanoic acid, CH_3COOH, contains hydrogen ions, H^+, and **ethanoate** ions, CH_3COO^-. When ethanoic acid reacts, it forms an **ethanoate**.

A **salt** is a compound formed when some or all of the hydrogen ions in an acid are replaced by metal ions or ammonium ions.

Exam tip

You can remove the H^+ ions from the acid molecule to work out the formula of the anion — for every H^+ you remove, give the anion one more negative charge.

The reactions of acids are summarised below.

Acids + metals

Dilute acids react with some metals to produce a salt and hydrogen gas. The general word equation is:

metal + acid → salt + hydrogen

Examples are:

- magnesium + hydrochloric acid → magnesium chloride + hydrogen

  $Mg + 2HCl \rightarrow MgCl_2 + H_2$

- zinc + sulfuric acid → zinc sulfate + hydrogen

  $Zn + H_2SO_4 \rightarrow ZnSO_4 + H_2$

Exam tip

Don't forget that hydrogen is diatomic. Also, remember that acids do not react with copper or any other metal below copper in the reactivity series.

The hydrogen produced in these reactions can be tested by applying a lit splint to the top of the test tube — if hydrogen is present a pop sound is heard. The pop is caused by the hydrogen burning explosively in air. It reacts with oxygen to produce water:

$2H_2 + O_2 \rightarrow 2H_2O$

Now test yourself

TESTED

5 What salt is produced when calcium reacts with ethanoic acid?
6 What is the test for hydrogen gas?
7 Write a balanced symbol equation for the reaction of calcium with nitric acid.

Acids + metal oxides/hydroxides

Dilute acids react with metal oxides and metal hydroxides to produce a salt and water. Metal oxides and hydroxides are bases. The general word equations are:

metal oxide + acid → salt + water

metal hydroxide + acid → salt + water

Examples are:

- copper(II) oxide + sulfuric acid → copper(II) sulfate + water

  $CuO + H_2SO_4 \rightarrow CuSO_4 + H_2O$
- sodium hydroxide + hydrochloric acid → sodium chloride + water

  $NaOH + HCl \rightarrow NaCl + H_2O$

Acids + metal carbonate/hydrogencarbonate

Dilute acids react with metal carbonates and metal hydrogencarbonates to produce a salt, carbon dioxide and water. Only group 1 elements form stable, solid hydrogencarbonates. Group 2 hydrogencarbonates can exist in solution but not as solids. The general word equations are:

metal carbonate + acid → salt + carbon dioxide + water

metal hydrogencarbonate + acid → salt + carbon dioxide + water

Examples are:

- copper(II) carbonate + sulfuric acid → copper(II) sulfate + carbon dioxide + water

  $CuCO_3 + H_2SO_4 \rightarrow CuSO_4 + CO_2 + H_2O$
- potassium hydrogencarbonate + nitric acid → potassium nitrate + carbon dioxide + water

  $KHCO_3 + HNO_3 \rightarrow KNO_3 + CO_2 + H_2O$

The carbon dioxide produced in these reactions can be tested by bubbling the gas through limewater (calcium hydroxide solution) using a delivery tube (Figure 19.2).

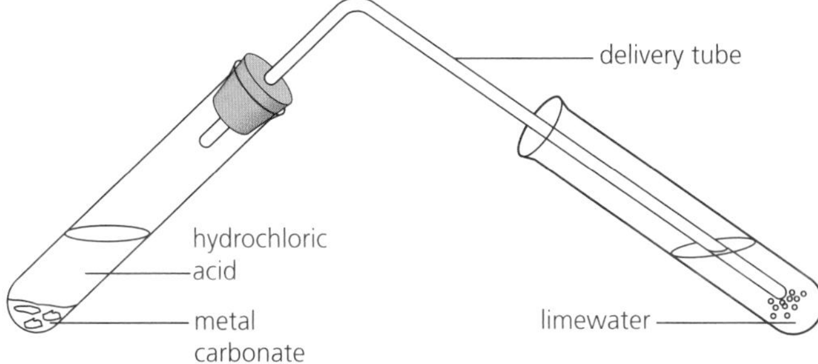

Figure 19.2 Testing for carbon dioxide

If carbon dioxide is present, the limewater changes from colourless to milky. The solution changing to milky is caused by the production of a white precipitate in the solution. A precipitate is a solid formed in a solution. The white precipitate is calcium carbonate, which is insoluble and is formed from the reaction of calcium hydroxide solution and carbon dioxide gas. The equation for the reaction, with state symbols, is:

$$Ca(OH)_2(aq) + CO_2(g) \rightarrow CaCO_3(s) + H_2O(l)$$

Now test yourself

TESTED

8 Name the products of the reaction between sodium carbonate and sulfuric acid.
9 What is used to test for carbon dioxide?
10 Name the salt formed in the reaction between sodium hydroxide and nitric acid.

Observations during acid reactions

REVISED

- If a solid metal or metal compound reacts with an excess of acid, the solid will disappear.
- Many acid reactions are exothermic — so the observation 'heat is released' applies to most reactions of acids.
- For metals that react with acids, and metal carbonates and metal hydrogencarbonates reacting with acids, a gas is produced — so the observation 'bubbles of a gas produced' applies.

All acids are colourless solutions, so you can work out if the colour of the solution changes during an acid reaction. For example:

magnesium + hydrochloric acid → magnesium chloride + hydrogen

- Magnesium is a grey solid.
- Hydrochloric acid is a colourless solution.
- Magnesium chloride is a salt and is soluble in water (*Data Leaflet*).
- Magnesium compounds are colourless in solution, so magnesium chloride solution is colourless.
- There is no change to the colour of the solution.

Observations: The solution remains colourless; bubbles of gas are produced; the grey solid disappears; heat is released.

Another example is:

copper(II) carbonate + sulfuric acid → copper(II) sulfate + water + carbon dioxide

- Copper(II) carbonate is a green solid.
- Sulfuric acid is colourless.
- Copper(II) sulfate is soluble in water (*Data Leaflet*) and is blue in solution.

Observations: The solution changes from colourless to blue; bubbles of gas are produced; the green solid disappears; heat is released.

Exam tip

All group 1 and 2, ammonium, aluminium and zinc compounds are white solids. If they dissolve in water they form colourless solutions. (When zinc oxide is heated it turns yellow, but it changes back to white on cooling.) Transition metal salts are generally coloured.

Exam tip

Copper(II) compounds vary in colour — copper(II) oxide is black; copper(II) carbonate is green; hydrated copper(II) sulfate is blue; anhydrous copper(II) sulfate is white. All copper(II) compounds that dissolve in water form blue solutions.

Prescribed practical C1

Investigate the reactions of acids, including temperature changes that occur

- Be able to carry out and describe practically how to set up a series of experiments to investigate the reactions of acids with metals, oxides, hydroxides, carbonates, hydrogencarbonates and ammonia.
- Be able to record and recall observations for the different reactions and temperature measurements where required.
- Be able to write balanced symbol, ionic and half-equations for the reactions where required.

Now test yourself

TESTED

11 Name and write the formula of the salt formed from the reaction between potassium carbonate and sulfuric acid.
12 Write a balanced symbol equation for the reaction of copper(II) oxide with nitric acid.

Exam practice

1 What is an alkali? [1 mark]
2 Name two compounds that would react with sulfuric acid to make a solution of copper(II) sulfate. [2 marks]
3 State the colour of universal indicator paper and the approximate pH for these solutions:
(a) sodium hydroxide
(b) hydrochloric acid
(c) ammonia
(d) ethanoic acid
(e) water [10 marks]
4 Write the formula of the ion that is present in all acids. [1 mark]
5 Name two bases. [2 marks]
H 6 Which of the following acid solutions would have the higher pH? [1 mark]
0.5 mol/dm^3 sulfuric acid   2.0 mol/dm^3 sulfuric acid
7 Write an ionic equation for neutralisation, including state symbols. [3 marks]
8 Write balanced symbol equations for each of these reactions of acids:
(a) magnesium + hydrochloric acid [3 marks]
(b) magnesium hydroxide + hydrochloric acid [3 marks]
(c) calcium carbonate + hydrochloric acid. [3 marks]
9 State what you would observe in these reactions:
(a) zinc and hydrochloric acid [3 marks]
(b) copper(II) carbonate and sulfuric acid. [3 marks]
10 Name the salt produced when sodium hydroxide reacts with hydrochloric acid. [1 mark]
11 Using a pH meter, a solution is found to have a pH of 3.84. State what colours you would expect to observe if this solution was tested with these indicator papers:
(a) red litmus
(b) blue litmus
(c) universal indicator
12 What gas is produced when zinc reacts with dilute sulfuric acid? How would you test for this gas? [3 marks]

Answers online

ONLINE

20 Chemical analysis

Pure substances, mixtures and formulations

Melting and boiling point

REVISED

When a substance melts or boils the temperature remains the same while this process is happening. Figure 20.1 shows how the temperature varies as a substance is heated.

- When a solid substance is heated, the temperature of the solid increases until it reaches its **melting point**.
- The temperature remains constant while the substance melts, even though it is still being heated.
- Once all the solid has changed to a liquid, the temperature of the liquid increases until it reaches its **boiling point**.
- The temperature remains constant while the substance boils, even though it is still being heated.
- Once all the liquid has changed to a gas, the temperature of the gas increases.

The **melting point** (or melting temperature) is the temperature at which a solid changes into a liquid.

The **boiling point** (or boiling temperature) is the temperature at which a liquid changes into a gas.

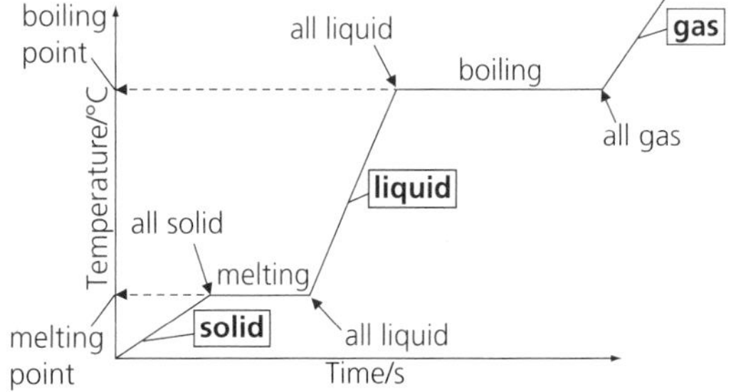

Figure 20.1 A heating curve

Assessing purity

REVISED

- **Pure substances** (elements and compounds) melt and boil at specific temperatures. These melting and boiling points can be used to distinguish substances from **mixtures**.
- An impure substance will have a lower melting point and will melt over a range of temperatures. The melting point is lower because impurities will disrupt the regular lattice arrangement that is present in most solids and so the bonds between the particles will be weaker. The greater the amount of impurity the lower the melting point. An example of this is when substances dissolved in water change the melting point:
  - The melting point of pure water is 0°C and its boiling point is 100°C.
  - Adding 10 g of sodium chloride to 100 cm^3 of water will reduce the melting point to −6°C. 20 g of sodium chloride would decrease the melting point to −16°C.

A **pure substance** is a single element or compound that is not mixed with any other substance.

A **mixture** is defined as two or more substances mixed together, which are usually easy to separate.

- The minimum melting point that a sodium chloride solution can have is −21°C. This is limited by the solubility of sodium chloride.
- This is why salt (with grit for grip) is added to roads in winter to lower the melting point of water so that ice cannot form at 0°C.
- At lower temperatures in colder countries, other compounds such as magnesium chloride are used.

- An impure liquid substance such as water with substances dissolved in it will have a higher boiling point. The greater the amount of impurity the higher the boiling point. An example of this is when substances such as ionic compounds dissolve in water and change the boiling point. The boiling point is higher because the water molecules are attracted to the ions and so more energy is required to make the water molecules escape from the water.
  - When sodium chloride is added to water the boiling point of water increases.
  - A solution containing 20 g of sodium chloride in 100 cm^3 of water would boil just below 102 °C.
  - The maximum boiling point that a sodium chloride solution can have is 108°C.
  - This is again limited by the solubility of sodium chloride in water.

Example

Which of the following would have the lowest melting point?

A water
B water containing 10 g of dissolved potassium chloride
C water containing 20 g of dissolved potassium chloride

Answer

C, because it contains the largest mass of impurity.

Determining state from melting and boiling points

REVISED

When the melting and boiling points of a substance are given, the state at any temperature can be determined as shown in the examples that follow.

Example 1

Bromine's melting point is −7°C and its boiling point is 59°C:

- At temperatures below −7°C, bromine is a solid.
- At temperatures between −7°C and 59°C, bromine is a liquid.
- At temperatures above 59°C, bromine is a gas.

On a temperature line, mark the melting point and the boiling point, with the lower value (or more negative) further to the left:

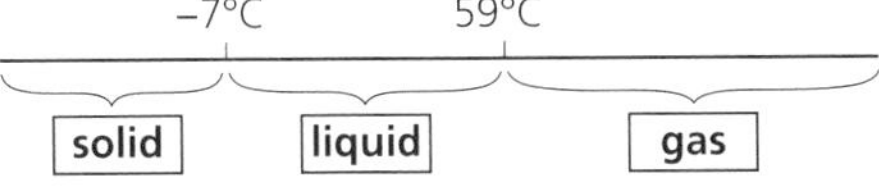

- At all temperatures *to the left* of (below) the melting point, the substance is a solid.
- At all temperatures *between* the melting and boiling points, the substance is a liquid.
- At all temperatures *to the right* of (above) the boiling point, the substance is a gas.

Example 2

Oxygen's melting point is −219°C and its boiling point is −183°C:

- At −191°C oxygen is a liquid.
- At 45°C oxygen is a gas.
- At −230°C oxygen is a solid.

−219 °C −183 °C

solid liquid gas

Now test yourself

TESTED

1 A substance melts at 114°C and boils at 444°C. What state is the substance in at 300°C?
2 Water boils at 100°C. How would the boiling point change if some sodium chloride were added?
3 What is meant by boiling point?

Formulations

REVISED

A **formulation** is a mixture that has been designed as a useful product and is formed by mixing together several different substances in carefully measured quantities to ensure the product has the required properties. Examples of formulations include alloys, medicines and fertilisers.

- 22 carat gold is an alloy containing 91.67% gold, 5% silver, 2% copper and 1.33% zinc. The properties of this gold, which is used in jewellery, make it harder and more durable than pure 24 carat (100%) gold.
- Some cold and flu tablets contain 500 mg of paracetamol (a painkiller), 6 mg of a decongestant and 25 mg of caffeine (a stimulant). These are formulated in this way to relieve pain, make your respiratory system feel less congested and give you a boost via the caffeine. If the formulation was incorrect it would not have the desired effect or could even be dangerous. Pharmacists are very conscious of formulations, especially when they prepare their own cough medicines for example.
- Fertilisers are formulated to provide plants with the correct minerals in suitable proportions. Often these are referred to as NPK fertilisers because they provide nitrogen, phosphorus and potassium compounds, which plants require for growth. Specific formulations are used for different plants.

Separating mixtures

Separation techniques

REVISED

Solids that dissolve in water are described as **soluble**; solids that do not dissolve in water are described as **insoluble**.

A solid that dissolves is called a **solute**; the liquid in which the solute dissolves (usually water) is called the **solvent**.

The resulting mixture of a solute dissolved in a solvent is called a **solution**.

Liquids that mix (for example alcohol and water) are described as **miscible**; liquids that do not mix (for example oil and water) are described as **immiscible**.

The method used to separate a mixture depends on the properties of the substances in the mixture. Different separation techniques are outlined below.

Filtration

- Filtration (Figure 20.2) separates an insoluble solid from a liquid.
- It can be used to separate a mixture of a soluble solid and an insoluble solid once added to a solvent.
- This method can be used to separate sand mixed with water — the sand would be the **residue** and the **filtrate** would be water.
- It can also be used to separate sand from salt. Add water to the mixture of salt and sand and then heat and stir to make sure all the salt dissolves. Filter to remove the sand from the mixture. The residue is the sand and the filtrate is the salt solution.

The filtered solution is called the **filtrate** and the solid remaining in the filter paper is called the **residue**.

residue
filter paper
filter funnel
conical flask
filtrate

Figure 20.2 Filtration

Now test yourself

4 What is the general name given to the solid trapped by the filter paper?
5 What are immiscible liquids?
6 What is the filtrate?

Exam tip

It is important that you can draw a fully labelled diagram of the assembled apparatus used to separate mixtures. These questions can be asked in different units throughout the specification.

Evaporation

Evaporation (Figure 20.3) separates a solute from a solution (the solution is usually the filtrate from filtration).

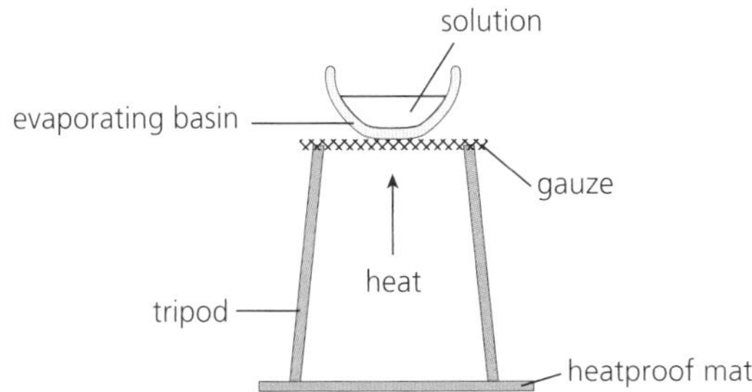

Figure 20.3 Evaporation

- The solution is heated gently in an evaporating basin — the water evaporates (or boils) off, leaving the solid solute behind in the evaporating basin.
- This method can be used to obtain salt from a salt solution, or simply to see if there are any dissolved solids in a liquid sample.
- If the solid to be obtained contains water of crystallisation, the solution is heated to evaporate to half volume and left aside to cool and crystallise. The crystals form in the solution and they can be filtered off. This method is known as **recrystallisation**.

Separating funnel

A separating funnel (Figure 20.4) can be used to separate immiscible liquids — for example, oil and water — based on a difference in their densities. The less dense liquid is the top layer and the more dense liquid is the bottom layer.

- The mixture of liquids is placed in a separating funnel. The separating funnel is clamped and the liquids separate into two layers. The stopper should then be removed from the top of the flask.
- The bottom layer (water in Figure 7.4) is run into a beaker below by opening the tap. When the junction between the layers reaches the tap, the tap is closed. The top layer can be then run out into another beaker.
- Water is a relatively dense liquid and is often the bottom layer, but not always.

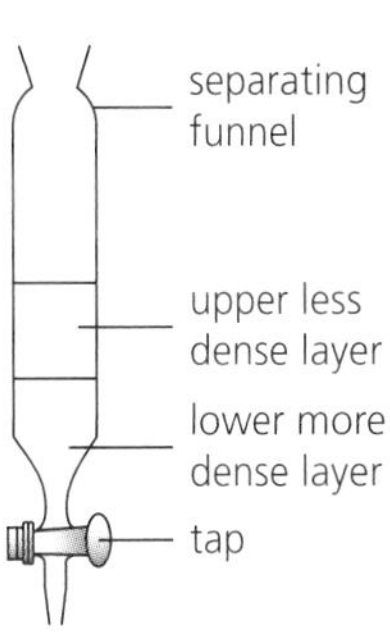

Figure 20.4 Separating funnel

Simple distillation

Simple distillation (Figure 20.5) separates a solvent from a solution (e.g. water from salt solution) or one of two miscible liquids (e.g. a mixture of ethanol and water).

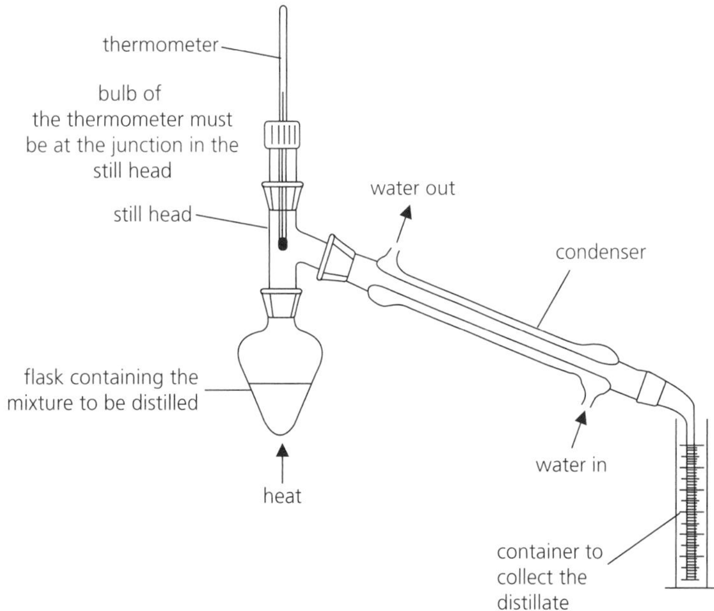

Figure 20.5 Simple distillation

- If a liquid in the mixture is flammable (e.g. ethanol) then a water bath or electric heating mantle is used to heat the flask rather than a Bunsen burner.
- Distillation separates substances because they have *different boiling points*. The temperature shown by the thermometer will remain the same while a particular liquid is boiling off.
- For salt solution, the solvent (water — boiling point 100°C) will boil off and collect as the **distillate**, and the thermometer will remain at 100°C until all the water is boiled off. The salt will remain in the flask.
- For a mixture of ethanol (boiling point 79°C) and water (boiling point 100°C) the ethanol will boil off first and the temperature will remain at 79°C until all the ethanol has boiled off.
- Even at exactly 79°C some water will evaporate and the ethanol collected will not be pure. Fractional distillation (see below) is a better method of separating liquids when the boiling points are reasonably close together.
- A condenser is a glass cylinder with another cylinder around it. Gases pass through the middle tube and cold water runs between the cylinders, providing a cold surface for the gases to condense on.
- The cooling water must go in at the bottom of the condenser and out at the top to ensure that the condenser is filled with water at all times.
- Anti-bumping granules are added to the contents of the flask to promote smooth boiling.

The liquid that is collected in distillation is called the **distillate**.

Fractional distillation

Fractional distillation (Figure 20.6) separates miscible liquids, such as ethanol and water, or the components of crude oil or of liquid air using *differences in their boiling points.*

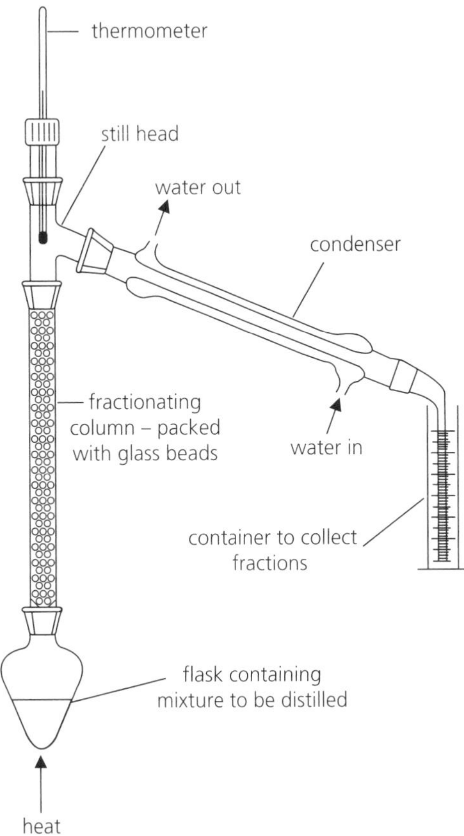

Figure 20.6 Fractional distillation

- Fractional distillation uses the same apparatus as simple distillation, with a fractionating column on top of the flask.
- This method provides better separation of miscible liquids than simple distillation because evaporated liquids below their boiling point do not reach the condenser — they condense on the glass beads in the fractionating column and return to the flask.
- The container collecting the distillate can be changed to collect distillates at different temperatures — distillates collected at different temperatures are called **fractions**.
- If a mixture of ethanol and water is fractionally distilled, a fraction would be collected around 79°C, which would be purer ethanol than the ethanol obtained by simple distillation.
- Fractional distillation should be used to separate miscible liquids with boiling points that are close together.

Now test yourself

TESTED

7 Why are anti-bumping granules added to the flask in distillation and fractional distillation?
8 Name the apparatus required to evaporate a solution of copper(II) sulfate.
9 Name the process used to separate miscible liquids with boiling points close together.

Paper chromatography

Paper chromatography (Figure 20.7) separates the components of a mixture in a solution. The components of the mixture are separated based on their solubility in the solvent (mobile phase) and their attraction to the paper (stationary phase).

Paper chromatography is commonly used. A pencil line is drawn about 1–2 cm from the bottom of a piece of chromatography paper. A pencil cross is drawn on the pencil line. Pencil is used because it will not dissolve in the solvent.

The mixture to be investigated is dissolved in a small volume of a solvent — water, ethanol or another solvent or a mixture of solvents. A capillary tube is used to spot a sample of the solution on the cross. The paper is then placed in a container with enough solvent in the bottom to reach the bottom of the paper (Figure 20.8).

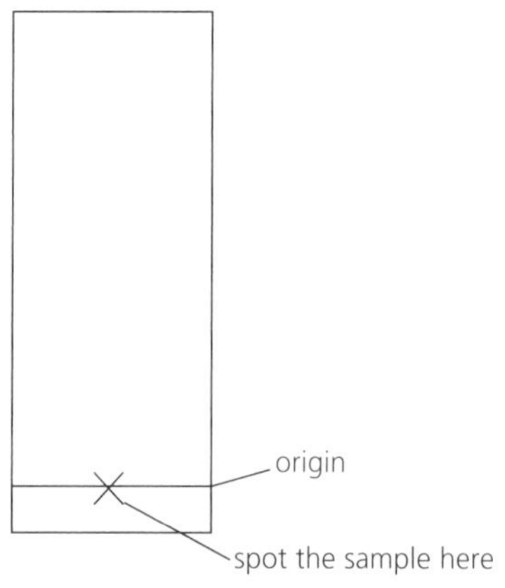

Figure 20.7 Setting up paper chromatography

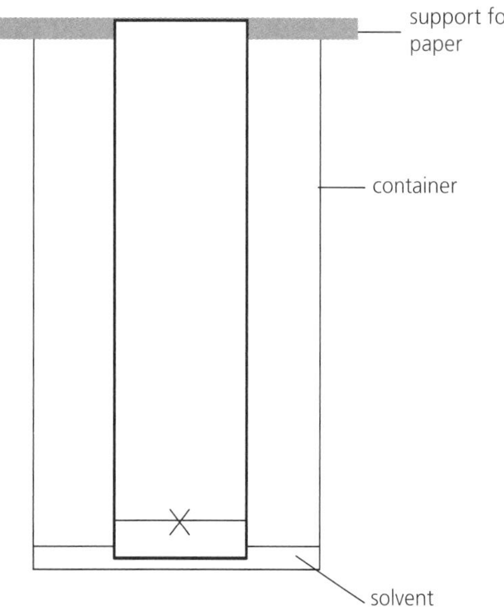

Figure 20.8 Chromatography

- The result of a chromatography experiment is called a **chromatogram**.
- The solvent is the mobile phase because it moves through the paper, taking some or all of the components in the mixture with it. The paper is the stationary phase because it does not move and substances that are attracted more to the paper do not move as far on the chromatogram.
- The spot that travels furthest on the chromatography paper is the component that is most soluble in the solvent (mobile phase).
- If a spot does not move up the chromatography paper during chromatography, it is not soluble in the solvent (mobile phase) used and is attracted to the paper (stationary phase). A different solvent, or sometimes a mixture of solvents, can be used to separate this type of component.
- The distance the solvent moves is marked with a pencil when the chromatogram is removed from the solvent and this is called the solvent front.
- The chromatogram shown in Figure 20.9 would indicate that there are three components in the mixture. However, there may be more but they did not separate well in the solvent used.

Exam tip

The mobile phase is the phase that is moving, in this case the solvent as it moves up the paper. The stationary phase is the one through which the mobile phase moves, in this case the paper.

- Allow the solvent to be drawn up the chromatography paper.
- When the solvent is close to the top of the paper, remove the paper from the solvent and mark how far the solvent has travelled with a pencil line. This is called the **solvent front.**
- The mixture should have separated into different components, which are seen as spots on the paper.

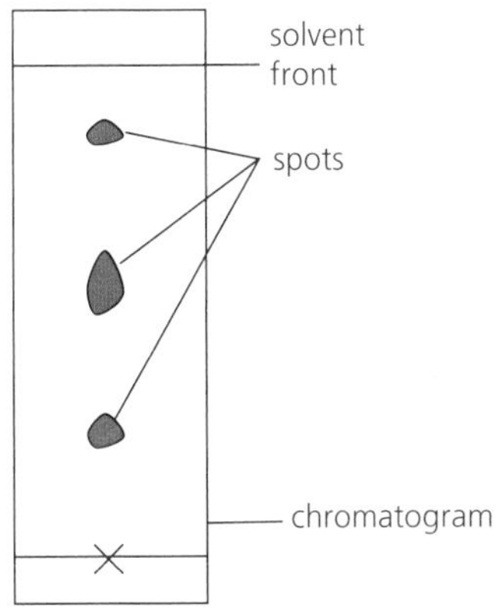

Figure 20.9 Example of a chromatogram

Chromatography can be used to identify specific components in dyes and food colourings. A sample of the unknown is run on a larger piece of chromatography paper with pure samples of the suspected dyes. Figure 20.10 shows such a chromatogram.

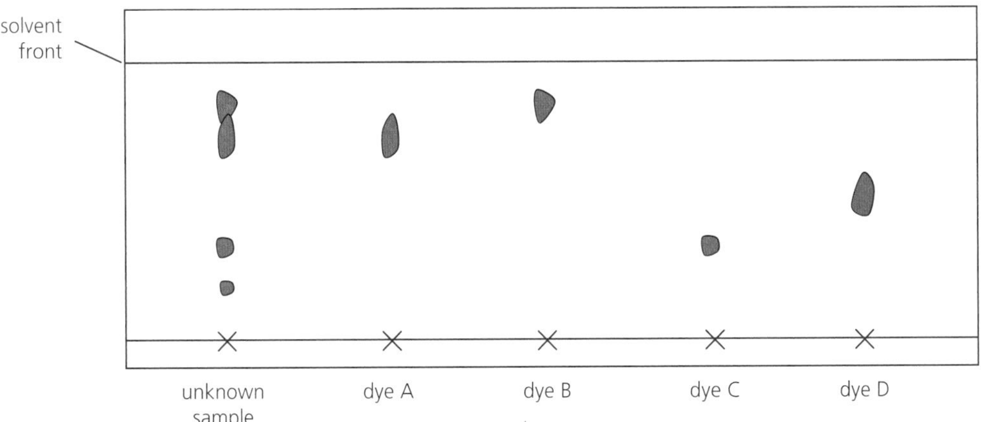

Figure 20.10 A typical chromatogram

The spots for the unknown sample can be compared with the spots for the pure dyes A, B, C and D.

The unknown sample appears to have four different components, though the separation of the spots nearest the solvent front is not complete. A different solvent, or a mixture of solvents, could be used or the solvent could have been allowed to run further up the paper to achieve better separation.

From the chromatogram it is possible to state the following:

- The unknown sample is made up of a minimum of four different components.
- The unknown sample is a mixture of dyes A, B and C but *not* dye D.
- Dye B is the most soluble in the solvent because it moves furthest up the chromatography paper.
- One component in the unknown sample has not been identified.

Paper chromatography can be used to separate the components in a dye, a food colouring or a chemical indicator. If the substances to be separated are not coloured then the spots can be viewed under ultraviolet light — the spots can be drawn round in pencil while under ultraviolet light. Alternatively, a chemical developing agent can be sprayed on the chromatogram to mark the position of the spots by making them coloured. If the chemical in the spot is required, the spot can be cut out and the substance extracted from the paper by placing it in the solvent. The solvent can then be evaporated.

R_f values

$R_f = \frac{\text{distance moved by spot}}{\text{distance moved by solvent}}$

R_f values (retardation factors) are calculated by dividing the distance moved by each spot by the distance moved by the solvent.

R_f values are values between zero and 1. An R_f value of zero means that the spot has not moved. The higher the value the further the spot has moved on the chromatogram.

The distance moved by the spot is taken from the origin to the centre of the spot. If the spot has an irregular appearance the approximate centre of the spot it taken.

For the three spots in Figure 20.11 the following R_f values are calculated:

- Spot 1: $R_f = \frac{\text{distance moved by spot}}{\text{distance moved by solvent}} = \frac{2.5}{8} = 0.3125$
- Spot 2: $R_f = \frac{\text{distance moved by spot}}{\text{distance moved by solvent}} = \frac{4}{8} = 0.5$
- Spot 3: $R_f = \frac{\text{distance moved by spot}}{\text{distance moved by solvent}} = \frac{7.2}{8} = 0.9$

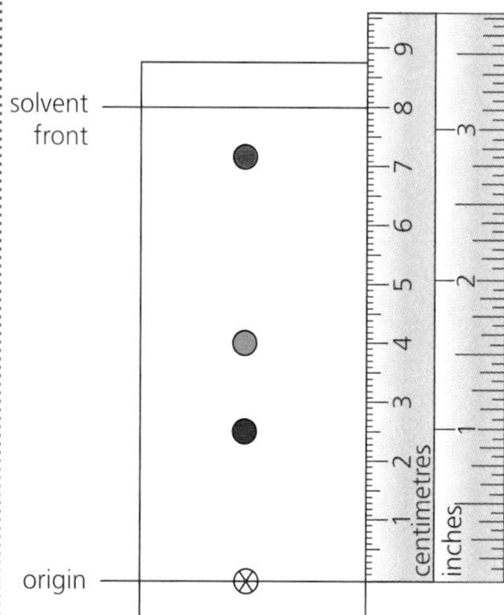

Figure 20.11 Calculating R_f values

The same substance will have the same R_f value in paper chromatography if the same solvent is used. This can be used to assess purity because an impure substance will have a different R_f value from the pure substance.

Now test yourself

TESTED

10 How is an R_f value calculated?
11 What does an R_f value of zero mean?
12 What is the mobile phase in paper chromatography using water as the solvent?

Planning methods of separation

REVISED

Examination questions give examples of mixtures and you will be expected to decide on the most appropriate method of separating the components. The first thing to do is decide on the type of mixture and then choose the appropriate method of separation (Table 20.1). Note that methods sometimes need to be combined to achieve a full separation.

Table 20.1 Methods of separating mixtures

Components	Most suitable method of separation	Examples	Separation based on difference in this physical property
Insoluble solid and a liquid	Filtration	Sand and water Sulfur and water	Solubility in water
Soluble solid and the liquid in which it is dissolved	Simple distillation to obtain liquid	Salt solution Sugar solution	Boiling points
Soluble solid and the liquid in which it is dissolved	Evaporation or recrystallisation to obtain the solid	Salt solution Sugar solution	Boiling points
Miscible liquids	Fractional distillation (can use simple distillation if boiling points differ substantially)	Ethanol and water Crude oil Liquid air	Boiling points
Immiscible liquids	Separating funnel	Oil and water	Density
Soluble substances	Chromatography	Dyes Food colourings Chemical indicators	Solubility in solvent
Two solids — one soluble in water, the other insoluble	Dissolve in water; filtration to obtain the insoluble solid; evaporation or recrystallisation to obtain the soluble solid from the filtrate	Sand and salt Sulfur and sugar	Solubility (filtration) Boiling points (evaporation or recrystallisation)

Test for water

The chemical test for water uses anhydrous copper(II) sulfate, which is a white solid that changes to blue when water is added to it.

Anhydrous copper(II) sulfate can be generated from hydrated copper(II) sulfate by heating to constant mass. Anhydrous copper(II) sulfate is $CuSO_4$ and it is a white solid. Hydrated copper(II) sulfate is $CuSO_4.5H_2O$ and it is a blue solid.

Now test yourself

TESTED

13 What colour is anhydrous copper(II) sulfate?

Tests for ions

Basic information

REVISED

H Remember that the positive ion in an ionic compound is called the cation; the negative ion is called the anion.

Cation tests

REVISED

Cations can be tested for using a flame test, sodium hydroxide solution and ammonia solution.

Flame test

A flame test is carried out as follows:

1 Dip nichrome wire in concentrated hydrochloric acid.
2 Dip the nichrome wire in the solid sample.
3 Place the nichrome wire in a blue Bunsen burner flame.
4 Observe the colour of the flame.

Many potassium compounds are contaminated with sodium compounds so it is difficult to see the lilac colour of the potassium ion because the sodium ion gives a strong golden yellow colour. Holding the sample at the edge of the flame can help in seeing the lilac colour.

Metal chlorides are usually used for flame tests (or the chloride can be made by adding a little concentrated hydrochloric acid to the solid) because the chloride ion will not affect the flame test colour in any way.

Results: Table 20.2 shows the results for the flame test of the ions you need to know.

Table 20.2

Colour	Ion present
Crimson	Li^+
Yellow/orange	Na^+
Lilac	K^+
Brick red	Ca^{2+}
Blue-green or green-blue	Cu^{2+}

Exam tip

If you are asked to describe how to carry out a flame test to identify a specific ion, remember to state the flame test colour you would expect.

Exam tip

Remember that you need to be able to identify an ion from the flame colour, and also to state the flame colour from the named ion present.

Prescribed practical C2

Identify the ions in an ionic compound using flame tests

- Be able to carry out and describe how to carry out a flame test on an ionic compound
- Interpret flame colours to identify the ions present

Exam practice

1 Explain what is meant by the following terms:
 (a) solute
 (b) solvent
 (c) solution. [3 marks]
2 Plan a practical method for the separation of a mixture of sand and water. [4 marks]
3 Name the separation technique that would provide the best separation of a mixture of ethanol and water. [1 mark]

4 The following chromatogram was obtained for different food colourings labelled 1, 2, 3 and 4.

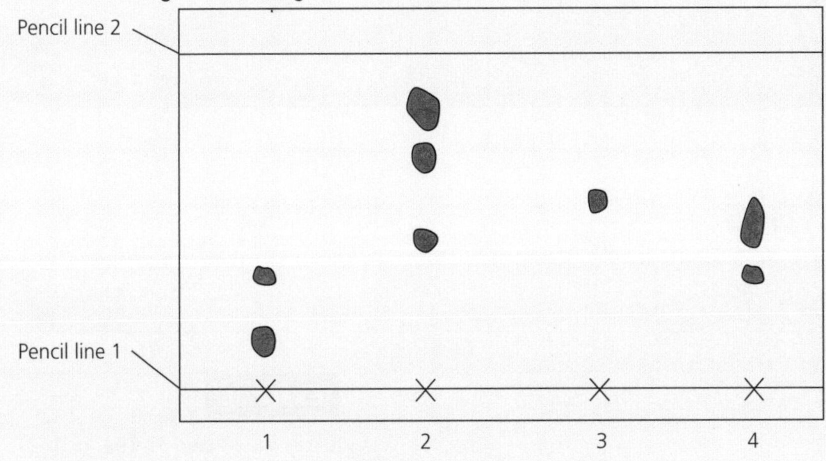

(a) Which of the food colourings is a pure substance? [1 mark]
(b) Name a suitable solvent for this chromatography. [1 mark]
(c) Which food colouring is composed of three components? [1 mark]
(d) State two food colourings that have a common component. [1 mark]
(e) Which food colouring has the component that is least soluble in the solvent? [1 mark]
(f) Suggest why a pencil is used to draw the horizontal lines on the chromatogram. [1 mark]
(g) What name is given to pencil line 2? [1 mark]

5 The diagram below shows a commonly used separation technique.

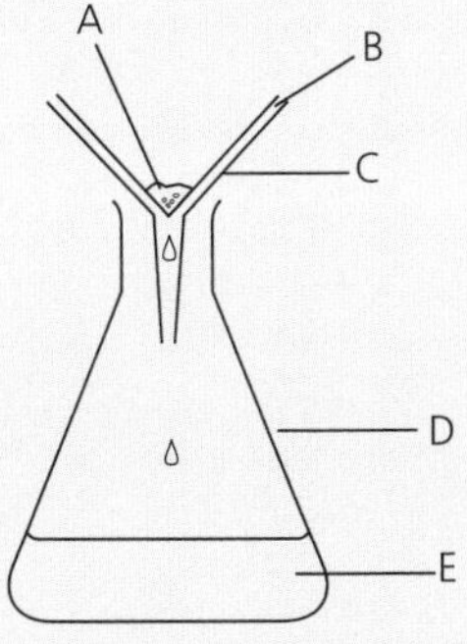

(a) What labels should be placed on the diagram at B, C and D? [3 marks]
(b) What is the general name for the solid obtained at A? [1 mark]
(c) What is the general name for the liquid obtained at E? [1 mark]
(d) Name one mixture that could be separated in this way. [1 mark]

6 Describe the colour change observed when water is added to anhydrous copper(II) sulfate. [1 mark]

7 Describe how to carry out a flame test on a sample of sodium chloride, and the observations you would make. [4 marks]

Answers online

ONLINE

21 Reactivity series of metals

Reactions of metals

Explaining the reactivity of metals

REVISED

The reactivity series for the common metals can be written as:

K Na Ca Mg Al Zn Fe Cu

Most reactive ⟶ Least reactive

Exam tip

You will need to learn this reactivity series and use it to predict whether or not a given reaction will occur. Precious metals like silver and gold are less reactive than copper.

Metals that are higher in the reactivity series react faster and more vigorously. The reactivity of a metal depends on the tendency of the metal to lose outer shell electron(s) and form its positive ion. The greater the tendency to lose outer shell electron(s), the more reactive the metal.

The reactivity increases down group 1. Potassium is more reactive than sodium. Both sodium atoms and potassium atoms lose one electron from their outer shell when they react to form the positive ion.

$Na \rightarrow Na^+ + e^-$

$K \rightarrow K^+ + e^-$

The outer electron lost from a potassium atom (2,8,8,**1**) is further from the nucleus than for the electron lost from a sodium atom (2,8,**1**), and the outer electron is shielded from the attraction of the positively charge nucleus by more inner electrons.

Group 2 elements are less reactive than group 1 elements. Magnesium is less reactive than sodium. Magnesium loses two electrons from its outer shell to form its positive ion when it reacts.

$Mg \rightarrow Mg^{2+} + 2e^-$

A magnesium atom is smaller than a sodium atom (atomic radius decreases across the periodic table) and so the outer electrons lost from a magnesium atom are closer to the positive nucleus. Also, two electrons are lost from a magnesium atom as opposed to one from a sodium atom.

Transition metal atoms are smaller than the atoms of group 1 or 2 in the same period and so the outer electrons are closer to the nucleus making them less reactive than group 1 or 2 metals.

Reactions of metals with air

REVISED

When metals react with air they gain mass because they form an oxide. Table 21.1 summarises the reactions of common metals with oxygen from the air.

Table 21.1 Reactions of some metals with air

Metal	Reaction when heated in air	Reaction with air under normal conditions	Balanced symbol equation and notes
K	Burns with lilac flame, forming a white solid	When freshly cut the shiny surface tarnishes (goes from shiny to dull) quickly	$4K + O_2 \rightarrow 2K_2O$
Na	Burns with a yellow/orange flame, forming a white solid		$4Na + O_2 \rightarrow 2Na_2O$
Ca	Burns with a brick red flame, forming a white solid	React slowly, forming an oxide layer on the surface	$2Ca + O_2 \rightarrow 2CaO$
Mg	Burns with a bright, white light, forming a white solid		$2Mg + O_2 \rightarrow 2MgO$
Al	Burns only when finely powdered, forming a white solid		$4Al + 3O_2 \rightarrow 2Al_2O_3$
Zn	Burns steadily, forming a yellow solid, which becomes white on cooling		$2Zn + O_2 \rightarrow 2ZnO$
Fe	Burns with orange sparks when in the form of filings, forming a black solid		$3Fe + 2O_2 \rightarrow Fe_3O_4$ Fe_3O_4 is a mixed oxide
Cu	Does not burn, but forms a black solid		$2Cu + O_2 \rightarrow 2CuO$

Sodium, potassium and calcium are only heated in air under very careful supervision and strict safety procedures. The reactions can be extremely dangerous.

All the other metals listed in the table can be heated in air in a crucible using the apparatus shown in Figure 21.1.

Usually the powder form of the metal is heated. The crucible lid is lifted occasionally during heating to allow more air to get into the crucible.

All of the metals reacts faster and more vigorously in pure oxygen compared with in air.

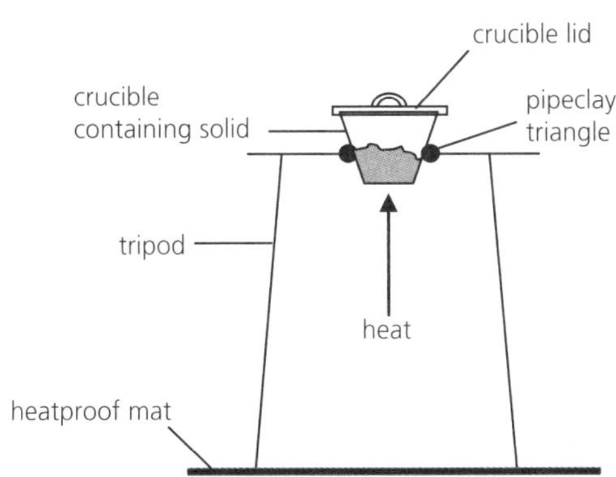

Figure 21.1 Reacting metals with air

Now test yourself

TESTED

1. What is the colour of the flame observed when potassium is heated in air?
2. Write a balanced symbol equation for the reaction of calcium with oxygen in air.
3. What is the colour of the product of the reaction of copper with oxygen?

Reaction of metals with water

REVISED

When sodium and potassium are reacted with water, a small piece of the metal is placed in a trough half-full of water. This reaction is carried out behind a safety screen. Tongs or tweezers are used to handle the metal.

Table 21.2 summarises the reactions of some of the more reactive metals with water.

Table 21.2 Reactions of some metals with water

Metal	Reaction with water	Balanced symbol equation and notes
K	Floats on the surface Moves around the surface Burns with a lilac flame Fizzes, giving off a gas Heat is released Small explosion/crackles Eventually disappears Forms a colourless solution	$2K + 2H_2O \rightarrow 2KOH + H_2$ Potassium is stored under oil to prevent it reacting with oxygen and moisture in the air
Na	Floats on the surface Moves about the surface Melts and forms a silvery ball Fizzes, giving off a gas Heat is released Eventually disappears Forms a colourless solution	$2Na + 2H_2O \rightarrow 2NaOH + H_2$ Sodium is stored under oil to prevent it reacting with oxygen and moisture in the air
Ca	Fizzes, giving off a gas Sinks, then rises Heat is released Eventually disappears Forms a colourless/milky solution	$Ca + 2H_2O \rightarrow Ca(OH)_2 + H_2$
Mg	A very slow reaction A few bubbles of gas given off	$Mg + 2H_2O \rightarrow Mg(OH)_2 + H_2$

To react calcium or magnesium with water, the metal is put in water in a beaker and an inverted filter funnel is placed over the metal. A boiling tube filled with water is used to collect the hydrogen produced. The apparatus is shown in Figure 21.2.

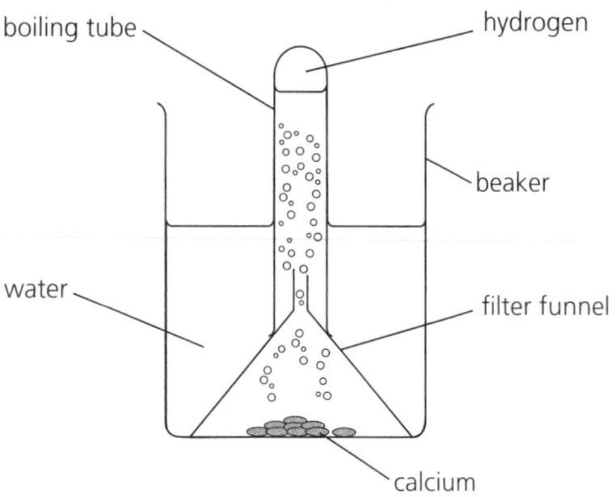

Figure 21.2 Reacting metals (not group 1) with cold water

When magnesium is used instead of calcium, only a few bubbles of gas are produced over a period of several days.

Now test yourself

TESTED

4 Rubidium reacts with water like sodium and potassium. Write a balanced symbol equation for the reaction of rubidium with water.
5 Name the products of the reaction of calcium with water.
6 What is observed when sodium reacts with water?

Exam tip

Copper (and metals below it in the reactivity series) do not react with water or steam.

Reaction of metals with steam

REVISED

Table 21.3 summarises the reactions of some metals with steam.

Table 21.3 Reactions of some metals with steam

Metal	Reaction with steam	Balanced symbol equation
Mg	The heated ribbon burns with a bright, white light Forms a white solid Heat is released	$Mg + H_2O \rightarrow MgO + H_2$
Al	No reaction in foil form, unless the protective layer of aluminium oxide is removed Powdered form burns to form a white solid Heat is released	$2Al + 3H_2O \rightarrow Al_2O_3 + 3H_2$
Zn	Powdered form glows to form a yellow solid, which changes to white on cooling Heat is released	$Zn + H_2O \rightarrow ZnO + H_2$
Fe	Powdered form glows at red heat, forming a black solid	$3Fe + 4H_2O \rightarrow Fe_3O_4 + 4H_2$

The apparatus in Figure 21.3 is used to allow steam to react with a heated metal. The apparatus is connected to a delivery tube and the gas produced is collected over water, using a beehive shelf and a gas jar. The gas produced is hydrogen.

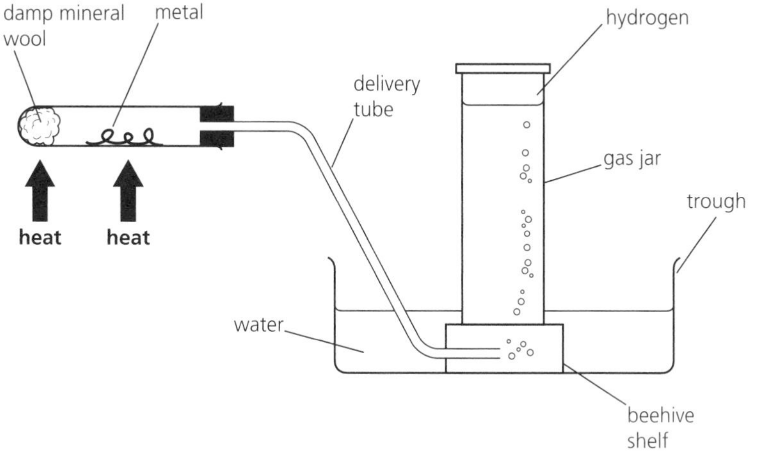

Figure 21.3 Reacting a metal with steam

Damp mineral wool is heated to generate steam. When the heating stops, there is a risk of 'suck back' occurring (the water in the trough being drawn back into the hot boiling tube). This can be prevented by removing the apparatus from the water in the trough, or taking the bung out of the boiling tube.

Exam tip

When drawing apparatus, remember that marks are awarded for the labels on an assembled and recognisable diagram — for example, you will *not* gain a mark for a beaker labelled as a crucible. Make the apparatus look like it should. The most common mistakes made when drawing the apparatus to react metals with steam is to put the delivery tube through the wall of the trough. Remember that the level of the water in the trough should always be above the beehive shelf.

Aluminium

The reactivity of aluminium is often 'hidden' because aluminium metal forms a **protective oxide layer** on its surface.

- A piece of aluminium foil does not react with air, water or steam because of the protective oxide layer.
- Powdered aluminium shows aluminium's true reactivity.
- Aluminium is reasonably reactive, but the protective aluminium oxide layer means that it can be used for saucepans without reacting with food.

Now test yourself

TESTED

7 Name the products of the reaction of zinc with steam.
8 Write a balanced symbol equation for the reaction of magnesium with steam.
9 What colour is magnesium oxide?

Displacement reactions

Displacement reactions involve the transfer of electrons (pages 174–175).

There are two main types of displacement reaction:

- a solid metal reacting with a solution of a metal ion
- a solid metal reacting with a solid metal oxide.

A **displacement reaction** is a reaction in which a more reactive metal takes the place of a less reactive metal in a compound.

Solutions

REVISED

The main observations to look out for are:

- any colour changes in the solution — copper(II) sulfate solution is frequently used because it gives a definite colour change
- the release of heat because displacement reactions are exothermic — the bigger the difference in the reactivity of the metals, the more heat will be released
- appearance of the solid metal that is being displaced from solution — remember to state its colour and the fact that a solid is produced.

Worked example

Magnesium metal reacts when placed in a solution of copper(II) sulfate.

(a) State the observations for the reaction. [3 marks]
(b) Write a balanced symbol equation for the reaction. [2 marks]
(c) Identify an ion that does not take part in the reaction (spectator ion). [1 mark]
H (d) Write an ionic equation for the reaction. [2 marks]
(e) Explain in terms of reactivity why a reaction occurs. [1 mark]

Answers

(a) Blue [1] solution fades/changes to colourless [1]; a red-pink/black [1] solid [1] appears; heat is released [1]. [max. 3]
(b) $Mg + CuSO_4 \rightarrow MgSO_4 + Cu$ [2]
(c) The sulfate ion/SO_4^{2-} (does not take part in the reaction and is the same in both reactants and products) [1]
H (d) $Mg + Cu^{2+} \rightarrow Mg^{2+} + Cu$ [2]
(e) Magnesium is more reactive than copper. [1]

Exam tip

Questions may be set using an unusual reaction, so you will need to use your knowledge of the colours of reactants and products to predict observations. Write down as many sensible observations as you can for the reaction — even if there are only 2 marks — because sometimes observations are combined in mark schemes.

Exam tip

You may be asked for balanced symbol equations, ionic equations and observations for these reactions. You may be asked to use information to determine a reactivity series.

A series of displacement reactions can be used to determine a reactivity series.

- A set of reactions between metals and their metal salt solutions is carried out.
- The metals are placed in a solution of the metal salt (usually the sulfate or nitrate because most sulfates are soluble in water and all nitrates are soluble in water). The results are recorded in a table like Table 21.4.

Table 21.4 Displacement reactions

Metal	Metal salt solution			
	Magnesium sulfate	Copper(II) sulfate	Iron(II) sulfate	Zinc sulfate
Magnesium		✓	✓	✓
Copper	✗		✗	✗
Iron	✗	✓		✗
Zinc	✗	✓	✓	

A tick (✓) indicates a reaction is occurring and a cross (✗) indicates no reaction.

- The parts of the table that are shaded show that the metal should not be placed in a solution of its own salt — for example, magnesium is not placed in magnesium sulfate solution.
- From the table it can be seen that magnesium displaces the other three metals from their solutions, indicating that it is the most reactive of the four metals.
- Zinc displaces copper and iron from their solutions, but does not displace magnesium, showing that it is the next most reactive metal.
- Iron displaces only copper from its solution, so iron is the third most reactive metal.
- Copper does not displace any of the other metals from their solutions, indicating that it is the least reactive metal in this investigation.
- The reactivity in order from most reactive to least reactive is magnesium, zinc, iron, copper.

Exam tip

When describing the reactivity of metals always state, for example, 'magnesium is more reactive than copper'. The most common mistake is to state simply 'magnesium is more reactive' — you must say what it is more reactive than. Under the same conditions, a more reactive metal will react faster than a less reactive metal.

Exam tip

Often an unfamiliar metal, such as chromium or cobalt, will be included and you may have to write a reactivity series based on the results from a table like the one above.

Balanced symbol equations, ionic equations and observations for the reactions can also be asked for. Remember that 'heat released' is the most common observation for a displacement reaction.

Exam tip

Questions often ask for reactivity in order from most reactive to least reactive, but be careful — some ask for the order from least reactive to most reactive.

Solids

REVISED

When solid metal compounds (usually oxides) are heated with a solid metal, a displacement reaction can occur (and it is a redox reaction).

- The reactions are carried out with both the metal oxide and the metal in powder form to increase the contact between the solids.
- The reactions are carried out in a crucible with the same apparatus as shown for heating a metal in air shown in Figure 21.1 (page 165).
- The reactions are exothermic.
- How exothermic a reaction is depends on how far apart the metals are in the reactivity series. The further apart the metals, the faster and more exothermic the reaction.

- If a mixture of magnesium powder and copper(II) oxide is heated strongly in a crucible, the crucible explodes. Powdered metals can be highly flammable.

Worked example

Zinc powder is mixed with black copper(II) oxide powder and heated in a crucible. The reaction produces a blue-green glow and a yellow solid, which changes to white on cooling.

(a) Write a balanced symbol equation for the reaction. [2 marks]
(b) Identify the ion that does not take part in the reaction (spectator ion). [1 mark]
H (c) Write an ionic equation for the reaction. [2 marks]

Answers

(a) $Zn + CuO \rightarrow ZnO + Cu$ [2]
(b) The oxide ion/O^{2-} [1]
H (c) $Zn + Cu^{2+} \rightarrow Zn^{2+} + Cu$ [2]

Exam tip

The observations for this reaction are important because the oxide formed is yellow when heated and changes to white on cooling — this indicates that the oxide formed is zinc oxide. The blue-green glow is caused by the presence of the copper(II) ions — this is similar to the flame test for copper(II) ions (page 162).

Now test yourself

TESTED

10 Write a balanced symbol equation for the reaction between magnesium and lead(II) nitrate, forming magnesium nitrate and lead.
11 Name the products of the reaction between zinc and copper(II) sulfate solution.
H 12 Write an ionic equation for the reaction between magnesium and zinc chloride solution.

Placing an unfamiliar metal in the reactivity series

REVISED

Reactions of an unfamiliar metal and its compounds may be used to place the metal in the reactivity series.

Example 1

Barium reacts vigorously when added to water. It sinks to the bottom and very rapid bubbling is observed and the reaction mixture releases a lot of heat. Place barium in the following reactivity series:

K Na Ca Mg Al Zn Fe Cu

Answer

K Na **Ba** Ca Mg Al Zn Fe Cu

Barium reacts more vigorously with water than calcium but less vigorously than sodium or potassium.

Example 2

The table below shows some displacement reactions of metals with solutions of their compounds.

Metal	Metal salt solution			
	Silver nitrate	Magnesium nitrate	Copper(II) nitrate	Iron(II) nitrate
Silver		✗	✗	✗
Magnesium	✓		✓	✓
Copper	✓	✗		✗
Iron	✓	✗	✓	

(a) Write the order of reactivity from most reactive to least reactive. [1 mark]

(b) Write a balanced symbol equation for the reaction of magnesium with silver nitrate solution. [3 marks]

Answers

(a) magnesium iron copper silver [1]

(b) $Mg + 2AgNO_3 \rightarrow Mg(NO_3)_2 + 2Ag$ [3]

Magnesium metal reacts with all of the other solutions so it is the most reactive. Iron reacts with two of the other solutions and copper with one of the solutions. Silver metal does not react with any of the solutions, so it is the least reactive of the metals here.

Exam tip

Metal nitrates are often used as the solution because all nitrates are soluble in water.

Exam tip

Remember to use your *Data Leaflet* — the ions are on the back, so you should know that silver ions are Ag^+. A common error is to assume a 2+ charge for silver ions.

Applications of metal reactivity

Extraction of metals from their ores

REVISED

Most metal ores are oxides of the metal — some are converted to oxides of the metal before the metal is extracted.

- The lowest reactivity metals are found uncombined in nature (also called native).
- Metals that are high in the reactivity series are extracted by **electrolysis**.
- Metals that are low in the reactivity series are extracted by **reduction** with carbon or carbon monoxide.
- For metals such as aluminium and those above it in the reactivity series, **electrolysis** must be used to extract the metal from its ore.
- All methods of extracting a metal from its ore require reduction, which can be explained either in terms of a gain of electrons or a loss of oxygen (page 173).

Reaction of carbon with metal oxides

Carbon is a non-metal but it can be included in the reactivity series of the metals because it can displace some metals from their compounds (Figure 21.4).

Carbon is used to extract metals below aluminium in the reactivity series.

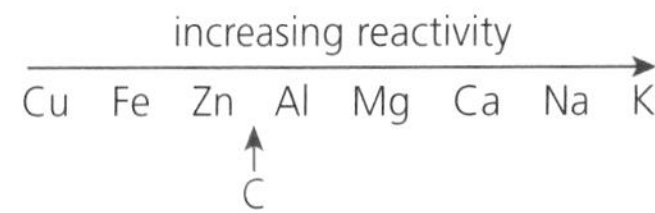

Figure 21.4 Carbon's place in the metal reactivity series

Prescribed practical C3

Investigate the reactivity of metals

- Be able to carry out and describe practically how to set up a series of experiments to form a reactivity series of metals.
- Be able to interpret given data on different reactions to form a reactivity series.
- Be able to record and recall observations for the different reactions of metals with air, water, steam, and for displacement reactions, where appropriate.
- Be able to write balanced symbol, **ionic** and **half-equations** for the reactions.

Exam practice

1 Name the gas produced when calcium reacts with water. [1 mark]
2 What is observed when potassium reacts with water? [4 marks]
3 Write a balanced symbol equation for the reaction of potassium with water. [3 marks]
4 What is observed when magnesium reacts with copper(II) sulfate solution? [3 marks]
5 In the reaction of magnesium with steam, how is the steam generated in the apparatus? [2 marks]
6 Copper metal reacts with silver(I) nitrate solution.
(a) Write a balanced symbol equation for the reaction. [3 marks]
(b) Which ion does not take part in the reaction? [1 mark]
H (c) Write an ionic equation for the reaction. [3 marks]
7 Explain why the reactivity of aluminium metal foil is lower than expected considering its position in the reactivity series. [1 mark]
8 What is observed when magnesium metal is heated in air? [3 marks]
9 For the metals calcium, copper, zinc and aluminium:
(a) Which one burns with a brick-red flame when heated in air? [1 mark]
(b) Which one would displace magnesium from magnesium sulfate solution? [1 mark]
(c) Place the metals in order of reactivity, from most reactive to least reactive. [1 mark]
10 Write a balanced symbol equation for the reaction of calcium with oxygen. [3 marks]

The table below gives information on displacement reactions when solid metals are added to solutions of the metal nitrates. Use the information in the table to answer the questions that follow.

Metal	Metal nitrate solution			
	Magnesium nitrate	Nickel(II) nitrate	Chromium(III) nitrate	Manganese(II) nitrate
Magnesium		✓	✓	✓
Nickel	✗		✗	✗
Chromium	✗	✓		✗
Manganese	✗	✓	✓	

A tick (✓) indicates a reaction occurring and a cross (✗) indicates no reaction.

11 Write the metals in order of reactivity — from most reactive to least reactive. [1 mark]
12 Write a balanced symbol equation for the reaction of chromium with nickel(II) nitrate solution, forming chromium(III) nitrate as one of the products. [3 marks]
H 13 Write an ionic equation for the reaction of magnesium with manganese(II) nitrate solution. [2 marks]
14 Suggest why metal nitrates are often used in solution displacement reactions. [1 mark]

Answers online

ONLINE

22 Redox, rusting and iron

Redox

A **redox reaction** is one in which **oxidation** and **reduction** occur at the same time.

Redox reactions

REVISED

Oxidation and reduction can be defined in one of three ways, as shown in Table 22.1.

Table 22.1

Oxidation	Reduction
Gain of oxygen	Loss of oxygen
Loss of hydrogen	Gain of hydrogen
Loss of electrons	Gain of electrons

- Many reactions can be simply described as oxidation or reduction in terms of the change in the oxygen or hydrogen content.
- Other reactions can only be described in terms of electrons lost or gained.
- Reduction is the reverse of oxidation.

Exam tip

The answer to a simple 'change in oxygen or hydrogen content' question is worth 2 marks. You should always give the following answer:

- Species (name) undergoes gain/loss in oxygen/hydrogen.
- Gain/loss of oxygen/hydrogen is oxidation/reduction.

Worked example

(a) Explain why this reaction is described as reduction:

$H_2 + Cl_2 \rightarrow 2HCl$ [2 marks]

(b) Magnesium burns in air according to the equation:

$2Mg + O_2 \rightarrow 2MgO$

Explain why this reaction is described as oxidation. [2 marks]

Answers

(a) Chlorine gains hydrogen [1]; gain of hydrogen is reduction [1].
(b) Magnesium gains oxygen [1]; gain of oxygen is oxidation [1].
or
Magnesium loses electrons [1]; loss of electrons is oxidation [1].

Now test yourself

TESTED

1 In the reaction $2ZnO + C \rightarrow 2Zn + CO_2$, what is oxidised?
2 Explain why the following reaction is an oxidation reaction:

$S + O_2 \rightarrow SO_2$

3 Explain why the following reaction is a reduction reaction:

$H_2 + I_2 \rightarrow 2HI$

Redox and displacement reactions

REVISED

A displacement reaction (page 168) is one in which a more reactive metal takes the place of a less reactive metal in a compound. This process involves the transfer of electrons. There are two main types of displacement reaction:

- a solid metal reacting with a solution containing metal ions
- a solid metal reacting with a solid metal oxide.

H One species will lose electrons and one will gain electrons:

- The loss of electrons is called **oxidation**.
- The gain of electrons is called **reduction**.

When both oxidation and reduction reactions occur in the same reaction, the overall reaction is described as a **redox reaction**.

Displacement reactions involving solutions

Example

Magnesium metal reacts when placed in a solution of copper(II) sulfate. Explain, in terms of electrons, why this reaction is described as a redox reaction.

Answer

To answer this question, use the following process:

Balanced symbol equation: $Mg + CuSO_4 \rightarrow MgSO_4 + Cu$

Ionic equation: $Mg + Cu^{2+} \rightarrow Mg^{2+} + Cu$

Spectator ion: SO_4^{2-} (does not take part in the reaction)

Half-equations:

$Mg \rightarrow Mg^{2+} + 2e^-$ oxidation

Magnesium atoms lose electrons, and loss of electrons is oxidation.

$Cu^{2+} + 2e^- \rightarrow Cu$ reduction

Copper(II) ions gain electrons, and gain of electrons is reduction.

The reaction is a redox reaction because both oxidation and reduction are occurring simultaneously.

Exam tip

Most questions are of this type. Observations are usually only asked for when it is a question involving copper(II) ions in solution, because this involves a colour change. There can be up to 7 marks for a question in which you are asked to explain why reaction is described as redox in terms of electrons if half-equations are asked for, and 5 marks for a written explanation:

The metal loses electrons [1] or $M \rightarrow M^{2+} + 2e^-$ [2], and the loss of electrons is oxidation [1].

The metal ion gains electrons [1] or $X^{2+} + 2e^- \rightarrow X$ [2], and the gain of electrons is reduction [1].

Redox is oxidation and reduction occurring simultaneously in the same reaction. [1]

A common error is failing to make your answer specific to the reaction in the question. You must state what loses electrons for oxidation and that oxidation is the loss of electrons. Similarly you must state what gains electrons for reduction and that reduction is gain of electrons.

Displacement reactions between solids

When solid metal compounds (often oxides) are heated with a solid metal, a displacement reaction can occur — it is a redox reaction (page 173).

Worked example

Zinc powder is mixed with black copper(II) oxide powder and heated in a crucible.

The reaction produces a blue-green glow and a yellow solid, which changes to white on cooling. Explain, in terms of oxygen content, why this is a redox reaction. [5 marks]

Answer

Balanced symbol equation:

$$Zn + CuO \rightarrow ZnO + Cu$$

Zinc gains oxygen [1], and the gain of oxygen is oxidation [1].

Copper(II) oxide (or copper) loses oxygen [1], and the loss of oxygen is reduction [1].

Redox is oxidation and reduction occurring simultaneously in the same reaction. [1]

Exam tip

This question could also have been asked in terms of electrons:

Zinc atoms lose electrons [1] /$Zn \rightarrow Zn^{2+} + 2e^-$ [2], and the loss of electrons is oxidation [1].

Copper(II) ions gain electrons [1] /$Cu^{2+} + 2e^- \rightarrow Cu$ [2], and the gain of electrons is reduction [1].

Redox is oxidation and reduction occurring simultaneously in the same reaction. [1]

Now test yourself

TESTED

4. For the half-equation $Fe \rightarrow Fe^{2+} + 2e^-$, explain why iron is oxidised.
5. Write half-equations for the reaction between copper and silver nitrate solution, and indicate which is oxidation and which is reduction.
6. What is reduced in the reaction between zinc and copper(II) chloride solution?

Rusting

The cause of rusting

REVISED

When iron is exposed to air (oxygen) and moisture (water in the air), the iron **rusts**.

Rust is hydrated iron(III) oxide, sometimes written $Fe_2O_3.xH_2O$.

Note: Steel is an alloy of iron containing between 0.2% and 2% carbon. Steel is stronger than iron. The iron in steel also rusts.

An investigation to determine the factors that cause rusting is shown in Figure 22.1.

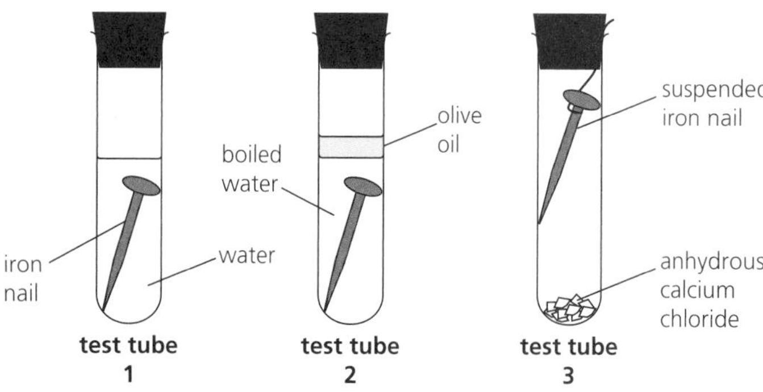

Figure 22.1 Investigating the factors that cause rusting

- Test tube 1 has air and water present.
- Test tube 2 has had the air removed from the water (by boiling), so only water is present. The olive oil prevents gases from the air dissolving in the water.
- Test tube 3 contains anhydrous calcium chloride, which removes the water vapour from the air, so only air is present. The nail is suspended to prevent contact between it and the calcium chloride.

The test tubes are left for several days and the iron nail rusts only in test tube 1. This indicates that both air and water are required for rusting.

Prevention of rusting

REVISED

Rust is unsightly and also dangerous because it weakens the metal. Iron and steel are used extensively in construction, so rusting is a major problem due to the cost of replacing the structures.

Rusting can be prevented in a variety of ways. Methods fall into three groups.

1 Preventing the surface of the iron coming into contact with water and air by using a barrier or protective layer:
   - Paint is used to protect cars, bridges and railings.
   - Oil or grease is used to protect tools and machinery.
   - Plastic coatings, such as those used to cover bicycle handlebars, garden chairs and dish racks. Car manufacturers are increasingly using plastic in cars to reduce the problem of rust.
   - The iron is plated with another metal.
   - Cans for food are made from steel and are coated on both sides with a thin layer of tin. Tin is unreactive and non-toxic. It is deposited on the steel by electrolysis.
   - Chromium is used to coat steel, giving it a shiny, attractive appearance. This is used for some vehicle bumpers and bicycle handlebars. Chromium can be applied by electrolysis.

H 2 Putting a more reactive metal in contact with the iron or steel. The more reactive metal reacts first, leaving the iron intact.
- Bars of magnesium are attached to the sides of ships, oil rigs and underwater pipes to prevent rusting. The magnesium **corrodes** instead of the iron or steel and must be replaced with fresh magnesium periodically. This method of rust prevention is called **sacrificial protection**.
- The experimental set-up in Figure 22.2 shows three iron nails in water but the nail in test tube 2 is wrapped in copper wire. The nail in test tube 3 is wrapped in magnesium ribbon. The nails in test tubes 1 and 2 will rust. The nail in test tube 3 will not rust. Copper is less reactive than iron so cannot prevent the iron from rusting, but magnesium is more reactive than iron so it offers sacrificial protection and reacts first before the iron.

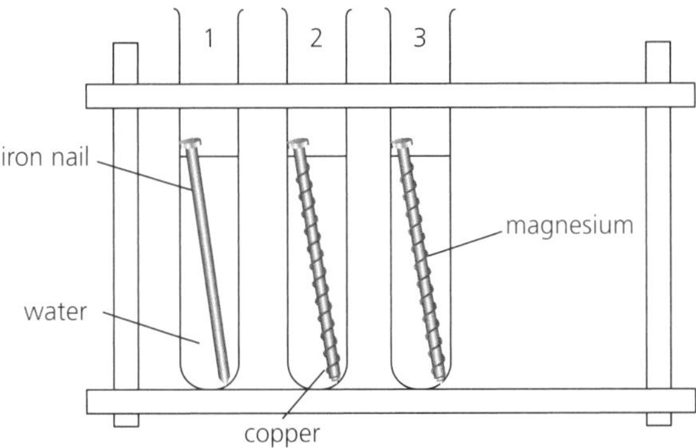

Figure 22.2 Demonstration of sacrificial protection

- Iron can be coated in zinc — this is called **galvanising**. Zinc is more reactive than iron and oxidises readily to form a layer of zinc oxide. Galvanising protects by sacrificial protection if the surface is scratched, and also the zinc oxide provides a barrier to air and water.

3 Alloying — an alloy is a mixture of two or more elements, at least one of which is a metal, and the resulting mixture has metallic properties. Alloys are often stronger and more resistant to corrosion than the pure metals they are made from. Stainless steel is an alloy that is resistant to corrosion.

Exam tip

Only iron and steel rust; other metals corrode.

Now test yourself

TESTED

7 What metal is used in galvanising?
8 What is the chemical name for rust?
H 9 Explain how magnesium can prevent iron from rusting by sacrificial protection.

Iron and its extraction

Extraction of iron in a blast furnace

REVISED

A simplified diagram of a blast furnace is shown in Figure 22.3.

The solid material put into the blast furnace is called the **charge**. It is made up of **iron ore** (**haematite**, Fe_2O_3), **limestone** (**calcium carbonate**, $CaCO_3$) and **coke** (**carbon**, C). Hot air is blasted in through pipes near the bottom of the blast furnace. Reduction of iron ore happens because the iron(III) oxide loses oxygen *or* Fe^{3+} ions gain electrons to form Fe atoms.

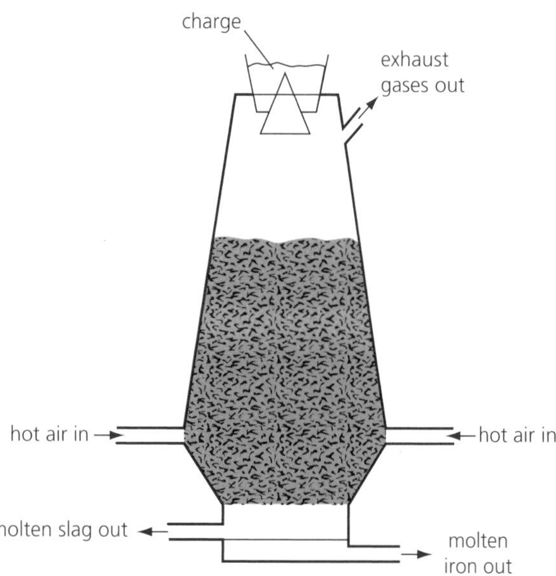

Figure 22.3 A blast furnace

There are five reactions involved in this process. The production of the reducing agent (carbon monoxide) occurs in two of them. The reduction of iron ore occurs in one reaction. Finally, removal of the acidic impurities occurs in two reactions.

Production of the reducing agent

1 Coke burns in oxygen (from hot air) to produce carbon dioxide:

$C + O_2 \rightarrow CO_2$

2 Carbon dioxide reacts with more coke (carbon) to produce the reducing agent, which is carbon monoxide:

$CO_2 + C \rightarrow 2CO$

Reduction of iron ore to iron

3 Iron(III) oxide reacts with carbon monoxide to produce molten iron and carbon dioxide:

$Fe_2O_3 + 3CO \rightarrow 2Fe + 3CO_2$

Note: This is the oxidation and reduction stage of the extraction of iron. Carbon (and carbon monoxide) will reduce iron(III) oxide to iron. Remember that the iron(III) oxide loses oxygen and loss of oxygen is reduction. The carbon monoxide is oxidised because it gains oxygen and gain of oxygen is oxidation.
For the iron, it may also be stated that iron(III) ions (Fe^{3+}) gain electrons ($Fe^{3+} + 3e^- \rightarrow Fe$), and gain of electrons is reduction.

Removal of acidic impurities

4 Calcium carbonate thermally decomposes to form calcium oxide and carbon dioxide:

$CaCO_3 \rightarrow CaO + CO_2$

5 Acidic silicon dioxide impurities react with calcium oxide to form molten slag (calcium silicate, $CaSiO_3$):

$SiO_2 + CaO \rightarrow CaSiO_3$

Exam tip

Many questions about iron involve the materials added to the blast furnace, the materials removed from it and the equations for the five reactions that occur during the process. The most common mistakes are to forget to write 'molten slag' and 'molten iron', and to get the order in which they are removed mixed up — remember that molten iron is denser than molten slag.

The molten slag and molten iron fall to the bottom of the furnace. Iron is denser than slag, so it sinks below the slag. They are tapped off separately at the base of the blast furnace.

Iron is a strong metal (particularly as the alloy steel, with between 0.2% and 2% carbon). This is why it is used in construction of buildings and bridges.

Now test yourself

TESTED

10 What is the chemical name and formula for slag?
11 Use balanced equations to explain how carbon monoxide is formed in the blast furnace.
12 Write a balanced symbol equation for the production of iron from iron(III) oxide.

Exam practice

1 In the reaction $N_2 + 3H_2 \rightarrow 2NH_3$, explain why nitrogen is reduced. [2 marks]
2 From the following gases:
hydrogen oxygen nitrogen carbon dioxide argon
(a) Which is gained in a reduction reaction? [1 mark]
(b) Which is gained in an oxidation reaction? [1 mark]
3 State the raw materials used in the extraction of iron from its ore. [4 marks]
4 Name one metal that can be used to prevent rusting by sacrificial protection. [1 mark]
5 What is meant by the following terms:
(a) rust [1 mark]
(b) redox [1 mark]
6 Complete the following paragraph using the terms below. [4 marks]
electrons oxygen hydrogen gain loss
Oxidation is the gain of _______ and the _______ of electrons and hydrogen. Reduction is the loss of _______ and the _______ of electrons and hydrogen.
7 When zinc reacts with copper(II) sulfate solution, a displacement reaction occurs. The zinc is converted to zinc ions and the copper ions are converted to copper atoms.
(a) Write a half-equation for the conversion of zinc atoms to zinc ions. [3 marks]
(b) Write a half-equation for the conversion of copper ions to copper atoms. [3 marks]
(c) State which of the above half-equations you have written is oxidation. [1 mark]
8 For the reaction $Mg + 2HCl \rightarrow MgCl_2 + H_2$, write half-equations for the oxidation and reduction processes and label them as oxidation and reduction. [8 marks]
9 In the ionic equation $Mg + Zn^{2+} \rightarrow Mg^{2+} + Zn$, which species is oxidised and which species is reduced? [2 marks]
10 What two conditions are necessary for rusting to occur? [2 marks]
11 State two barrier methods used to prevent iron from rusting. [2 marks]
12 Write a balanced symbol equation for the reduction of iron(III) oxide in the blast furnace using carbon monoxide. [3 marks]
13 Explain, using two balanced symbol equations, how acidic impurities are removed from iron ore in the blast furnace. [6 marks]
14 The following reactions occur in the blast furnace:
Reaction A carbon + oxygen → carbon dioxide
Reaction B carbon dioxide + carbon → carbon monoxide
(a) Explain why reaction A is described as oxidation. [2 marks]
(b) Write a balanced symbol equation for reaction B. [3 marks]
15 A piece of magnesium ribbon was wrapped around an iron nail to prevent it from rusting. Explain how the magnesium protects the iron from rusting. [2 marks]

Answers online

ONLINE

23 Rates of reaction

Factors affecting rate of reaction

REVISED

The **rate of reaction** is the rate of change of the reactants into the products in a chemical reaction. The rate depends on:

- the surface area or size of solid particles
- the concentration of solutions
- the temperature
- the presence of a catalyst.

Each of these factors can be studied experimentally. There are different methods of measuring rate — all of them measure a quantity against time, for example a change in mass or a change in gas volume. Other factors, such as light and pressure, may affect the rate of some chemical reactions, but these are not studied practically at GCSE.

Reactions used in rate of reaction experiments

REVISED

The following reactions are the most often used:

- A metal reacting with a dilute acid — for example, zinc or magnesium with dilute hydrochloric acid or dilute sulfuric acid. These reactions produce hydrogen gas:

  $Mg + 2HCl \rightarrow MgCl_2 + H_2$ *or* $Mg + H_2SO_4 \rightarrow MgSO_4 + H_2$

  $Zn + 2HCl \rightarrow ZnCl_2 + H_2$ *or* $Zn + H_2SO_4 \rightarrow ZnSO_4 + H_2$

- Calcium carbonate (often in the form of marble chips) reacting with dilute hydrochloric acid. This reaction produces carbon dioxide gas:

  $CaCO_3 + 2HCl \rightarrow CaCl_2 + H_2O + CO_2$

- The catalytic decomposition of hydrogen peroxide solution (H_2O_2). This reaction produces oxygen gas:

  $2H_2O_2 \rightarrow 2H_2O + O_2$

- The catalyst is manganese(IV) oxide, MnO_2 (also called manganese dioxide).
- The reaction between sodium thiosulfate solution and hydrochloric acid, which produces a precipitate of sulfur, making the solution go cloudy. The equation for the reaction is not required but you should know that sulfur causes the cloudiness.

Exam tip

The production of gases in these reactions allows the rate to be monitored in several ways. The mass of the reaction mixture will decrease as the gas is released. The volume of gas produced could be measured. The time for the reaction to stop (fizzing) could also be measured.

Measuring rate of reaction

Measuring a change in mass

REVISED

The reaction between marble chips (calcium carbonate, $CaCO_3$) and hydrochloric acid can be used to investigate how the size of solid particles affects the rate of the reaction.

Mass is 'lost' during this reaction because carbon dioxide escapes from the reaction vessel. Recording the loss in mass over a certain period of time at regular intervals, using the apparatus shown in Figure 23.1, gives an indication of the rate of the reaction. The cotton wool stops any liquid loss from the flask — there is **effervescence** (bubbling), which can cause the solution to splash out.

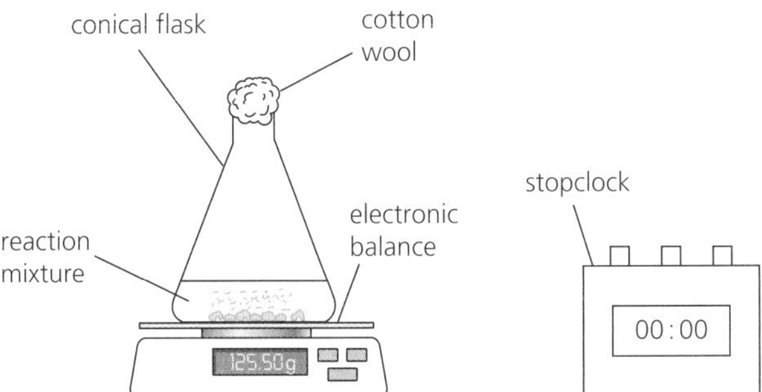

Figure 23.1 Investigating the effect of particle size on reaction rate

A measured mass of large marble chips was used first; and then the experiment was repeated using the same mass of smaller chips. The results are shown in a graph of mass plotted against time (Figure 23.2).

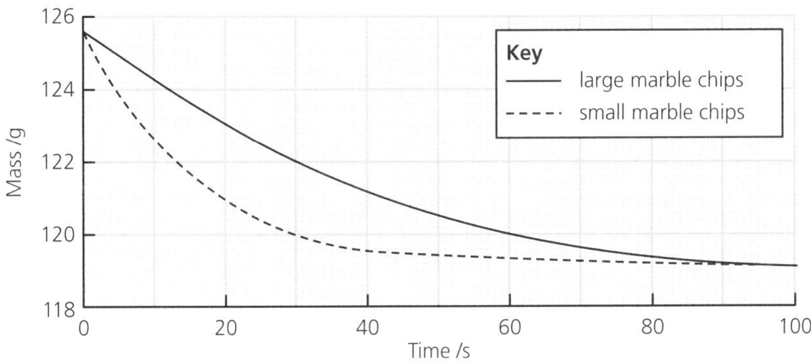

Figure 23.2 The effect of particle size on reaction rate

- From the graph you can see that the reaction occurred more rapidly with smaller marble chips. The smaller marble chips have a much larger surface area-to-volume ratio. There is more contact between the surface of the smaller marble chips and the acid. This causes a greater rate of reaction.
- As the same mass of marble chips and the same volume and concentration of acid were used, the graphs for both reactions start and level off at the same masses.
- The steeper initial slope of the curve for small marble chips indicates that the mass is decreasing more quickly. So the **rate of reaction** is higher with smaller solid particles.

Now test yourself

TESTED

1. State the effect on the rate of reaction between marble chips and hydrochloric acid of using smaller marble chips.
2. Write a balanced symbol equation for the reaction of magnesium with sulfuric acid.
3. Name the catalyst used to decompose hydrogen peroxide.

Exam tip

As hydrogen is the lightest gas, it is not practical to measure the mass or change in mass when hydrogen is the gas produced.

Exam tip

The reaction mixture here is labelled as this is a general diagram — you should label any chemicals in the flask and make sure they are in contact so that the reaction occurs. For example, the marble chips and the hydrochloric acid should both be labelled.

Measuring gas volume

REVISED

If a reaction produces a gas, a good method of measuring a rate of reaction is by collecting the gas in a gas syringe over a period of time. Figure 23.3 shows the apparatus used to measure the volume of a gas produced from a reaction. The volume of gas is measured by taking readings from the gas syringe at various time intervals.

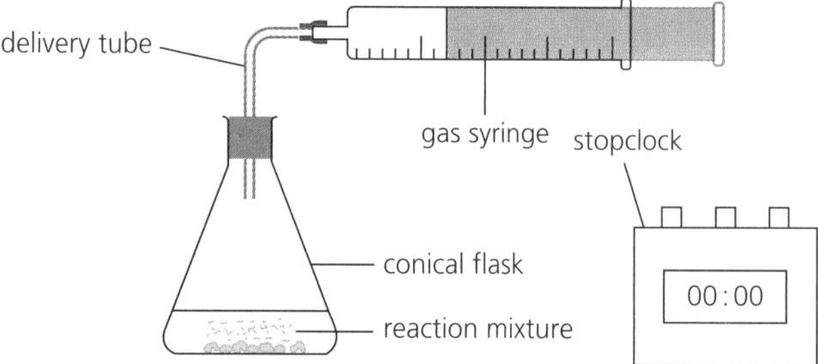

Figure 23.3 Investigating rate of reaction by measuring the volume of a gas produced

Exam tip

The marks for such a diagram are for the labels and are usually divided into four sections (the items *italicised* are the standard labels required):

- Preparation of the gas — you need to draw the reactants in contact with each other in the *conical flask*. If the reactants in the *reaction mixture* are named they should be labelled and they must be in contact.
- Connection — you need to draw a correctly fitted *delivery tube*. It may be a side arm tube or the arrangement shown in Figure 23.3.
- Collection of the gas — you need to draw a *gas syringe* correctly connected to the end of the delivery tube.
- Timing — a *stopclock* or *stopwatch* should be included with every rate experiment diagram.

Note: An inverted measuring cylinder filled with water over a beehive shelf in a trough of water could also be used to measure the volume of gas produced.

A typical graph obtained in experiments such as this is shown in Figure 23.4.

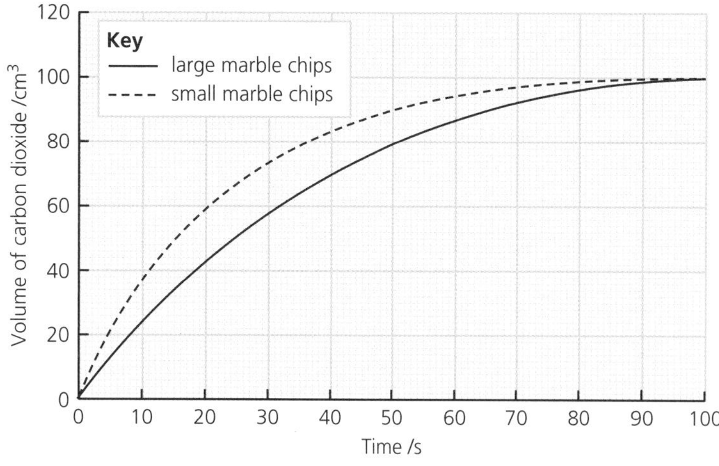

Figure 23.4 Graph showing how particle size affects the volume of gas produced

- The volume of gas produced increases more rapidly initially when small marble chips are used.
- The curves on the graph start and end at the same gas volume because the same mass of marble chips and the same volume and concentration of hydrochloric acid are used.

- For both curves on the graph note that the curve is steepest at the start, which indicates the reaction is fastest at the start and gradually gets slower as the reaction proceeds. This is because the reactants are being used up, and so the rate of reaction decreases as the reaction proceeds.

Measuring the time for a reaction to be completed

REVISED

In reactions that produce a gas or in which a solid disappears, the time until the reaction stops can be measured. For a reaction that produces a gas, the time for the reaction to finish can be determined by:

- reading the time when the mass first reaches its minimum on a graph of mass against time
- reading the time when the gas volume first reaches its maximum on a graph of gas volume against time
- timing the reaction until you see that the production of gas has stopped.

For a reaction in which a solid disappears, the reaction can be timed until no more solid can be seen in the reaction vessel.

Measuring production of a solid precipitate

REVISED

Sodium thiosulfate solution reacts with dilute hydrochloric acid and the reaction produces solid sulfur. If the solutions are mixed in a conical flask placed on a piece of white paper with a black cross on it, then as the solid sulfur is produced during the reaction it obscures the cross (Figure 23.5). The time is measured until the cross can no longer be seen when viewed from above. The higher the concentration of the sodium thiosulfate solution or hydrochloric acid, the less time is taken for the cross to be obscured, which means a higher rate of reaction.

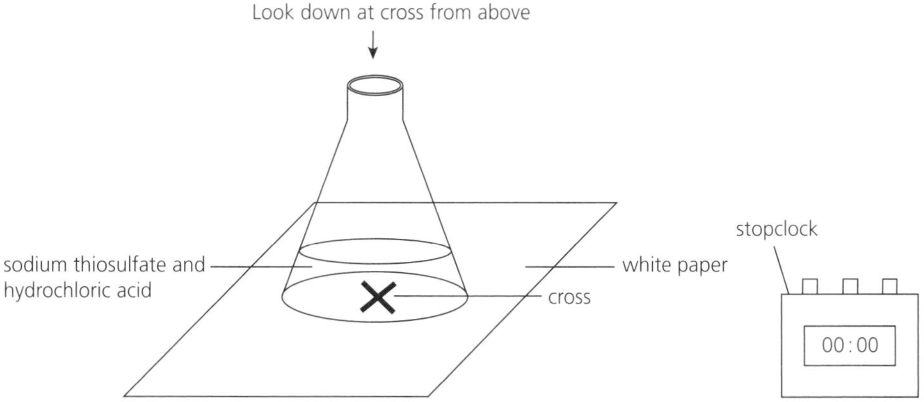

Figure 23.5 Measuring production of a solid precipitate

Calculating rate from time

REVISED

A value for rate can be calculated from the time for the reaction to be completed:

$$\text{rate} = \frac{1}{\text{time}}$$

The units of rate are s^{-1}.

Investigative work

Planning an investigation

REVISED

Planning an investigation involves several steps, but making an initial prediction and ensuring that the investigation is a fair test are both important.

Predictions about the outcome of an investigation are often written in the form 'as *quantity one* increases, *quantity two* increases/decreases'. (Quantity two will either increase or decrease as quantity one increases.)

For an investigation into the effects of changing the concentration of hydrochloric acid on the rate of reaction with magnesium ribbon, the prediction could read: as the *concentration of the acid* increases, the *rate of the reaction* increases.

A 'fair test' is the phrase used for an investigation in which one variable is changed (the independent variable), one variable is measured (the dependent variable) and all other variables are kept the same (controlled variables).

Validity and reliability of data collection

Validity is part of the overall design of the experiment. Suppose an experiment uses two concentrations of acid (0.2 mol/dm^3 and 0.25 mol/dm^3) to determine the link between concentration and rate of reaction. The results obtained from this may not be valid because of the limited range of the concentrations. The validity could be improved by using a greater number and range of concentrations of acid.

Reliability depends on whether or not the same result could be obtained again if the experiment were to be repeated. Measuring the time for a reaction to reach completion by simply observing the reaction may not be reliable because the results are judged by the observer. Measuring the gas volume or the change in mass can produce more reliable results because the time when the reaction finishes is determined by the apparatus used in the experiment — i.e. a gas syringe or an electronic balance.

Any **anomalies** are easily spotted from a graph — they are indicated by points that do not fall on or close to the line of best fit.

An **anomaly** is a piece of data that does not match the pattern shown in the rest of the investigation.

H Explaining the effects of temperature and concentration on reaction rate

REVISED

Reactions happen when reacting atoms, molecules or ions collide with each other. Only *some* of the collisions result in a reaction, and these are called successful collisions. The minimum energy that the reacting particles require in a collision in order for a collision to cause a reaction is called the **activation energy**. An increase in temperature increases the rate of most reactions. Activation energy is considered again on page 224.

Activation energy is the minimum energy required for a reaction to occur.

At a *higher temperature*:

- the particles have more energy and move faster
- this leads to more collisions between particles
- and so more successful collisions in a given period of time — more particles have more energy than the activation energy
- which increases the rate of the reaction.

H At a *higher concentration* of solution:

- there are more particles present in the same volume
- this leads to more collisions between particles
- and so more successful collisions in a given period of time
- which increases the rate of the reaction.

Worked example

An increase in the concentration of hydrochloric acid increases the rate of reaction of the acid with magnesium ribbon. Explain, in terms of particles, how this occurs. [4 marks]

Answer

As the concentration of hydrochloric acid increases, the number of hydrogen ions increases. [1] This increases the number of collisions [1], which increases the number of successful collisions [1] in a given period of time [1].

Exam tip

The link between particles and rate is important. You should always be able to explain why an increase in concentration causes an increase in rate, and also why an increase in temperature causes an increase in rate. The last part of the answer is often left out — you must indicate that an increase in rate is an increase in the number of successful collisions *per second* or in a *given period of time*.

Now test yourself

TESTED

4 What is the activation energy?

H 5 Explain how an increase in temperature increases the rate of reaction.

6 Name the precipitate formed in the reaction between sodium thiosulfate solution and hydrochloric acid.

Investigating the effect of the presence of a catalyst

REVISED

A reaction commonly used for the study of the effect of a catalyst is the decomposition of hydrogen peroxide using manganese(IV) oxide, MnO_2, as a catalyst. The reaction can be monitored using a gas syringe to collect and measure the volume of oxygen gas produced. Graphs of gas volume against time are then drawn (Figure 23.6).

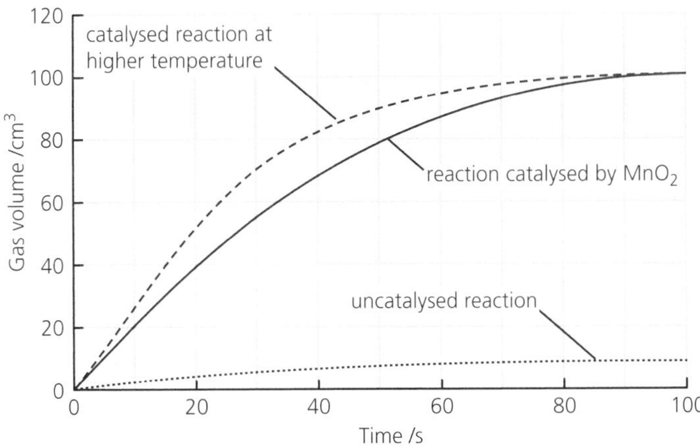

Figure 23.6 Graph showing volume of oxygen collected from the decomposition of hydrogen peroxide

From the graph, the catalysed reaction happens at a faster rate than the uncatalysed reaction, and the catalysed reaction at a higher temperature occurs at the highest rate of the reactions studied.

Points to note for the graph at higher temperature, when all other factors are kept the same (concentration, volume etc.), are:

- The gas volume starts at zero.
- The gas volume is higher at every time.
- The line on the graph levels off earlier.
- The line ends at same final gas volume.

Exam tip

Learn these four bullet points. They are important because they form the basis of many questions about gas volume as a measure of rate of reaction. At a lower temperature, the gas volume would start at zero, stay lower and level off later, but at the same final gas volume.

A change in the concentration of the hydrogen peroxide solution would change the volume at which the graph levels off. This is usually simple — for example, if the concentration is doubled, the volume at which the graph levels off will also double.

Metal with acid reactions

REVISED

The effects of changes in both temperature and concentration can be studied using two other reactions:

- zinc with dilute acid
- magnesium with dilute acid

Both reactions produce hydrogen gas, so gas volume can be used as a measure of the rate of the reaction.

- A change of temperature will alter the shape of the gas volume against time graph — it does not make any difference whether the acid or the metal is in excess.
- A change of concentration of the acid will alter the shape of the gas volume graph, but the final gas volume will be changed only if all the acid is used up (i.e. the metal is in excess).

Graphs for the reaction of magnesium with dilute hydrochloric acid are shown in Figure 23.7 for three different temperatures of the acid.

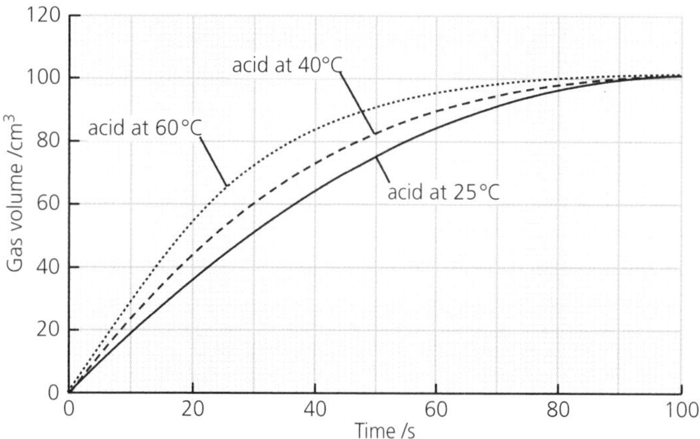

Figure 23.7 Gas volume against time graphs at different temperatures

Catalysts

REVISED

Catalysts work by providing an alternative reaction pathway of lower activation energy, which increases the number of successful collisions in a given period of time — this increases the rate of a reaction. (Remember that activation energy is the minimum energy required for a reaction to occur.)

A **catalyst** is a substance that increases the rate of a chemical reaction without being used up.

The mass of catalyst will be the same at the end of the reaction as at the start because a catalyst is not used up.

Transition metals and their compounds are often used as catalysts, for example manganese(IV) oxide in the catalytic decomposition of hydrogen peroxide and iron in the Haber process.

Now test yourself

TESTED

7 What is a catalyst?

8 Explain how a catalyst increases the rate of reaction.

9 1.0 g of manganese(IV) oxide is added to 25.0 cm^3 of 0.2 mol/dm^3 hydrogen peroxide solution. Once the decomposition of the hydrogen peroxide is complete, what mass of manganese(IV) oxide remains?

Prescribed practical C4

Investigate how changing a variable changes the rate of reaction

- Be able to carry out and describe practically how to set up rate experiments measuring gas volume, mass or time for the reaction to stop/obscure an X on a piece of paper.
- Be able to draw and label the individual and assembled apparatus required for any of the methods of measuring rate of reaction.
- Be able to record data and understand the need for repeating for reliability.
- Be able to calculate rate from time measurements using 1/time.
- Be able to plot graphs of gas volume against time, mass (or loss in mass) against time.
- Be able to explain and sketch graphs showing how changes in factors (concentration, temperature, particle size and presence of a catalyst) would affect the shape of the graphs.

Exam practice

1 Name the catalyst used in the decomposition of hydrogen peroxide. [1 mark]

2 Name the product which makes the solution become cloudy in the reaction between sodium thiosulfate solution and hydrochloric acid. [1 mark]

3 What piece of apparatus is used to measure gas volume? [1 mark]

4 Write a balanced symbol equation for the decomposition of hydrogen peroxide. [3 marks]

5 State the effect of an increase in temperature on the rate of a chemical reaction. [1 mark]

6 What is meant by the term 'catalyst'? [2 marks]

7 What gas is produced when magnesium reacts with dilute hydrochloric acid? [1 mark]

8 Apart from temperature and concentration, name one factor that will affect the rate of a chemical reaction. [1 mark]

9 Briefly describe a method you could use to determine the rate of reaction between marble chips and dilute hydrochloric acid. [2 marks]

H 10 Explain, in terms of particles, how increasing the temperature of dilute hydrochloric acid would affect the rate of reaction with zinc metal. [4 marks]

11 The graph shows the volume of hydrogen gas produced against time for the reaction of zinc with $25.0\,cm^3$ of $0.5\,mol/dm^3$ hydrochloric acid at 25°C. All the zinc was used up in the reaction.

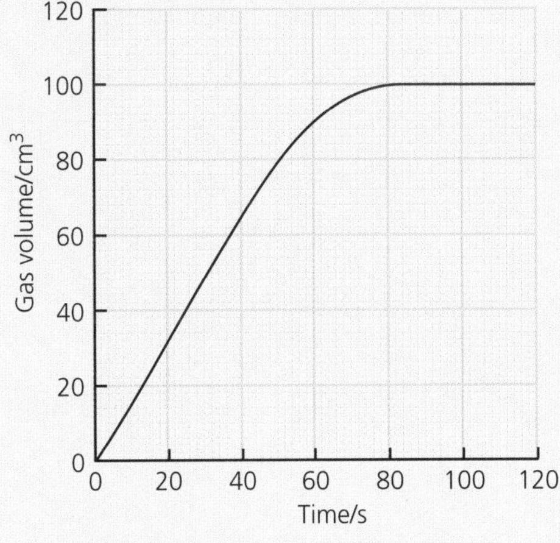

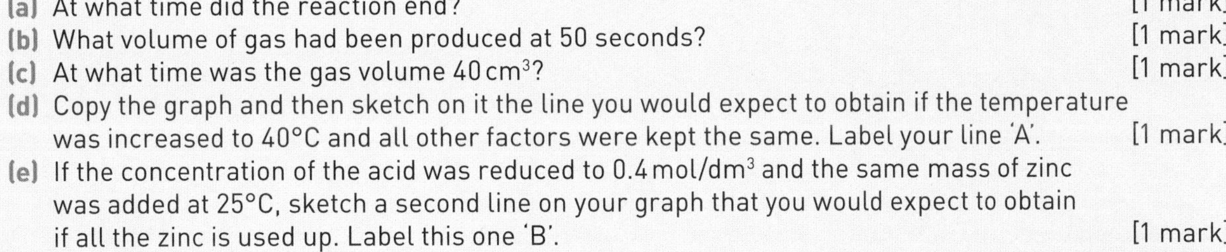

(a) At what time did the reaction end? [1 mark]

(b) What volume of gas had been produced at 50 seconds? [1 mark]

(c) At what time was the gas volume $40\,cm^3$? [1 mark]

(d) Copy the graph and then sketch on it the line you would expect to obtain if the temperature was increased to 40°C and all other factors were kept the same. Label your line 'A'. [1 mark]

(e) If the concentration of the acid was reduced to $0.4\,mol/dm^3$ and the same mass of zinc was added at 25°C, sketch a second line on your graph that you would expect to obtain if all the zinc is used up. Label this one 'B'. [1 mark]

Answers online

ONLINE

24 Equilibrium

Reversible reactions and equilibrium

Reversible reactions

REVISED

- Many reactions are **reversible**.
- The reactions start with the reactants and, as the products form, some products react to form reactants again.
- Equilibrium is achieved when the amount of reactants and products remains constant.
- An equilibrium is described as a **dynamic equilibrium** when the rate of the forward reaction is equal to the rate of the reverse reaction, resulting in the amount of reactants and products remaining constant.
- Where only the reactants and products are present, this is known as a **closed system**. Heat energy can be exchanged with the surroundings and external factors such as pressure can change.

A **reversible reaction** is one where reactants can change into products and the products can change back into the reactants.

Dynamic equilibrium occurs when the rate of the forward reaction is equal to the rate of the reverse reaction and the amounts of reactants and products remain constant.

In a **closed system** only the reactants and products are present.

Exam tip

The two points for dynamic equilibrium are that the amount of reactants and products remains constant *and* that the rate of the forward reaction is equal to the reverse reaction.

Understanding dynamic equilibrium

REVISED

- Imagine 100 year-11 pupils (A) and 100 year-12 pupils (B) moving randomly around an assembly hall. The assembly hall is the closed system.
- When they collide with enough energy they may pair up to form C. This is a reaction.
- Not all collisions may result in a reaction if there is not enough energy (the minimum energy required is the activation energy).
- If a pair collides with another pupil, they may split up again.
- Over time, the numbers of pairs (C) and singles (A and B) remain constant as the rate of formation of pairs is equal to the rate of break-up of pairs.
- Initially there are only singles (A and B, which are the reactants) and at equilibrium there is a mixture of pairs (C) and singles (A and B).

Dynamic equilibrium reactions

REVISED

For a general reaction:

$$A(g) + B(g) \rightleftharpoons C(g)$$

A reacts with B and forms C *and* C breaks down to form A and B. Imagine capital letter As and capital letter Bs moving around randomly. When they collide they may react if they collide with sufficient energy to form C. However C can also break up to release one A and one B.

The reaction starts with A and B only present and as the reaction starts some C appears. The reaction is described as having reached equilibrium when the number of A, B and C remains constant. However the A, B and C present are not always the same ones even though the number of each remains the same. They are constantly being formed and broken up. This is called a dynamic equilibrium, in which you have a constant amount of the reactants and products and the rates of the forward reaction and the reverse reaction are the same (Figure 24.1).

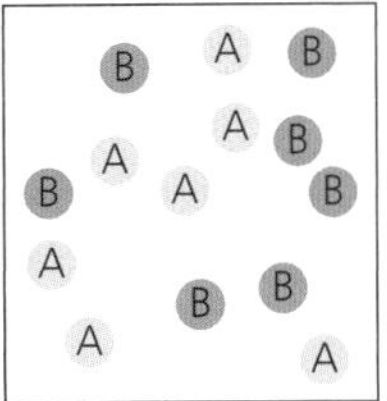
Initially only reactants

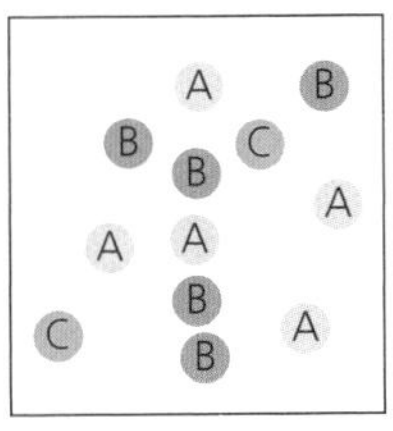
Forward reaction proceeds, making product C

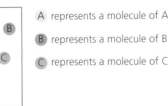

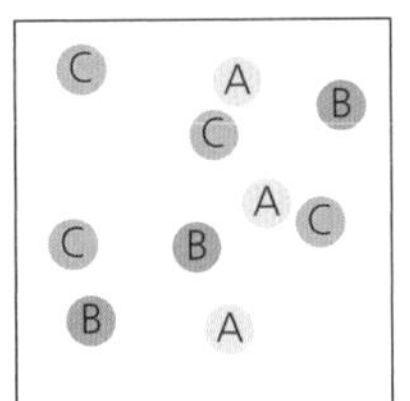
Equilibrium has been reached (4C, 3A and 3B)

A represents a molecule of A
B represents a molecule of B
C represents a molecule of C

Figure 24.1 Dynamic equilibrium

Example

Hydrogen reacts with iodine to form hydrogen iodide in a reversible reaction. Write an equation for the reversible reaction.

Answer

$H_2 + I_2 \rightleftharpoons 2HI$

The balanced symbol equation for a reversible reaction is balanced in the same way as for a standard reaction, but a reversible arrow is used.

Features of reversible reactions

REVISED

- A reversible reaction is shown with a $\rightleftharpoons$ arrow as opposed to a traditional $\rightarrow$ arrow. The double two-way arrow indicates that the reaction is reversible.
- The direction of a reversible reaction can be changed by altering the reaction conditions, such as temperature and pressure.

Now test yourself

TESTED

1 What is a reversible reaction?
2 What is meant by a dynamic equilibrium?
3 Write an equation for the reversible reaction between hydrogen and nitrogen forming ammonia.
4 For the equilibrium reaction $N_2O_4(g) \rightleftharpoons 2NO_2(g)$ how does the concentration of N_2O_4 change in the equilibrium if the pressure is increased?
5 What effect does a catalyst have on the position of equilibrium?

Exam practice

1 Write a balanced symbol equation for the reversible reaction between hydrogen and bromine, forming hydrogen bromide. [3 marks]
2 State two conditions which could affect the direction of a reversible reaction. [2 marks]

Answers online

ONLINE

25 Organic chemistry

Homologous series

Fossil fuels and living things are based on the element **carbon**. Organic chemicals are mostly obtained from crude oil. Carbon atoms can form four covalent bonds. Chains of carbon atoms can form when carbon atoms covalently bond with other carbon atoms. This means that there are a large number of carbon compounds — these are grouped into **homologous series** to simplify their study.

A **homologous series** is a family of organic compounds where members:

- have the same general formula
- show similar chemical properties
- show a gradation in their physical properties
- differ successively by a CH_2 unit.

Exam tip

The definition of a homologous series is often asked for in the exam. Often three out of the four points listed are required, but also two of the points could be given and the other two asked for, so make sure you know all of them.

General formulae for organic molecules

REVISED

A general formula is a one involving a variable number, *n*, which allows the molecular formula of any compound in a particular homologous series to be determined.

There are four homologous series required for GCSE chemistry: alkanes, alkenes, alcohols and carboxylic acids.

- **Alkanes** are relatively unreactive. They have the general formula C_nH_{2n+2}.
- **Alkenes** are more reactive than alkanes. They have the general formula C_nH_{2n}.
- Alcohols and carboxylic acids are reactive organic compounds.
- Alcohols have the general formula $C_nH_{2n+1}OH$.
- Carboxylic acids have the general formula $C_nH_{2n}O_2$.

Alcohols and carboxylic acids will be discussed later.

Table 25.1 summarises the general formulae for the four homologous series studied at this level.

Table 25.1 Formulae of some organic compounds

Homologous series	General formula	Values of *n* required
Alkanes	C_nH_{2n+2}	1, 2, 3 and 4
Alkenes	C_nH_{2n}	2, 3 and 4
Alcohols	$C_nH_{2n+1}OH$	1, 2 and 3
Carboxylic acids	$C_nH_{2n}O_2$	1, 2, 3 and 4

In terms of naming organic compounds:

- For organic compounds containing one carbon atom, the prefix **meth-** is used.
- For organic compounds containing two carbon atoms, the prefix **eth-** is used.
- For three carbon atoms, **prop-** is used
- For four carbon atoms, **but-** is used.

The general formula of the alkanes is C_nH_{2n+2}. If $n = 3$ then the formula of the alkane with three carbon atoms is $C_3H_{(2\times3+2)} = C_3H_8$. This is the molecular formula of propane.

For alkenes the general formula is C_nH_{2n}. This means that the number of hydrogen atoms in an alkene is twice the number of carbon atoms. An alkene with three carbon atoms is C_3H_6. This is the molecular formula of propene.

Sometimes numbers are required within the name of an organic compound; this is due to the position of a group, such as OH and C=C, in the carbon chain. An alkene with four carbon atoms can have the C=C starting at carbon 1 (but-1-ene) or carbon 2 (but-2-ene). The C=C starting at carbon 3 would be the same as the C=C starting at carbon 1 counting from the other end.

H For alcohols the general formula is $C_nH_{2n+1}OH$. An alcohol with two carbon atoms is C_2H_5OH. This is the molecular formula of ethanol.

An alcohol with three carbon atoms can have the OH group at carbon 1 in the chain or carbon 2. This means we have propan-1-ol (meaning the OH group is bonded to carbon 1) and propan-2-ol (meaning the OH group is bonded to carbon 2). The OH bonded to carbon 3 is the same as the OH group bonded to carbon 1.

Now test yourself

TESTED

H 1 C_2H_5COOH is an organic compound. State the homologous series to which it belongs.
2 What is a homologous series?
3 What is the general formula of the alkenes?

Drawing structural formulae for organic molecules

REVISED

- A structural formula shows the covalent bonds between the atoms in a molecule.
- Atoms are represented as the symbol for that atom.
- A line (—) is used to represent a single covalent bond.
- A double line (=) is used to represent a double covalent bond.

It is important to count the number of bonds around the atoms in any organic molecules you draw. This will mean that you will not draw extra bonds or miss some.

Exam tip

Look back at pages 100–101 to revise the formation of covalent bonds and single and double covalent bonds.

Carbon atoms can form four covalent bonds. A double bond, for example C=C, counts as two bonds to each C atom.

Hydrogen atoms can only form one covalent bond, so there should only be one line to an H atom.

Oxygen atoms can form two covalent bonds, so there should be two lines to an O atom. A double bond, for example C=O, counts as two bonds to the O atom (and also two bonds to the C atom).

The four examples in Figure 25.1 come from the four different homologous series (alkanes, alkenes, alcohols and carboxylic acids), so make sure that you can see that each C atom forms four covalent bonds (count four lines to each pink C atom), each H atom forms one covalent bond (count one line to each blue H atom) and each O atom forms two covalent bonds (count two lines to each green O atom).

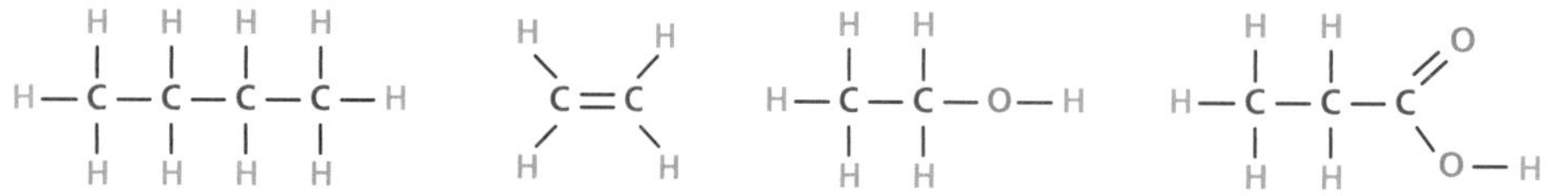

> **Exam tip**
>
> All covalent bonds are usually shown, but the –O–H group in alcohols and carboxylic acids can often be shown as –OH for simplicity. There is still a bond between the O and H atom. If you are asked to draw a structure showing all bonds it should include the bond between the O and H.

an alkane (butane)

an alkene (ethene)

an alcohol (ethanol)

a carboxylic acid (propanoic acid)

Figure 25.1 Structural formulae

Hydrocarbons

All alkanes and alkenes are **hydrocarbons**.

> A **hydrocarbon** is a compound (or molecule) consisting of hydrogen and carbon only.

> **Exam tip**
>
> The definition of a hydrocarbon is a very common question. The key points are that a hydrocarbon is a *compound (or molecule)* consisting of *only carbon and hydrogen*.

Alkanes

REVISED

Alkanes have the general formula C_nH_{2n+2}. They are saturated hydrocarbons. Saturated means that the molecules do not contain any carbon–carbon double bonds (C=C). Unsaturated hydrocarbons contain at least one C=C.

> **Exam tip**
>
> The general formula for an alkane means that, with *n* for number of carbon atoms, an alkane will have 2*n* + 2 hydrogen atoms. For example, if *n* = 4 (four carbon atoms), 2*n* + 2 = 10, so the alkane has four carbon atoms and 10 hydrogen atoms. Its molecular formula is therefore C_4H_{10}. You may be expected to do this for a higher number of carbon atoms. Try *n* = 5 (C_5H_{12}), *n* = 6 (C_6H_{14}) etc.

Table 25.2 names the first four alkanes, with their molecular formulae, structural formulae and states at room temperature and pressure. The *n* value is based on *n* in the general formula C_nH_{2n+2}.

Table 25.2 Alkanes

n	Name	Molecular formula	Structural formula	State at room temperature and pressure
1	Methane	CH_4	H–C(H)(H)–H	Gas
2	Ethane	C_2H_6	H–C(H)(H)–C(H)(H)–H	Gas
3	Propane	C_3H_8	H–C(H)(H)–C(H)(H)–C(H)(H)–H	Gas
4	Butane	C_4H_{10}	H–C(H)(H)–C(H)(H)–C(H)(H)–C(H)(H)–H	Gas

Now test yourself

TESTED

4 State the name of C_4H_{10}.
5 What is the molecular formula of ethane?
6 Write the molecular formula for the alkane with seven carbon atoms.

Fractional distillation of crude oil

REVISED

Fractional distillation separates crude oil into simpler mixtures of hydrocarbons called fractions. Fractional distillation is carried out in a fractionating column (Figure 25.2).

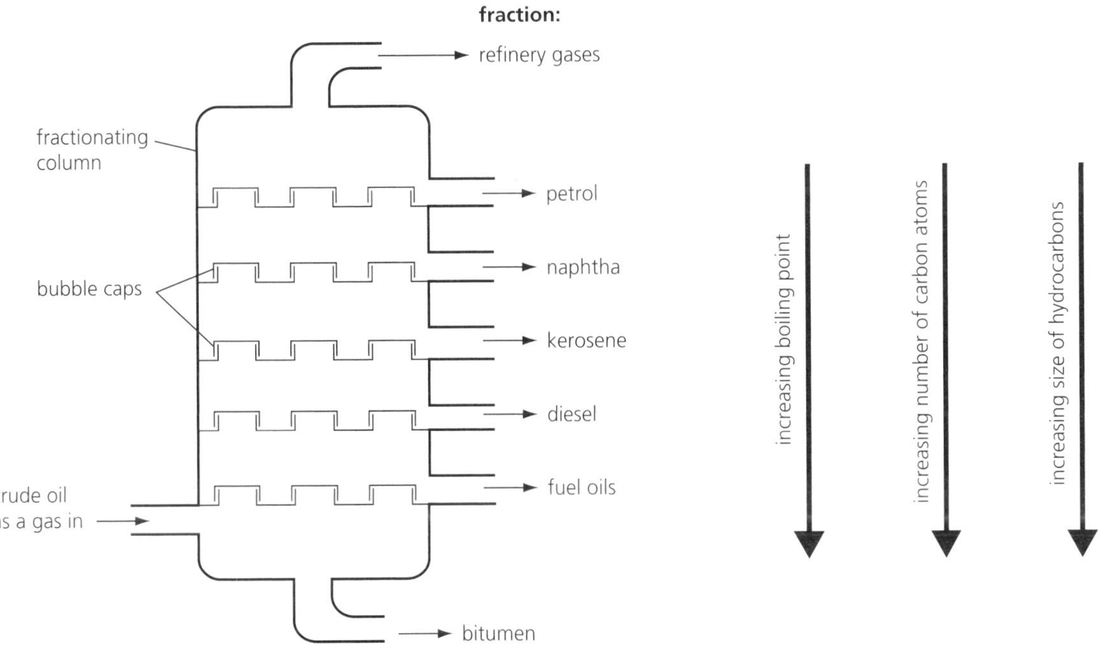

Figure 25.2 A fractionating column is used to separate crude oil into fractions

- The crude oil enters at the bottom as a hot, gaseous mixture.
- The fractionating column has bubble caps, which allow gases to move upwards.
- The temperature decreases *up* the column.
- As the gases move up the column, hydrocarbons condense when the temperature of the column is the same as their boiling point.

Table 25.3 shows the major fractions obtained from crude oil in order of increasing size of molecules and increasing boiling point, along with some of their uses.

Table 25.3 Uses of some fractions from crude oil

Fraction	Use
Refinery gases	Bottled gases
Petrol	Fuel for cars
Naphtha	Chemicals and plastics
Kerosene	Fuel for aircraft
Diesel	Fuel for cars and trains
Fuel oil	Fuel for ships
Bitumen	Surfacing roads as road tar, and sealing roofs

Crude oil is a finite resource, meaning that it will run out. The fractions obtained from crude oil contain hydrocarbons, many of which are used as fuels but also act as a feedstock for the petrochemical industry. This means they are used as a starting point for the synthesis of other chemicals, such as polymers and pharmaceutical drugs.

Cracking

Larger hydrocarbon molecules (such as those from fuel oils and bitumen) are not as useful as smaller ones. This is because they are not as useful as fuels. They are therefore broken down by **cracking**.

- Cracking uses heat in the absence of air or a catalyst. If heat is used it is called thermal cracking; if a catalyst is used it is called catalytic cracking.
- The equations for cracking only involve hydrocarbons and must have the same number of carbon atoms and hydrogen atoms on the left and right.
- The products of cracking must include at least one alkene (an unsaturated hydrocarbon containing a C=C), which have the general formula C_nH_{2n}, because there are not enough hydrogen atoms for all the products to be alkanes.

Cracking is the breakdown of larger/longer less useful saturated hydrocarbon molecules into smaller/shorter ones that are more useful, some of which are unsaturated (contain at least one C=C).

Example 1

Octane (C_8H_{18}) can be cracked to form butane and ethene (C_2H_4). Write a balanced symbol equation for this reaction.

Answer

The reactants and products as detailed in the question are as follows:

$$C_8H_{18} \rightarrow C_4H_{10} + C_2H_4$$

This is not balanced because the number of carbon atoms and hydrogen atoms are not the same on both sides of the equation. To balance the atoms, $2C_2H_4$ are required, so the balanced symbol equation is:

$$C_8H_{18} \rightarrow C_4H_{10} + 2C_2H_4$$

Exam tip

You are given the formula of ethene here, but once you have studied alkenes the formula would not be given.

Exam tip

Always balance the carbon atoms and hydrogen atoms on each side. Remember that there should be at least one alkene on the right-hand side.

Example 2

Nonane (C_9H_{20}) is cracked to form hexane (C_6H_{14}) and one other product. Write a balanced symbol equation for the reaction.

Answer

$$C_9H_{20} \rightarrow C_6H_{14} + C_xH_y$$

C_xH_y is the other product, and you can see that $x = 3$ and $y = 6$ to make the number of carbon and hydrogen atoms balance in the equation. This means that C_xH_y is C_3H_6. The equation is:

$$C_9H_{20} \rightarrow C_6H_{14} + C_3H_6$$

Exam tip

You might be expected to name the alkene product, which you will be able to do once you have studied alkenes. Remember that to identify an alkane the general formula C_nH_{2n+2} must be followed, and for alkene it must be C_nH_{2n}.

The alkenes produced from cracking can be used by the petrochemical industry to synthesise polymers and pharmaceutical drugs.

Now test yourself

TESTED

7 What process is used to separate the components of crude oil?
8 What is cracking?
9 Cracking of pentane (C_5H_{12}) produces ethane. Write the formula for the one other product.

Alkenes

REVISED

Table 25.4 names four alkenes, with their molecular formulae, structural formulae and states at room temperature and pressure. The *n* value is based on *n* in the general formula C_nH_{2n}.

Table 25.4 Alkenes

n	Name	Molecular formula	Structural formula	State at room temperature and pressure
2	Ethene	C_2H_4	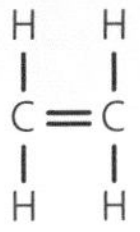 $CH_2{=}CH_2$	Gas
3	Propene	C_3H_6	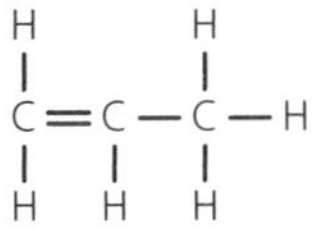 $CH_2{=}CH{-}CH_3$	Gas
4	But-1-ene	C_4H_8	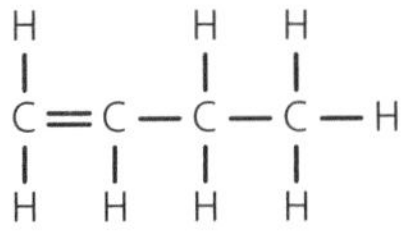 $CH_2{=}CH{-}CH_2{-}CH_3$	Gas
4	But-2-ene	C_4H_8	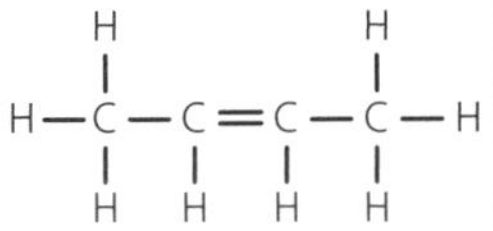 $CH_3{-}CH{=}CH{-}CH_3$	Gas

But-1-ene and but-2-ene have the same molecular formula but they have different structural formulae. The name but-1-ene is used because the C=C starts from the first carbon in the chain shown in purple in Figure 25.3.

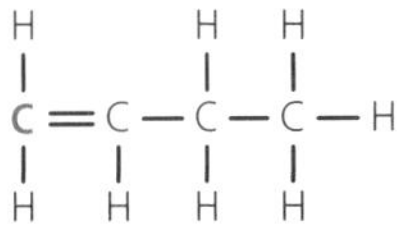

$CH_2{=}CH{-}CH_2{-}CH_3$

Figure 25.3 But-1-ene

In but-2-ene (Figure 25.4) C=C starts at the second carbon atom (counting from either end).

The lowest number is taken counting from either end.

```
    H               H
    |               |
H — C — C = C — C — H
    |   |   |   |
    H   H   H   H
```

Figure 25.4 But-2-ene

Example

Compounds with six carbon atoms begin with the prefix hex-. Suggest a chemical name for this organic molecule.

```
    H           H   H   H
    |           |   |   |
H — C — C = C — C — C — C — H
    |   |   |   |   |   |
    H   H   H   H   H   H
```

It is an alkene because it has a C=C, so it is hexene.

The C=C starts at carbon 2 from the left and carbon 4 from the right, so the lowest number is used. This is hex-2-ene.

Now test yourself

TESTED

10 Name C_3H_6.
11 What is the molecular formula of ethene?
12 Name the two alkenes that have the molecular formula C_4H_8.

Alcohols and carboxylic acids

Alcohols

REVISED

Table 25.5 names the first four alcohols, with their molecular formulae, structural formulae and states at room temperature and pressure. The *n* value is based on *n* in the general formula $C_nH_{2n+1}OH$.

Table 25.5 Alcohols

n	Name	Molecular formula	Structural formula	State at room temperature and pressure
1	Methanol	CH_3OH	H | H — C — OH | H	Liquid
2	Ethanol	C_2H_5OH	H H | | H — C — C — OH | | H H	Liquid
3	Propan-1-ol	C_3H_7OH	H H H | | | H — C — C — C — OH | | | H H H	Liquid
3	Propan-2-ol	C_3H_7OH	H H H | | | H — C — C — C — H | | | H OH H	Liquid

The presence of the bond between the O and H atom depends on the question. If a question asks for all bonds to be shown there should be a bond shown between the O and the H, as O–H, but normally it is shown as –OH.

Propan-1-ol and propan-2-ol have the same molecular formula but different structural formulae. The name propan-1-ol is used because the OH group is bonded to the first carbon in the chain labelled purple in Figure 25.5.

In propan-2-ol (Figure 25.6) the OH group is bonded to the second carbon atom (counting from either end), again labelled purple.

The lowest number is taken counting from either end.

```
   H   H   H
   |   |   |
H— C — C — C — OH
   |   |   |
   H   H   H
```

Figure 25.5 Propan-1-ol

```
   H   H   H
   |   |   |
H— C — C — C — H
   |   |   |
   H   OH  H
```

Figure 25.6 Propan-2-ol

Example

Organic compounds with five carbon atoms begin with pent-. Suggest a chemical name for this compound.

```
   H   H   H   H   H
   |   |   |   |   |
H— C — C — C — C — C — H
   |   |   |   |   |
   H   H   H   OH  H
```

This is pentan-2-ol. The OH group is bonded to carbon 2, counting from the right-hand side.

Carboxylic acids

REVISED

Table 25.6 names the first four carboxylic acids, with their molecular formulae, structural formulae and states at room temperature and pressure. The *n* value is based on *n* in the general formula $C_nH_{2n}O_2$, but they are often written showing the functional group as COOH. So a carboxylic acid with three carbon atoms has the molecular formula $C_3H_6O_2$, but this would most often be written C_2H_5COOH (check that the atoms add up to $C_3H_6O_2$).

Table 25.6 Carboxylic acids

n	Name	Molecular formula	Structural formula	State at room temperature and pressure
1	Methanoic acid	HCOOH	H–C(=O)–OH	Liquid
2	Ethanoic acid	CH_3COOH	CH_3–C(=O)–OH (all bonds shown)	Liquid
3	Propanoic acid	C_2H_5COOH	CH_3CH_2–C(=O)–OH (all bonds shown)	Liquid
4	Butanoic acid	C_3H_7COOH	$CH_3CH_2CH_2$–C(=O)–OH (all bonds shown)	Liquid

The presence of the bond between the O and H atom depends on the question. If a questions asks for all bonds to be shown, there should be a bond shown between the O and H atom, as O–H, but normally it is shown as –OH.

Now test yourself

TESTED

13 Name the carboxylic acid with four carbon atoms.
14 What is the molecular formula of ethanoic acid?
15 What is the name of C_2H_5COOH?

Functional groups

REVISED

Many organic compounds have a **functional group**.

Alkanes do not have a functional group and so are a less reactive homologous series. However, they do undergo combustion. Natural gas is mainly methane and bottled gas is a mixture of propane and butane.

- The functional group of alkenes is C=C
- The functional group of alcohols is −OH
- The functional group of carboxylic acids is:

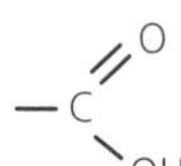

> A **functional group** is a reactive group in a molecule. It causes many of the reactions of the compounds in the homologous series.

Combustion of organic compounds

> **Combustion** is the reaction in which a fuel reacts with oxygen, producing oxides and releasing heat.

The higher the percentage carbon content by mass of an organic compound, the more orange the flame with which it burns in air. A lower percentage carbon content causes the flame to be blue, as opposed to orange.

Complete and incomplete combustion

REVISED

Complete combustion occurs when a fuel burns in a plentiful supply of oxygen/air to form carbon dioxide and water, and release heat. For example:

CH_4	+	$2O_2$	→	CO_2	+	$2H_2O$
methane		oxygen		carbon dioxide		water
$2C_2H_6$	+	$7O_2$	→	$4CO_2$	+	$6H_2O$
ethane		oxygen		carbon dioxide		water

Incomplete combustion occurs when a fuel burns in a limited supply of oxygen/air to form carbon monoxide (and sometimes soot — carbon) and water, and release heat. For example:

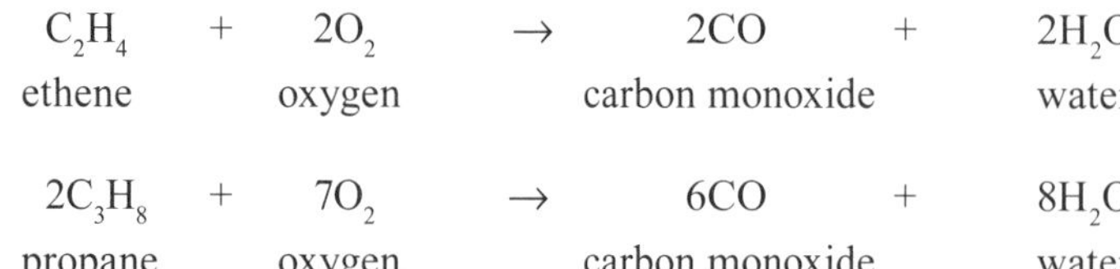

C_2H_4	+	$2O_2$	→	$2CO$	+	$2H_2O$
ethene		oxygen		carbon monoxide		water
$2C_3H_8$	+	$7O_2$	→	$6CO$	+	$8H_2O$
propane		oxygen		carbon monoxide		water

> **Exam tip**
>
> Equations for combustion are common in organic chemistry questions. Equations are not required for the production of soot, but you may be asked to write a balanced symbol equation for incomplete combustion, forming carbon monoxide and water.

Carbon monoxide is a colourless and odourless gas. It is toxic because it binds to haemoglobin in blood and reduces the blood's capacity to carry oxygen. This is one of the reasons why boilers (particularly inside properties) have to be serviced regularly to prevent the formation of carbon monoxide. Also, carbon monoxide sensors are often fitted in properties to detect the presence of the gas.

Flueless gas fires are fitted with catalytic convertors, which convert carbon monoxide to carbon dioxide, which is not toxic.

Balancing equations for combustion of hydrocarbons

REVISED

To balance a complete combustion equation:

1. The number of carbon atoms in the hydrocarbon and the balancing number in front of the CO_2 are the same.
2. The number of hydrogen atoms in the hydrocarbon is divided by 2 to get the number in front of the H_2O.
3. Count the *total number of oxygen atoms* in the CO_2 (remember that CO_2 has 2) and the H_2O, and divide by 2 to get the number in front of O_2.
4. If the number in front of O_2 has a half (e.g. 2½) multiply all the balancing numbers by 2 to get whole numbers.

Exam tip

If you are balancing an incomplete combustion equation, replace carbon dioxide with carbon monoxide in step 1 and step 3. Remember that carbon monoxide, CO, has only one oxygen atom per molecule.

Ⓗ Combustion of alcohols

REVISED

Alcohols also undergo combustion. Alcohols have a lower percentage carbon content than alkanes or alkenes, and so burn with a blue flame (sometimes with an orange tip). Alcohols normally undergo complete combustion, but in a very limited supply of oxygen, carbon monoxide and even soot (carbon) may be formed.

Here are the word equations and balanced symbol equations for complete combustion:

$2CH_3OH$	+	$3O_2$	→	$2CO_2$	+	$4H_2O$
methanol		oxygen		carbon dioxide		water
C_2H_5OH	+	$3O_2$	→	$2CO_2$	+	$3H_2O$
ethanol		oxygen		carbon dioxide		water

For incomplete combustion:

C_2H_5OH	+	$2O_2$	→	$2CO$	+	$3H_2O$
ethanol		oxygen		carbon monoxide		water

Carboxylic acids do not undergo combustion easily.

Exam tip

Balancing the equations for the combustion of alcohols follows the same process given for hydrocarbons — but remember to take the oxygen atom in the alcohol into account when counting the total oxygen atoms on both sides of the equation in step 3.

Testing the products of combustion

REVISED

The main products of combustion are carbon dioxide and water vapour. The presence of these two products can be detected using two chemical tests:

- Cool the gases and a colourless liquid will condense — add this liquid to anhydrous copper(II) sulfate, any water present will change the anhydrous copper(II) sulfate from white to blue.
- Bubble the gases through limewater — any carbon dioxide present changes the limewater from colourless to milky.

Now test yourself

TESTED

Ⓗ 16 Name the products of incomplete combustion of ethene.

17 Explain why incomplete combustion is hazardous.

Ⓗ 18 Write a balanced symbol equation for the complete combustion of methanol.

Chemistry of hydrocarbons

Chemistry of alkanes

REVISED

Alkanes are relatively unreactive organic molecules. They do undergo combustion.

Observations: Alkanes burn with an orange flame, releasing heat and producing colourless gases. For example:

$$C_3H_8 + 5O_2 \rightarrow 3CO_2 + 4H_2O$$

Chemistry of alkenes

REVISED

H Combustion of alkenes

Observations: Alkenes burn with an orange flame, releasing heat and producing colourless gases. For example:

$$C_2H_4 + 3O_2 \rightarrow 2CO_2 + 2H_2O$$

Addition reactions of alkenes

All other reactions of alkenes are addition reactions. An addition reaction is a reaction in which a molecule is added across a double bond, such as the C=C double bond in alkenes. For example:

$$>C=C< + X{-}Y \longrightarrow -\overset{|}{\underset{X}{C}}-\overset{|}{\underset{Y}{C}}-$$

Addition reactions only produce one product. The molecule X–Y can be bromine (Br–Br), hydrogen (H–H) or water (H–OH).

Reaction of alkenes with bromine

Ethene (and other alkenes) react with bromine (or the bromine in bromine water). For example:

$$C_2H_4 + Br_2 \rightarrow C_2H_4Br_2$$

$$\underset{H}{\overset{H}{C}}=\underset{H}{\overset{H}{C}} + Br_2 \longrightarrow H-\underset{Br}{\overset{H}{C}}-\underset{Br}{\overset{H}{C}}-H$$

Exam tip

You do not need to name the product of the reaction of an alkene with bromine.

Saturation and unsaturation

Alkanes have no C=C double bonds — they are **saturated**. Alkenes have one C=C double bond per molecule — they are **unsaturated**.

The test for a C=C double bond or unsaturation is to add the substance to bromine water and mix well. An unsaturated substance/any alkene will cause a colour change from orange/brown to colourless; a saturated substance/any alkane will cause no change and the bromine water will remain orange/brown.

Note: Alcohols and carboxylic acids do not contain the C=C functional group and so are saturated — when added to bromine water the colour will remain orange/brown. Bromine water is often used to distinguish between alkanes and alkenes.

A **saturated** compound contains no C=C bonds.

An **unsaturated** compound contains at least one C=C bond.

Reaction of ethene with steam

Ethene reacts with steam, producing ethanol:

$$C_2H_4 + H_2O \rightarrow C_2H_5OH$$

```
H   H                      H   H
|   |                      |   |
C = C   +  H2O  ───►   H — C — C — H
|   |                      |   |
H   H                      H   OH
```

Reaction of ethene with hydrogen

Alkenes react with hydrogen to produce an alkane. For example:

C_2H_4 +	H_2	$\rightarrow$	C_2H_6
ethene			ethane

Addition polymerisation

REVISED

Addition **polymerisation** is the process of adding molecules together to form a **polymer** as the only product — the long molecule is the polymer.

- The simple molecule from which a polymer is formed is called the **monomer**.
- The monomer has a C=C double bond.
- The monomer is an **alkene**.
- The polymer is shown as the monomer with only a single bond in a square bracket.
- '*n*' molecules of monomer must be at the beginning of the equation.
- The polymer structure has '*n*' after it to show that the polymer repeats *n* times.

Polymerisation is the process of creating a long chain molecule from small molecules. The small molecules form the repeating unit in the polymer.

Figure 25.7 shows the general equation for addition polymerisation.

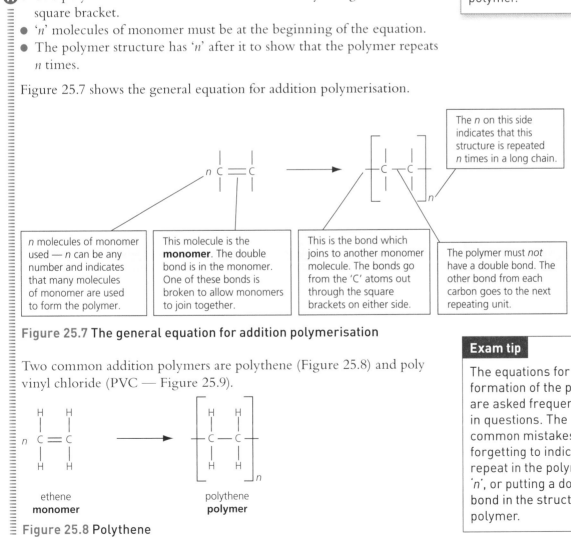

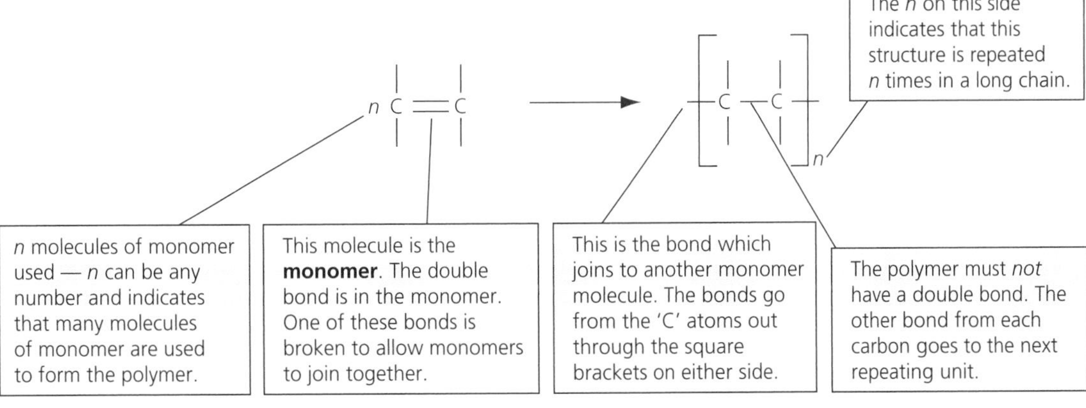

n C=C
n
The n on this side indicates that this structure is repeated n times in a long chain.
n molecules of monomer used — n can be any number and indicates that many molecules of monomer are used to form the polymer.
This molecule is the monomer. The double bond is in the monomer. One of these bonds is broken to allow monomers to join together.
This is the bond which joins to another monomer molecule. The bonds go from the 'C' atoms out through the square brackets on either side.
The polymer must not have a double bond. The other bond from each carbon goes to the next repeating unit.

Figure 25.7 The general equation for addition polymerisation

Two common addition polymers are polythene (Figure 25.8) and poly vinyl chloride (PVC — Figure 25.9).

```
  H   H             [ H   H ]
  |   |             [ |   | ]
n C = C   ───►    —[— C — C —]—
  |   |             [ |   | ]
  H   H             [ H   H ]n
```

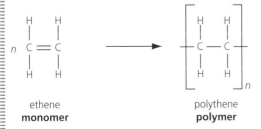

ethene **monomer** — polythene **polymer**

Figure 25.8 Polythene

Exam tip

The equations for the formation of the polymers are asked frequently in questions. The most common mistakes are forgetting to indicate the repeat in the polymer using '*n*', or putting a double bond in the structure of the polymer.

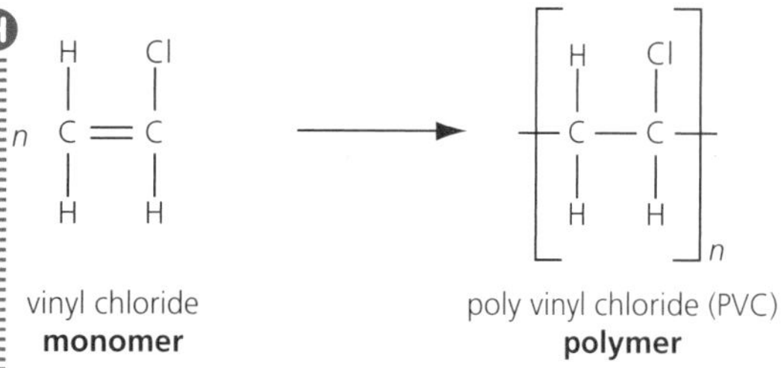

H Cl | | n C=C | | H H → [–C(H)(H)–C(Cl)(H)–]n

vinyl chloride
monomer

poly vinyl chloride (PVC)
polymer

Figure 25.9 PVC

Exam tip

Vinyl chloride is also called chloroethene. PVC can also be called poly(chloroethene).

Example 1

The following alkene is called phenylethene. It is commonly known as styrene.

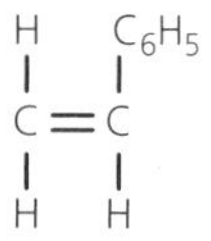

H C_6H_5 | | C=C | | H H

Draw the structure of the polymer formed when phenylethene undergoes addition polymerisation, and suggest the common name of the polymer.

Answer

The structure of the polymer is shown as:

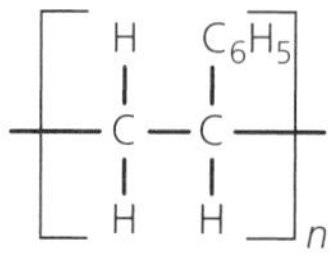

[–C(H)(H)–C(C_6H_5)(H)–]n

The common name of the polymer is polystyrene.

Example 2

The following structure shows part of a polymer.

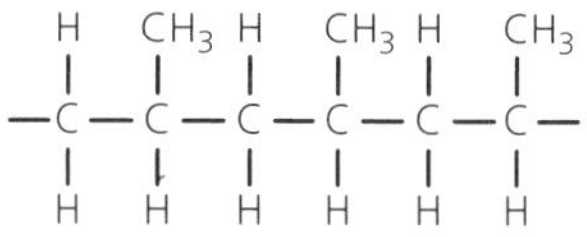

–C(H)(H)–C(CH_3)(H)–C(H)(H)–C(CH_3)(H)–C(H)(H)–C(CH_3)(H)–

Draw the structure of the monomer from which this addition polymer is formed.

Answer

The repeating unit of the monomer is shown in the square, and this repeating unit is formed from the monomer.

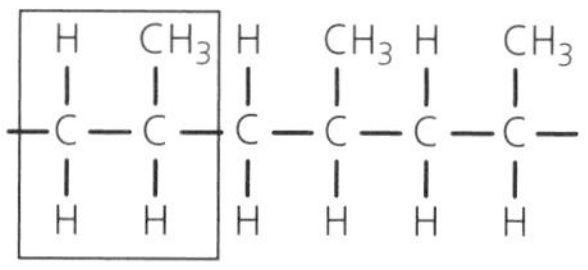

[–C(H)(H)–C(CH_3)(H)–] –C(H)(H)–C(CH_3)(H)–C(H)(H)–C(CH_3)(H)–

This means the monomer is an alkene with the following structure:

H CH_3 | | C=C | | H H

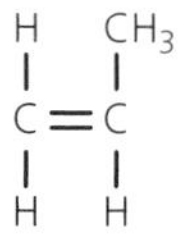

Now test yourself

TESTED

19 Name the product of the reaction of ethene with steam.

20 What chemical is used to test for a C=C bond and what is observed if it is present?

21 Name the type of polymerisation that forms polythene.

Chemistry of alcohols and carboxylic acids

Chemistry of alcohols

REVISED

Combustion

Observations: The four alcohols (methanol, ethanol, propan-1-ol and propan-2-ol) burn with a clean blue flame and heat is released. For example:

C_2H_5OH	+	$3O_2$	→	$2CO_2$	+	$3H_2O$
ethanol		oxygen		carbon dioxide		water

Fermentation

Ethanol can be produced by fermentation of sugars. Sugar solution is mixed with yeast in warm conditions in the absence of oxygen. The reaction produces carbon dioxide and ethanol.

Fermentation can produce a solution of ethanol in water of approximately 12–15%. This concentration of ethanol kills the yeast, so a higher concentration is not possible simply by fermentation.

Chemistry of carboxylic acids

REVISED

Carboxylic acids are weak acids because they are only partially ionised in water.

- Methanoic acid forms methanoate salts — the methanoate ion is $HCOO^-$.
- Ethanoic acid forms ethanoate salts — the ethanoate ion is CH_3COO^-.
- Propanoic acid forms propanoate salts — the propanoate ion is $C_2H_5COO^-$.
- Butanoic acid forms butanoate salts — the butanoate ion is $C_3H_7COO^-$.

Reactions of carboxylic acids with sodium carbonate

Observations: Bubbles of gas are formed, the white solid disappears and a colourless solution is formed.

General equation:

carboxylic acid + sodium carbonate → sodium salt + water + carbon dioxide

Example:

$2CH_3COOH$	+	Na_2CO_3	→	$2CH_3COONa$	+	H_2O	+	CO_2
ethanoic acid				sodium ethanoate				

Of the organic compounds in this course, only carboxylic acids will release carbon dioxide when sodium carbonate is added to a sample. This is a way of distinguishing a carboxylic acid from other organic compounds. The gas can be identified as carbon dioxide by bubbling it through limewater, which changes from colourless to milky if carbon dioxide is present (pages 149, 199 and 238).

Exam tip

In the formula of the salt, 'methanoate' or 'ethanoate' is usually written first, for example, magnesium methanoate is $(HCOO)_2Mg$. However, it is acceptable to write it the other way round as $Mg(HCOO)_2$.

Reactions of carboxylic acids with magnesium

Observations: Bubbles of a gas are produced, heat is released, the grey solid disappears and a colourless solution is formed.

General equation:

carboxylic acid + magnesium → magnesium salt + hydrogen

Example:

H $2C_2H_5COOH + Mg \rightarrow (C_2H_5COO)_2Mg + H_2$

propanoic acid → magnesium propanoate

The gas can be tested with a lit splint — it should pop if it is hydrogen.

Reactions of carboxylic acids with sodium hydroxide

Observations: Heat is released and the solution remains colourless.

General equation:

carboxylic acid + sodium hydroxide → sodium salt + water

Example:

H $C_3H_7COOH + NaOH \rightarrow C_3H_7COONa + H_2O$

butanoic acid → sodium butanoate

Now test yourself

TESTED ☐

22 Name the salt formed in the reaction of ethanoic acid with sodium hydroxide.
23 What is observed when magnesium reacts with ethanoic acid?
H 24 Write a balanced symbol equation for the reaction of propanoic acid with calcium carbonate.

Properties of organic compounds

- The alkanes (methane, ethane, propane and butane) are colourless gases.
- The alkenes (ethene, propene, but-1-ene and but-2-ene) are colourless gases.
- All polymers are white solids — coloured dyes can be added to make coloured plastics.
- The alcohols (methanol, ethanol, propan-1-ol and propan-2-ol) are colourless liquids with an alcohol-like smell.
- Methanoic acid is a colourless liquid with a pungent odour.
- Ethanoic acid is a colourless liquid with a vinegar-like smell.
- Propanoic acid is a colourless liquid with a sharp, irritating odour.
- Butanoic acid is a colourless liquid with a rancid odour.
- All the alcohols and carboxylic acids will mix with water, forming one layer (liquid alkenes and alkanes do not mix with water and will form two layers).
- The higher the carbon content of an organic compound, the sootier (and more orange) the flame will be when it burns.
- The lower the carbon content of an organic compound, the less sooty (and more clean and blue) the flame will be when it burns.
- Alcohols burn with a cleaner, blue flame than the alkanes or alkenes.

Atmospheric pollution due to combustion of hydrocarbon fuels

Greenhouse effect

REVISED

Burning hydrocarbon fuels for heat and for the production of electrical energy leads to increased levels of gases such as carbon dioxide and sulfur dioxide in the atmosphere. These gases are considered to be polluting because they cause adverse environmental problems.

Scientists think that increased levels of carbon dioxide in the atmosphere are leading to the greenhouse effect, which causes:

- global warming
- melting of the polar ice caps
- rising sea levels
- flooding of low-lying areas
- climate change.

Carbon burns in a limited supply of oxygen to form carbon monoxide and soot (unburnt carbon). Carbon monoxide is a colourless, odourless, toxic gas. It can cause death by inhalation and can be the cause of deaths in house fires and when boilers and other appliances that burn fossil fuels are faulty. Soot in the atmosphere can lead to lung damage and also contributes to global dimming (less sunlight caused by solid particles like soot in the atmosphere).

Acid rain

REVISED

Fossil fuels often contain sulfur impurities. When the fossil fuel burns, the sulfur reacts to form sulfur dioxide:

$$S + O_2 \rightarrow SO_2$$

Sulfur dioxide is an acidic gas that dissolves in rainwater, forming sulfurous acid:

$$SO_2 + H_2O \rightarrow H_2SO_3$$

Rain containing sulfurous acid falls as acid rain, which has several effects:

- It destroys limestone statues, buildings and natural limestone features.
- It kills fish in rivers and lakes.
- It defoliates trees.

Now test yourself

TESTED

25 State one effect of increased carbon dioxide levels in the atmosphere.
26 What gas causes acid rain to form?
27 State one effect of acid rain.

Exam tip

There are cost implications for developing new and less polluting ways of producing energy. Wind turbines, and wave and tidal generators are expensive to build and they require maintenance. Nuclear power has health and safety issues for those working in and living near a nuclear power station. Delivery and storage of hydrogen presents technological problems, which scientists are working to solve.

Exam practice

1 Write the molecular formula for each of these organic substances:
(a) ethane (c) butane
(b) ethene (d) methane [4 marks]

2 What is observed when ethene is bubbled through bromine water? [1 mark]

3 What is meant by the term hydrocarbon? [1 mark]

4 Name these molecules:

(a)
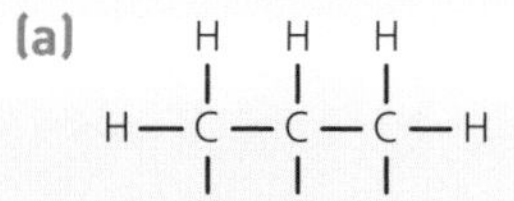

(b)
```
   H               H
   |               |
H— C — C = C — C — H
   |   |   |   |
   H   H   H   H
```

H (c)
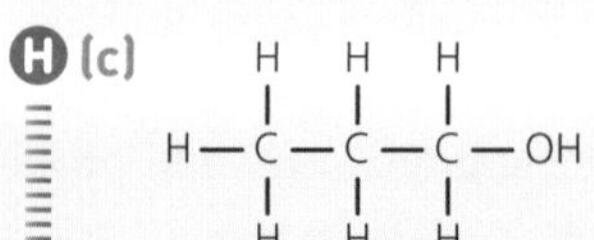

[3 marks]

H 5 Write an equation for the complete combustion of ethanol. [3 marks]

6 What is produced when ethene reacts with hydrogen? [1 mark]

7 What process is used to separate the hydrocarbons in crude oil? [1 mark]

H 8 Write an equation for the reaction of magnesium with ethanoic acid. [3 marks]

9 What is meant by the term 'homologous series'? [3 marks]

H 10 Name these polymers:

(a)
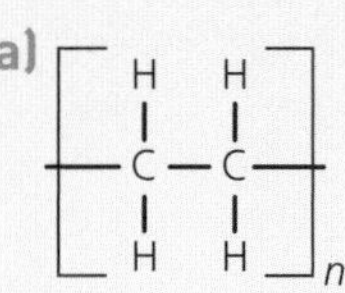

(b)
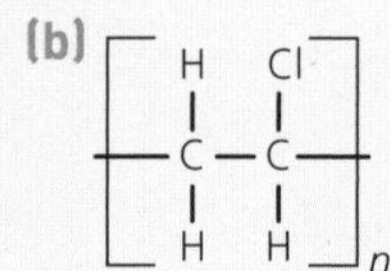

[2 marks]

11 State three effects of increased carbon dioxide in the atmosphere. [3 marks]

12 What are the products of incomplete combustion of butane? [3 marks]

H 13 Write a balanced symbol equation for the reaction of sodium carbonate with butanoic acid and name the products. [4 marks]

14 What is the general formula for the alkanes? [1 mark]

Answers online

ONLINE

26 Quantitative chemistry II

Water of crystallisation and degree of hydration

REVISED

Note: The rest of this chapter is in Unit 2 for Double Award Science.

Water that is chemically bonded into the crystal structure of a compound is known as **water of crystallisation**. The **degree of hydration** of a compound refers to the number of moles of water of crystallisation chemically bonded in 1 mole of that compound. The degree of hydration of **hydrated** copper(II) sulfate, $CuSO_4.5H_2O$, is 5. It is sometimes called copper(II) sulfate-5-water or copper(II) sulfate pentahydrate. The '$5H_2O$' in $CuSO_4.5H_2O$ is part of the mass of the solid.

The degree of hydration of hydrated substances can be determined by heating to constant mass, taking mass measurements before and after, or by titration.

Another example of a hydrated substance is $BaCl_2.2H_2O$, which is hydrated barium chloride but also called barium chloride-2-water or barium chloride dihydrate. The '2' is the degree of hydration. You will need to recognise all of these ways of expressing the degree of hydration. **Anhydrous** barium chloride would have the formula $BaCl_2$ (no water of crystallisation present).

Water of crystallisation is water that is chemically bonded into the crystal structure.

Hydrated means that solid crystals contain water of crystallisation.

Dehydration means removal of water of crystallisation.

An **anhydrous** substance does not contain water of crystallisation.

H Empirical formula and molecular formula

REVISED

- The formula that is determined from experimental mass (or percentage) data is called the **empirical formula**.
- The **molecular formula** is a simple multiple of the empirical formula.
- Remember you can use the M_r to determine the molecular formula from the empirical formula.
- Remember to cancel down the number of each type of atom to the lowest number when determining an empirical formula.

The **empirical formula** is the simplest whole-number ratio of the atoms of each element in a compound.

The **molecular formula** is the actual number of atoms of each element in a molecule.

Example 1

The empirical formula of a compound has been determined to be CH_3 and the M_r of the compound is 30. Determine the molecular formula of the compound.

Answer

The M_r of CH_3 is 12 + (1 × 3) = 15.

So 2 × CH_3 must be present in the compound as 2 × 15 = 30.
Its molecular formula is therefore C_2H_6.

Example 2

The empirical formula of a compound is CH_2O, and its M_r is 180. Determine the molecular formula of the compound.

Answer

The M_r of CH_2O is 12 + (1 × 2) + 16 = 30.

So 6 × CH_2O must be present in the compound as 6 × 30 = 180.
Its molecular formula is therefore $C_6H_{12}O_6$.

H **Example 3**

The molecular formula of a compound is $Na_2S_4O_6$. What is its empirical formula?

Answer

The empirical formula is found by simplifying the ratio of atoms in the formula to the simplest possible, in this case by dividing each number by 2. So the empirical formula is NaS_2O_3.

Determining formulae of simple compounds

REVISED

Note: The calculation of the formula is higher tier, but at foundation tier the practical processes and the mass and calculation of moles could appear in the exam without the final step of determining the formula.

Simple compounds are formed from just two elements — for example, sodium chloride and magnesium oxide. You can already work out the formula of a simple compound using valency values (page 109). However, you must also be able to use information about percentage composition or mass to determine the formula of a simple compound.

- You can calculate the numbers of moles of the atoms of each element by dividing the mass of each element by its A_r (always use A_r values for these calculations with elements).
- The moles are then converted to a simple ratio — this is best done by making the lowest mole value equal to 1 and reducing the others in the same scale. You divide them all by the lowest mole value.
- In some examples, you may be given the masses of the elements that combine; in others you may be given the mass of one element and the mass of the compound (a simple subtraction will calculate the mass of the second element).
- You also need to be able to plan how to carry out these experiments practically to determine the formula of a simple compound. Most of the experiments involve heating to constant mass, but full practical details and apparatus required may be expected.

Example 1

1.5 g of magnesium combines with oxygen to give 2.5 g of magnesium oxide. Find the formula of the oxide of magnesium.

Answer

First, work out the mass of oxygen combining with the magnesium using 2.5 – 1.5 = 1.0 g.

You can then calculate the formula of the oxide of magnesium:

Element	Magnesium	Oxygen
Mass (g)	1.5	1.0
A_r	24	16
Moles	$\frac{1.5}{24} = 0.0625$	$\frac{1.0}{16} = 0.0625$
Ratio (÷ 0.0625)	1	1
Empirical formula	MgO	

Exam tip

Often the masses are given to two or three decimal places, so the moles may not be exactly the same. Use at least three significant figures for the number of moles.

Heating a solid to determine the empirical formula

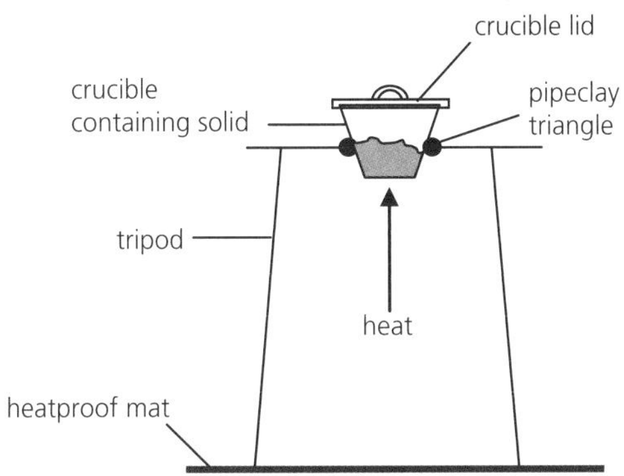

Figure 26.1 Heating a solid

This experiment is carried out as follows (Figure 26.1):

1. Measure the mass of the empty crucible and lid (mass = *a*).
2. Measure the mass of the lid and crucible containing some magnesium powder (mass = *b*).
3. Heat the contents of the crucible with the lid on a pipeclay triangle on a tripod over a Bunsen burner.
4. Raise the lid a little now and again to let more air in.
5. Allow the crucible and lid to cool and then measure the mass.
6. Heat again and measure the mass again; repeat until the mass is constant (mass = *c*).
7. The results would not be reliable without these measures in place. Any lack of reliability in the measurements would be because not all of the magnesium reacted, or some product was lost from the crucible.

The mass of the magnesium is $b - a$; the mass of oxygen gained is $c - b$. The data may be presented in this way and you have to calculate the masses needed to determine the empirical formula.

Exam tip

Powdered magnesium is used because it has a larger surface area, so it reacts more readily. The sample is heated to constant mass to ensure that all of the magnesium reacts. The lid is used to prevent loss of solid during the reaction but it must be lifted now and again to allow more air in.

Example 2

An oxide of manganese contains 63.2% manganese. Determine the empirical formula of the oxide.

Answer

If the compound only contains manganese and oxygen, 100 – 63.2 = 36.8% must be oxygen.

Using 100 g of the compound means that 63.2 g are manganese and 36.8 g are oxygen. The calculation is carried out as in the previous example:

Element	Manganese	Oxygen
Mass (g)	63.2	36.8
A_r	55	16
Moles	$\frac{63.2}{55} = 1.15$	$\frac{36.8}{16} = 2.3$
Ratio (÷ 1.15)	1	2
Empirical formula	MnO_2	

The empirical formula is MnO_2. This is manganese(IV) oxide or manganese dioxide.

H **Worked example**

A sample of solid phosphorus was burned in excess oxygen. 0.775 g of phosphorus reacted with 1.0 g of oxygen.

(a) Determine the empirical formula of the oxide of phosphorus formed. [3 marks]

(b) Given that the M_r of the oxide of phosphorus is 284, determine the molecular formula of the oxide. [1 mark]

Answer

Element	Phosphorus	Oxygen
Mass (g)	0.775	1
A_r	31	16
Moles	$\frac{0.775}{31} = 0.025$	$\frac{1}{16} = 0.0625$
Ratio (÷ 0.025)	1	2.5
	The ratio works out at 1:2.5, so both are multiplied by 2 to give whole numbers	
	2	5
Empirical formula	P_2O_5	

(a) The empirical formula is P_2O_5. [1]

(b) The M_r of P_2O_5 = (31 × 2) + (16 × 5) = 142; and the M_r of the oxide is 284. So 2 × P_2O_5 must be present in the compound, giving the molecular formula P_4O_{10}. [1]

Now test yourself

TESTED

1 An oxide of silver contains 93.1% silver. Determine its empirical formula.

2 An organic compound containing only carbon and hydrogen has 80% carbon. Determine its empirical formula.

3 Calculate the percentage yield if 3.8 g of magnesium chloride were obtained from the reaction of 1.2 g of magnesium with excess hydrochloric acid:

$$Mg + 2HCl \rightarrow MgCl_2 + H_2$$

Determining degree of hydration by heating to constant mass

REVISED

The method of determining empirical formulae can also be applied to hydrated compounds. The apparatus used to heat hydrated compounds is shown in Figure 26.2.

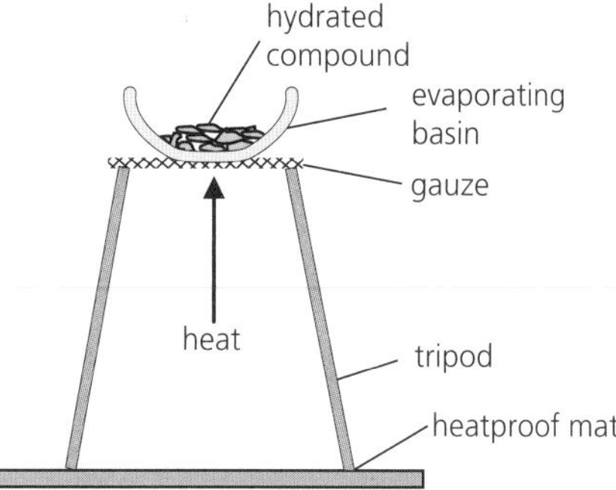

Figure 26.2 Apparatus for heating a hydrated compound

- If hydrated compounds are heated, they lose water of crystallisation and their mass decreases as the anhydrous compound is formed.
- When the compound has been heated to constant mass, the decrease in mass is the mass of water lost.
- By using the mass of the anhydrous compound and the mass of water lost, the degree of hydration can be determined.

Example 1

Given that 4.0 g of hydrated copper(II) sulfate, $CuSO_4.nH_2O$, produces 2.56 g of the anhydrous copper(II) sulfate, $CuSO_4$, on heating to constant mass, find the value of *n* in the formula of the hydrated salt.

Answer

The two pieces of mass data required here are the mass of the anhydrous salt and the mass of water lost on heating to constant mass.

mass of anhydrous salt ($CuSO_4$) = 2.56 g

mass of water lost (H_2O) = 4.0 − 2.56 = 1.44 g

Compound	Copper(II) sulfate	Water
Formula	$CuSO_4$	H_2O
Mass (g)	2.56	1.44
M_r	160	18
Moles	$\frac{2.56}{160} = 0.016$	$\frac{1.44}{18} = 0.08$
Ratio (÷ 0.016)	1	5
Empirical formula	$CuSO_4.5H_2O$	

You can see from the empirical formula that the value of *n* = 5.

Example 2

The following mass measurements were taken when a sample of hydrated aluminium nitrate, $Al(NO_3)_3.nH_2O$, was heated to constant mass in an evaporating basin in a low temperature oven:

mass of evaporating basin = 54.13 g

mass of evaporating basin and hydrated salt = 61.63 g

mass of evaporating basin and contents after heating to constant mass = 58.39 g

Find the degree of hydration in $Al(NO_3)_3.nH_2O$.

Answer

Mass of anhydrous salt = 58.39 − 54.13 = 4.26 g

Mass of water lost = 61.63 − 58.39 = 3.24 g

Compound	Aluminium nitrate	Water
Formula	$Al(NO_3)_3$	H_2O
Mass (g)	4.26	3.24
M_r	213	18
Moles	$\frac{4.26}{213} = 0.02$	$\frac{3.24}{18} = 0.18$
Ratio (÷ 0.02)	1	9
Empirical formula	$Al(NO_3)_3.9H_2O$	

You can see from the empirical formula that the value of *n* = 9.

Exam tip

The solid remaining after heating to constant mass is the anhydrous compound. The difference in mass before and after heating is the mass of water lost.

Prescribed practical C5

Determine the mass of water present in hydrated crystals

- Be able to carry out and describe practically heating hydrated crystals to constant mass, including the mass measurements that must be taken.
- Be able to draw and label the individual and assembled apparatus required.
- Be able to manipulate given mass data to calculate the mass of water in a sample of hydrated crystals.

Heating to constant mass

REVISED

Note: 'Heating to constant mass' is in Unit 2 for Double Award Science.

- When heated, some substances produce gases (including water vapour), which are released to the atmosphere — these substances lose mass on heating.
- When heated, some substances react with gases in the air — these substances gain mass on heating.
- When a solid produces a gas on heating, or reacts with a gas from the air, you should **heat to constant mass** to ensure that the reaction has gone to completion.
- Heating to constant mass means that you heat the substance, allow it to cool and record the mass of the substance — this is repeated until the mass no longer changes.
- The process involves measuring the mass of the empty container, for example a crucible or evaporating basin. The substance is then put in the container and the total mass is measured. The substance is heated in the container and the mass measured on several occasions, after cooling, until there is no further change.
- Data involving heating to constant mass may be presented to you as follows:

  mass of crucible = 21.12 g

  mass of crucible and substance = 21.60 g

  mass of crucible and contents after heating for 2 minutes = 21.87 g

  mass of crucible and contents after heating for 4 minutes = 21.92 g

  mass of crucible and contents after heating for 6 minutes = 21.92 g

  - Heating to constant mass has been achieved.
  - This substance gains mass on heating so it is reacting with a gas in the air.
  - The gas in the air that reacts is usually oxygen, though some substances may react with carbon dioxide and even nitrogen.
  - Mole calculations using these masses can be carried out if you know the M_r of the substance.

4 Calculate the percentage of water in hydrated zinc sulfate, $ZnSO_4.7H_2O$.
5 Explain the process of heating to constant mass.
6 11.07 g of hydrated magnesium sulfate, $MgSO_4.nH_2O$, were heated to constant mass and 5.4 g of anhydrous magnesium sulfate remained. Calculate the value of n in $MgSO_4.nH_2O$.

Working with solutions

Solution calculations

REVISED

Solutions are slightly more complicated than solids because the number of moles depends on the volume of the solution (in cm^3) and on the concentration of the solution (in mol/dm^3).

A solution that has a concentration of 1 mol/dm^3 has 1 mole of the solute dissolved in 1 dm^3 of solution. If you had 500 cm^3 of a solution of concentration 1 mol/dm^3, you would have 0.5 moles of the solute in that volume.

When making a solution, a certain mass of a solid is dissolved in a certain volume of water. The *number of moles* of a solute in a solution can be calculated using:

$$\text{moles} = \frac{\text{volume (cm}^3) \times \text{concentration (mol/dm}^3)}{1000}$$

Exam tip

1 dm^3 is the same as 1 litre. 1 cm^3 is the same as 1 ml. There are 1000 cm^3 in 1 dm^3 (1000 ml in 1 litre). Mole is abbreviated to mol, so you will see mol after any number of moles.

Example 1

1.4 g of KOH (potassium hydroxide) were dissolved completely in 100 cm^3 of water. Calculate the concentration of the solution formed in mol/dm^3.

Answer

M_r of KOH is 39 + 16 + 1 = 56

$$1.4\text{ g of solid KOH} = \frac{1.4}{56} = 0.025\text{ mol}$$

0.025 mol of KOH are present in 100 cm^3.

For a solution:

$$\text{moles} = \frac{\text{volume (cm}^3) \times \text{concentration (mol/dm}^3)}{1000}$$

$$0.025 = \frac{100 \times \text{concentration}}{1000}$$

So:

$$\text{concentration} = \frac{0.025 \times 1000}{100} = 0.25\text{ mol/dm}^3$$

More simply, this is 10 × the number of moles in 100 cm^3 because 1 dm^3 is 1000 cm^3.

H

Example 2

Calculate the concentration, in mol/dm^3, of a solution formed when 3.75 g of hydrated copper(II) sulfate, $CuSO_4.5H_2O$ are dissolved in 75 cm^3 of deionised water.

Answer

$$M_r \text{ of } CuSO_4.5H_2O = 64 + 32 + (4 \times 16) + 5 \times (2 \times 1 + 16) = 250$$

$$3.75\,\text{g of } CuSO_4.5H_2O = \frac{3.75}{250} = 0.015\,\text{mol}$$

0.015 mol of $CuSO_4.5H_2O$ are present in 75 cm^3.

For a solution:

$$\text{moles} = \frac{\text{volume (cm}^3) \times \text{concentration (mol/dm}^3)}{1000}$$

$$0.015 = \frac{75 \times \text{concentration}}{1000}$$

So:

$$\text{concentration} = \frac{0.015 \times 1000}{75} = 0.2\,\text{mol/dm}^3$$

Now test yourself

TESTED

7 Calculate the concentration of a solution, in mol/dm^3, that contains 0.24 moles in 250 cm^3.

8 Calculate the concentration of a solution, in mol/dm^3, which contains 249 g of potassium iodide in 2 dm^3.

9 Calculate the concentration of a solution of sodium chloride, in mol/dm^3, that has 4.68 g of sodium chloride dissolved in 50 cm^3 of water.

Determining moles of solute in a solution

REVISED

The same equation is also used to calculate the number of moles of a **solute** in a certain volume of solution.

$$\text{moles} = \frac{\text{volume (cm}^3) \times \text{concentration (mol/dm}^3)}{1000}$$

Example

45.0 cm^3 of a 0.1 mol/dm^3 solution of hydrochloric acid were used in a titration. Calculate the number of moles of hydrochloric acid used.

Answer

For a solution:

$$\text{moles} = \frac{\text{volume (cm}^3) \times \text{concentration (mol/dm}^3)}{1000}$$

$$\text{moles of HCl} = \frac{45.0 \times 0.1}{1000} = 0.0045\,\text{mol}$$

The mass of HCl in the solution could be determined by multiplying the number of moles by the M_r of HCl, in this case $0.0045 \times 36.5 = 0.164$ g.

H

Now test yourself

TESTED

10 Calculate the number of moles present in 12.5 cm^3 of a 0.02 mol/dm^3 solution of sodium hydroxide.

11 Calculate the mass of iron(II) nitrate, $Fe(NO_3)_2$, that is present in 25.0 cm^3 of a 1.12 mol/dm^3 solution.

Atom economy

Atom economies

REVISED

The atom economy in a chemical reaction is calculated as a percentage atom economy using:

$$\text{atom economy} = \frac{\text{mass of desired product}}{\text{total mass of products}} \times 100$$

- A high atom economy indicates low waste in a reaction.
- A low atom economy indicates a substantial amount of waste in the chemical reaction.
- Chemists aim for a high atom economy to avoid waste and so that development of new materials is sustainable.
- Economically speaking a greater mass of reactants is needed to form the same mass of products if the atom economy is low, so it costs companies more to produce chemicals when the atom economy is low.
- If a use can be found for the waste products (for example, the companies may be able to recycle them or sell them on to other companies), then the atom economy is increased.

Example

Iron(III) oxide can be reduced to iron using carbon. The following reaction occurs:

$$2Fe_2O_3 + 3C \rightarrow 4Fe + 3CO_2$$

The useful product is iron. Calculate the atom economy to one decimal place.

Answer

$$\text{atom economy} = \frac{\text{mass of desired product}}{\text{total mass of products}} \times 100$$

In the calculation the balancing numbers have to be used to calculate the mass of all atoms.

mass of desired product = $4 \times 56 = 224$

total mass of products = $(4 \times 56) + (3 \times (12 + 32)) = 356$

$$\text{atom economy} = \frac{224}{356} \times 100 = 62.9\%$$

Now test yourself

TESTED

12 Calculate the atom economy, to one decimal place, for the production of hydrogen from methane. Hydrogen is the useful product.

$$CH_4(g) + 2H_2O(g) \rightarrow CO_2(g) + 4H_2(g)$$

Exam practice

1 Calculate the M_r of these hydrated compounds:
  (a) nickel sulfate-7-water [1 mark]
  (b) $CoCl_2.6H_2O$ [1 mark]

H 2 100 g of an oxide of iron contains 27.6 g of oxygen. Determine the empirical formula of the oxide. [4 marks]

3 3.2 g of copper turnings are heated in air. The copper reacts with oxygen to form 4.0 g of an oxide of copper. Determine the empirical formula of the oxide of copper. [3 marks]

4 Which one of these represents an empirical formula? [1 mark]
  $C_2H_4O_2$ $C_{18}H_{36}O_2$ CH_2O

5 A compound of sulfur was found to contain 40 g of sulfur and 60 g of oxygen. Determine the empirical formula of the compound. [3 marks]

6 A sample of hydrated sodium carbonate, $Na_2CO_3.nH_2O$, was heated to constant mass in an evaporating basin. The measurements below were taken at 5-minute intervals:

  mass of evaporating basin = 122.400 g

  mass of evaporating basin and hydrated sample = 122.900 g

  mass of evaporating basin and sample after 5 minutes heating = 122.714 g

  mass of evaporating basin and sample after 10 minutes heating = 122.612 g

  mass of evaporating basin and sample after 15 minutes heating = 122.612 g

  (a) Calculate the mass of anhydrous sodium carbonate present at the end of the experiment. [1 mark]
  (b) Calculate the number of moles of anhydrous sodium carbonate present at the end of the experiment. [1 mark]
  (c) Calculate the mass of water lost by heating. [1 mark]
  (d) Calculate the number of moles of water lost by heating. [1 mark]
  (e) Using your answer to parts (b) and (d), determine the value of n in $Na_2CO_3.nH_2O$. [1 mark]

7 For the following reaction:

$$4NH_3(g) + 3O_2(g) \rightarrow 2N_2(g) + 6H_2O(g)$$

  Calculate the atom economy if nitrogen is the useful product. [3 marks]

Answers online

ONLINE

27 Electrochemistry

Electrolysis

What is electrolysis?

REVISED

Electrolysis is the decomposition of a liquid **electrolyte** using a direct current of electricity.

Two graphite rods, placed in a liquid and connected externally to a power supply such as a battery or a power pack, can be used to test if a liquid conducts electricity. If a liquid conducts electricity (Figure 27.1) and is **decomposed** by it, then electrolysis is taking place.

- The graphite rods used in electrolysis are called **electrodes**. They are called **inert electrodes** as they do not take part in the reactions.
- Graphite is the main material used to make electrodes because it is a good conductor of electricity and is unreactive.
- The negative electrode is called the **cathode** and the positive electrode is called the **anode**.

The **electrolyte** is the liquid or solution that conducts electricity and is decomposed by it.

Inert electrodes are electrodes that do not take part in the reactions.

The **cathode** is the negative electrode.

The **anode** is the positive electrode.

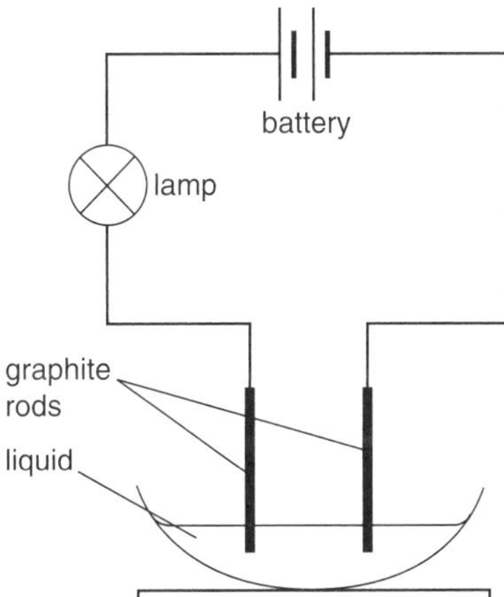

Figure 27.1 Testing to see if a liquid conducts electricity

Exam tip

The definitions of the terms used in electrolysis are asked in almost every question on this topic. Learn them exactly as they are written above. The reasons why graphite is used for the electrodes are: 'it is a *good conductor of electricity* and it is *unreactive*'. This is a frequently asked question.

Now test yourself

TESTED

1 What is electrolysis?
2 What name is used for the negative electrode?
3 What name is used for the positive electrode?

How electrolysis works

REVISED

All electrolytes conduct electricity because they have **free ions** that can move and carry charge. When positive and negative ions are free to move, the positive ions (called cations) move to the negative electrode (called the cathode), and the negative ions (called anions) move to the positive electrode (called the anode).

- The positive ions at the negative electrode (cathode) gain electrons to become atoms.
- The gain of electrons is called reduction — reduction happens at the cathode.
- The negative ions at the positive electrode (anode) lose electrons to become atoms, which may combine to form diatomic molecules for elements such as chlorine, Cl_2, and bromine, Br_2.
- The loss of electrons is called oxidation — oxidation happens at the anode.
- Molten ionic compounds and aqueous ionic compounds (including acids) are the most common electrolytes.

Examples of electrolysis

Electrolysis of molten ionic compounds

REVISED

A molten ionic compound contains ions that are free to move and carry charge.

Typical apparatus used to electrolyse a molten electrolyte is shown in Figure 27.2.

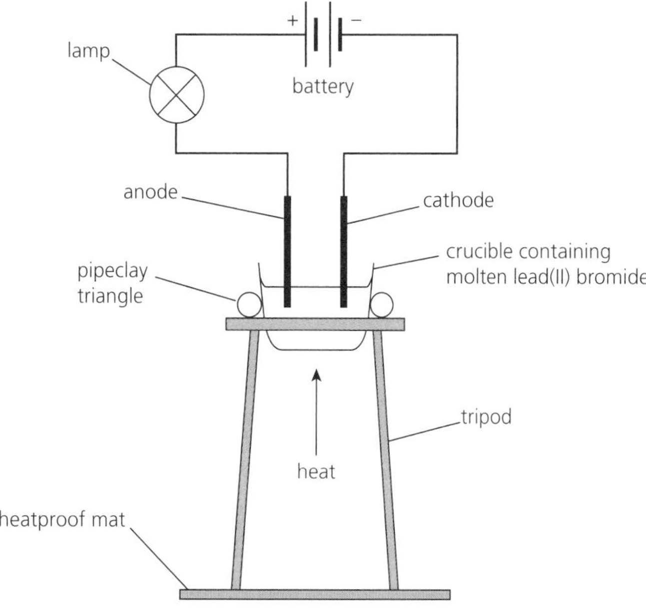

Figure 27.2 Apparatus used to electrolyse a molten electrolyte

Exam tip

Be careful to label every piece of apparatus in the diagram and ensure that the apparatus looks the way it should. You will not gain a mark for a labelled crucible that looks like a beaker.

Exam tip

Do not confuse the conduction of electricity in metals and graphite (caused by delocalised electrons that can move and carry charge) with the conduction of electricity in electrolytes (which is caused by free ions that can move and carry charge).

Observations and half-equations

For each of the following electrolytes, you must be able to describe what is observed at both electrodes. You must also be able to write **half-equations** to represent the reactions at each of the electrodes in terms of loss or gain of electrons.

- First, identify the positive metal ions and the negative non-metal ions present in the ionic compound.
- The positive metal ions in the molten compound are attracted to the negative cathode, where the ions gain electrons to form atoms.
- The negative non-metal ions in the molten compound are attracted to the positive anode, where the ions lose electrons to form atoms. For diatomic elements, two atoms combine together to form a molecule.
- It is often easier to write the half-equations first, and then work out what would be observed from the products of the electrolysis.

Molten lead(II) bromide, $PbBr_2$

Lead(II) bromide, $PbBr_2$, contains the following ions:

- Cation — lead(II), Pb^{2+}
- Anion — bromide, Br^-

Cathode: positive ion, Pb^{2+}, attracted to cathode

H *Half-equation:* $Pb^{2+} + 2e^- \rightarrow Pb$

Observations: Silvery grey liquid formed, which sinks to the bottom (can only be seen when the molten electrolyte is poured off)

Anode: negative ion, Br^-, attracted to anode

H *Half-equation:* $2Br^- \rightarrow Br_2 + 2e^-$

Observations: red-brown pungent gas evolved

Molten lithium chloride, LiCl

Lithium chloride, LiCl, contains the following ions:

- Cation — lithium ion, Li^+
- Anion — chloride ion, Cl^-

Cathode: positive ion, Li^+, attracted to cathode

H *Half-equation:* $Li^+ + e^- \rightarrow Li$

Observations: Silvery grey liquid formed

Anode: negative ion, Cl^-, attracted to anode

H *Half-equation:* $2Cl^- \rightarrow Cl_2 + 2e^-$

Observations: yellow-green, pungent gas evolved

Molten sodium iodide, NaI

This is an example of applying your knowledge to an unfamiliar metal halide.

Sodium iodide, NaI, contains the following ions:

- Cation — sodium ion, Na^+
- Anion — iodide ion, I^-

Exam tip

Because these are all molten ionic compounds, heat is being applied and so a low melting point and low density metal such as lithium, sodium or potassium will appear as a grey liquid around the cathode. Lead has a relatively low melting point compared with the transition metals, but is dense and so will sink to the bottom.

Exam tip

The halogens will all appear as gases if they are produced at the anode during molten electrolysis.

If a molten metal oxide is electrolysed, oxygen gas is produced at the anode and the observations would include a colourless, odourless gas being evolved (though often this can be difficult to see apart from a few bubbles around the electrode).

Cathode: positive ion, Na^+, attracted to cathode

H *Half-equation:* $Na^+ + e^- \rightarrow Na$

Observations: Silvery grey liquid formed

Anode: negative ion, I^-, attracted to anode

H *Half-equation:* $2I^- \rightarrow I_2 + 2e^-$

Observations: Purple, pungent gas evolved

H Now test yourself

TESTED

4 Name the products of the electrolysis of molten zinc chloride.

Extraction of aluminium from its ore

REVISED

Aluminium metal is extracted from its ore using electrolysis — aluminium ore is called **bauxite**.

Bauxite is the ore from which aluminium is extracted.

Bauxite is purified to form aluminium oxide (called alumina). The alumina is dissolved in molten cryolite to reduce the operating temperature and increase its conductivity (Figure 27.3).

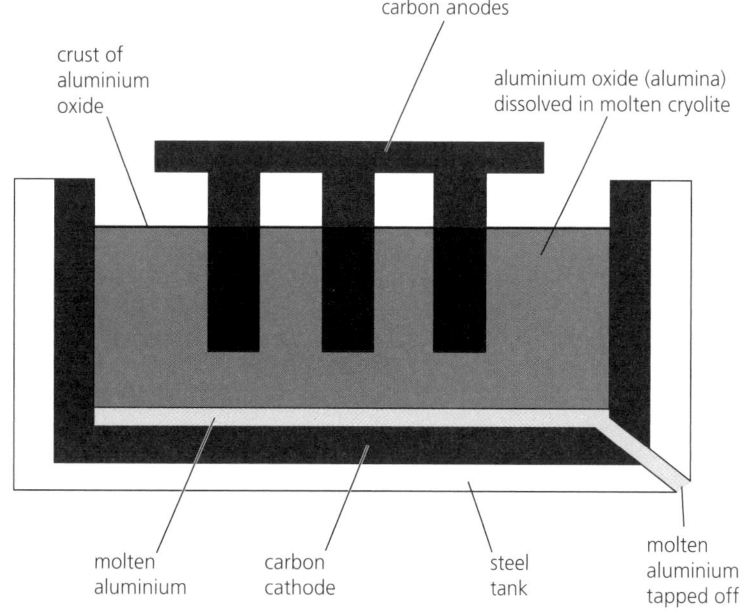

Figure 27.3 Apparatus used to extract aluminium from its ore

Exam tip

The bauxite is purified by mixing it with sodium hydroxide solution. The alumina in the ore dissolves but the other chemicals do not. It is filtered and acid is added to the filtrate which is then heated to form alumina.

- A crust of aluminium oxide keeps heat in. The operating temperature is between 900 and 1000°C.
- The cathode and anode are both made of carbon.
- H The reaction at the cathode is:

  $Al^{3+} + 3e^- \rightarrow Al$
- The reaction at the anode is:

  $2O^{2-} \rightarrow O_2 + 4e^-$
- The carbon anode has to be replaced periodically because it wears away as it reacts with oxygen:

  $C + O_2 \rightarrow CO_2$
- The reaction at the cathode is reduction because aluminium ions are gaining electrons, and the gain of electrons is reduction.

Exam tip

When explaining oxidation at the anode or reduction at the cathode during electrolysis, always include the names of the ions gaining or losing electrons and then state the definition of either oxidation or reduction.

- The reaction at the anode is oxidation because oxide ions are losing electrons, and the loss of electrons is oxidation.
- The extraction of aluminium is expensive because the cost of electricity is high and a high temperature is needed to keep the aluminium oxide molten. The use of cryolite increases the conductivity and reduces the operating temperature, saving money. The aluminium oxide crust keeps some heat in, again saving money.
- The expense of recycling aluminium is only a fraction of the cost of producing new aluminium from bauxite. This is why it is important to recycle materials such as aluminium — it saves resources, saves energy, prevents waste going to landfill and costs less.

Now test yourself

TESTED

5 What is the purified aluminium ore dissolved in before electrolysis takes place?
H 6 Write a half-equation for the production of oxygen at the anode.
7 Write a half-equation for the production of aluminium at the cathode.

Exam practice

1 What is meant by the term 'electrolysis'? [1 mark]
2 What is observed at the anode during the electrolysis of molten lead(II) bromide? [2 marks]
H 3 Write a half-equation for the reaction at the anode during the electrolysis of molten lithium chloride. [3 marks]
4 What is:
(a) an anode
(b) a cathode? [2 marks]
5 What does the term 'electrolyte' mean? [2 marks]
6 State two reasons why graphite is used for the electrodes during the electrolysis of molten lead(II) bromide. [2 marks]
7 Explain why molten lithium chloride conducts electricity. [2 marks]
8 What is the name of aluminium ore? [1 mark]
9 Give two reasons why the purified aluminium ore is dissolved in molten cryolite. [2 marks]
10 What material is used for the electrodes during the extraction of aluminium? [1 mark]
11 Explain why the anode has to be replaced during the electrolytic extraction of aluminium. You may use an equation to help answer this question. [2 marks]
H 12 Write a half-equation for the discharge of aluminium ions at the cathode. [3 marks]
13 Write a half-equation for the production of oxygen at the anode during the extraction of aluminium from its ore and explain why this is described as an oxidation reaction. [5 marks]

Answers online

ONLINE

28 Energy changes in chemistry

Exothermic and endothermic reactions

Chemical reactions are either **exothermic** (give out heat to the surroundings) or **endothermic** (take in heat from the surroundings).

Table 28.1 shows important exothermic and endothermic reactions.

Table 28.1

Exothermic	Endothermic
Neutralisation	Thermal decomposition
Displacement	Electrolysis
Combustion	Dehydration of hydrated salts
Hydration of anhydrous salts	
Rusting	

Reaction profile diagrams

REVISED

A reaction profile for a chemical reaction shows how the energy of the chemicals changes as the reaction proceeds.

A reaction profile is drawn as a graph sketch of energy on the vertical axis (of the chemicals in the reaction) against the progress of the reaction from reactants to products on the horizontal axis. The chemical energy of the reactants is shown as one line labelled reactants and the chemical energy of the products is shown as a line further to the right labelled products.

Exothermic reaction profile

In an exothermic reaction, chemical energy in the reactants is converted into heat energy, which is released. This means that the chemical energy going from reactants to products has decreased.

As the products are at a lower energy value than the reactants, the change in energy (which = energy of products − energy of reactants) in an exothermic reaction from reactants to products is written as a negative number.

The reaction profile for an exothermic reaction is shown in Figure 28.1.

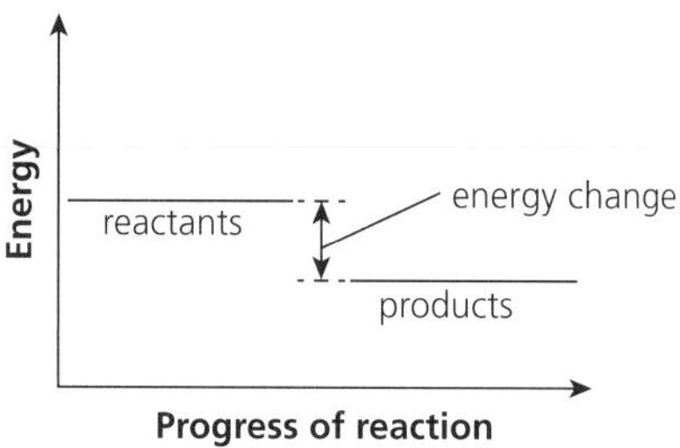

Figure 28.1 Reaction profile for an exothermic reaction

Endothermic reaction profile

In an endothermic reaction, heat energy taken in from the surroundings is converted into chemical energy in the products. This means that the chemical energy going from reactants to products has increased.

As the products are at a higher energy value than the reactants, the change in energy (which = energy of products − energy of reactants) in an endothermic reaction from reactants to products is written as a positive number.

The reaction profile for an endothermic reaction is shown in Figure 28.2.

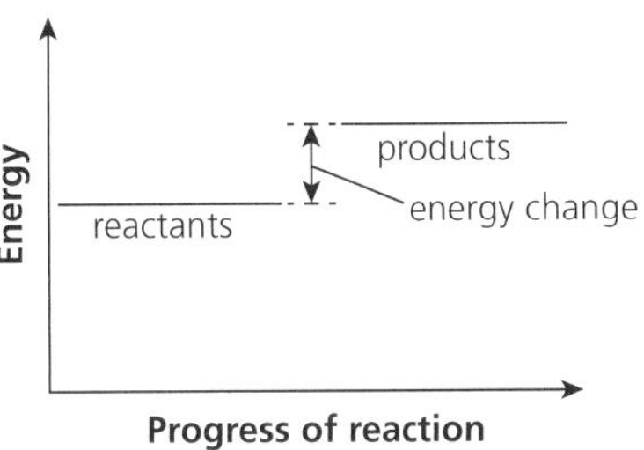

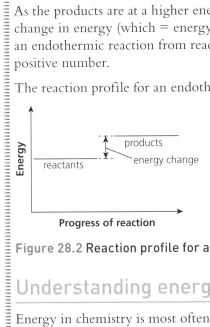

Figure 28.2 Reaction profile for an endothermic reaction

Understanding energy change values

REVISED

Energy in chemistry is most often measured in kilojoules (kJ). Energy changes in a chemical reaction are also measured in kilojoules.

For a hypothetical exothermic reaction, assume that the energy of the reactants is 150 kJ and the energy of the products is 120 kJ.

energy change = energy of the products − energy of the reactants

So in this hypothetical example energy change = 120 −150 = −30 kJ. A negative value for energy change indicates an exothermic reaction. 30 kJ of energy is released to the surroundings as heat for the reaction as written in the balanced symbol equation.

For a hypothetical endothermic reaction, assume that the energy of the reactants is 100 kJ and the energy of the products is 140 kJ.

energy change = energy of the products − energy of the reactants

So in this hypothetical example energy change = 140 − 100 = +40 kJ. A positive value for energy change indicates an endothermic reaction. 40 kJ of energy is taken in from the surroundings as heat for the reaction as written in the balanced symbol equation.

Reversible reactions

If the reaction is reversible (as met in Chapter 24) the reverse reaction has the opposite energy change. For a reversible reaction where the forward reaction is exothermic, the reverse reaction is endothermic. If the energy change for the forward reaction is endothermic, the reverse reaction is exothermic.

If the forward reaction has an energy change of −30 kJ the reverse reaction will have an energy change of +30 kJ.

Exam tip

The + and – signs for energy changes should always be included.

H

Worked example

For the reversible reaction:

$$N_2(g) + 3H_2(g) \rightleftharpoons 2NH_3(g)$$

The energy change is −92 kJ.

(a) Explain whether the forward reaction is endothermic or exothermic. [1 mark]

(b) What is the energy change for the reverse reaction? [1 mark]

Answer

(a) Exothermic, because the energy change value is negative. [1]

(b) +92 kJ [1]

Activation energy

REVISED

The reacting particles collide and not all collisions cause a reaction to occur. Only those collisions with the **activation energy** or greater will cause a reaction.

The **activation energy** is the minimum energy required for a reaction to occur.

On a reaction profile the reaction pathway is shown from the reactants to the products as shown in Figure 28.3. The energy difference between the energy of the **reactants** and the top of the pathway is the activation energy. This applies to both exothermic and endothermic reaction profiles.

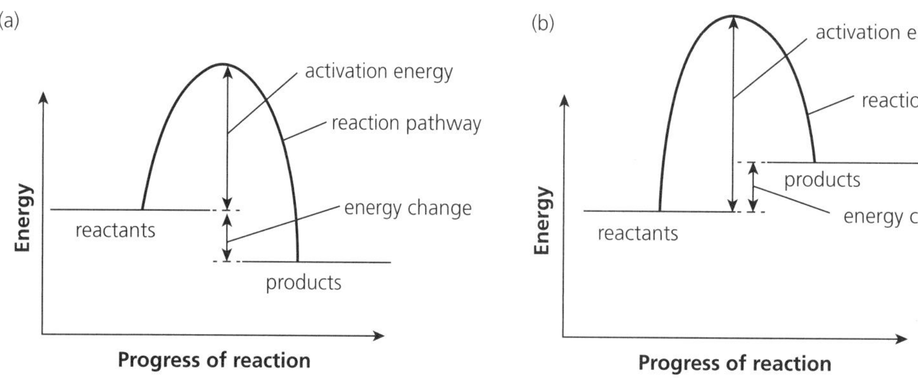

Figure 28.3 Reaction profiles for (a) an exothermic reaction and (b) an endothermic reaction

Catalyst

As stated in Chapter 23, a catalyst works by providing an alternative reaction pathway of lower activation energy. This can also be shown on a reaction profile such as the one in Figure 28.4, which is for an exothermic reaction.

The reaction profile shows that the catalysed reaction has a different or alternative reaction pathway and that this pathway has a lower activation energy.

Exam tip

You would be expected to be able to draw a reaction profile or to recognise or label the features, including the reaction pathway, the energy change and the activation energy.

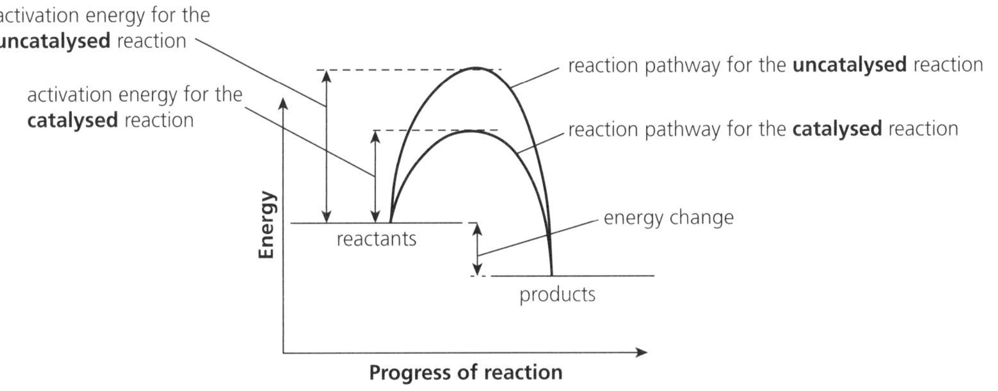

Figure 28.4 Effect of a catalyst on the reaction pathway

Worked example

A reaction profile is shown below.

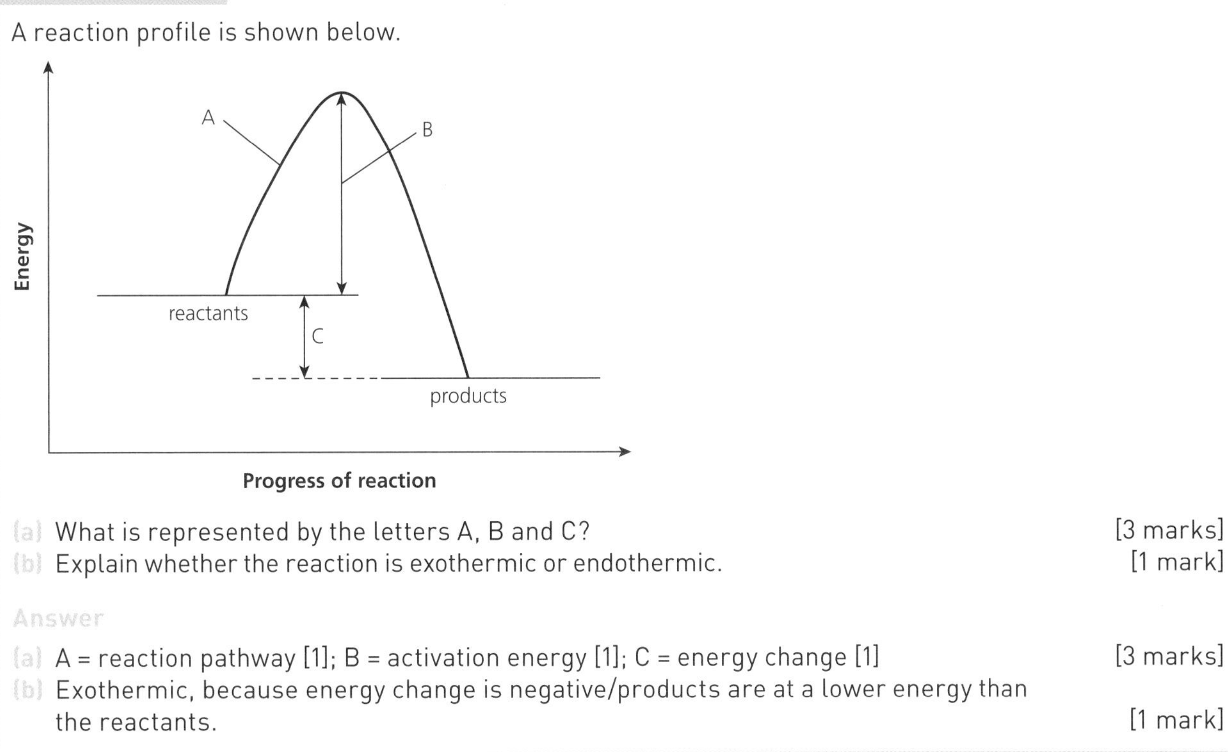

(a) What is represented by the letters A, B and C? [3 marks]
(b) Explain whether the reaction is exothermic or endothermic. [1 mark]

Answer

(a) A = reaction pathway [1]; B = activation energy [1]; C = energy change [1] [3 marks]
(b) Exothermic, because energy change is negative/products are at a lower energy than the reactants. [1 mark]

Now test yourself

TESTED

1 A reaction has a negative energy change. Is the reaction exothermic or endothermic?
2 What name is given to the minimum energy required for a reaction to occur?
3 What are the axes labels on a reaction profile diagram?

Bond energies

Explaining energy changes in terms of bonds

REVISED

All chemicals possess **internal energy** in their bonds. Energy is required to break all types of bond, and energy is released when all types of bond are formed. This means that bond breaking is endothermic and bond making is exothermic.

H

Worked example

(a) Methane burns in oxygen, releasing energy to the surroundings:

$CH_4 + 2O_2 \rightarrow CO_2 + 2H_2O$

Explain why the combustion of methane is exothermic in terms of the energy of the bonds. [3 marks]

(b) The reaction of hydrogen with iodine to form hydrogen iodide is endothermic:

$H_2 + I_2 \rightarrow 2HI$

Explain why this reaction is endothermic in terms of the energy of the bonds. [3 marks]

Answers

(a) The energy required to break bonds in the reactants, methane and oxygen [1], is less [1] than the energy released when bonds are formed in the products, carbon dioxide and water [1].

(b) The energy required to break bonds in the reactants, hydrogen and iodine [1], is more [1] than the energy released when bonds are formed in the product, hydrogen iodide [1].

Exam tip

In an energetics question, your answer needs to be specific to the reaction given. So your answer needs to follow this model:

'The energy required to break the bonds in the reactants (name them) is (comparison — state 'less than' for exothermic, or 'more than' for endothermic) the energy released when bonds are made in the products (name them).'

Calculating energy changes from bond energies

REVISED

A bond energy is the energy required to break one mole of a covalent bond, measured in kilojoules, kJ. For example, the bond energy of a C−H bond is 412 kJ.

Bond breaking is endothermic (takes in energy) and bond making is exothermic (releases energy). Breaking one mole of C−H requires 412 kJ of energy. Making one mole of C−H releases 412 kJ of energy.

General points for bond energies

Bond energy relates to the strength of a covalent bond. A higher bond energy value means a stronger covalent bond.

Triple covalent bonds generally have a higher bond energy value than double covalent bonds, and double covalent bonds generally have a higher bond energy value than single covalent bonds. For example, Table 28.2 shows the bond energies of some carbon–carbon covalent bonds.

Table 28.2

Covalent bond	Bond energy (kJ)
C−C	348
C=C	611
C≡C	838

Exam tip

Bond enthalpies should always be quoted as a number. They do not have a positive or negative sign when written as they may be endothermic or exothermic depending on whether the bond is being broken or made, respectively.

Calculating energy changes

The best way to calculate an energy change from bond energy values is to add up the bond energies of all the bonds broken in the reaction and also add up all the bond energies of the bonds made in the reaction. Remember that bonds are broken in the reactants and bonds are made in the products.

The energy change is calculated using:

$$\text{energy change} = \begin{matrix}\text{total of bond energies} \\ \text{of all bonds broken}\end{matrix} - \begin{matrix}\text{total of bond energies} \\ \text{of all bonds made}\end{matrix}$$

Example 1

Methane reacts with oxygen to form carbon dioxide and water. The equation for the reaction is given below, with the structures showing the covalent bonds:

$$CH_4(g) + 2O_2(g) \rightarrow CO_2(g) + 2H_2O(l)$$

H–C(–H)(–H)–H + O=O, O=O → O=C=O + H–O–H, H–O–H

The bond energy values are given in the table below:

Bond	Bond energy (kJ)
C–H	412
O=O	496
C=O	803
O–H	463

Calculate the energy change in this reaction using the bond energy values.

Answer

The following bonds are broken:

4 C–H bonds = 4 × 412

2 O=O bonds = 2 × 496

And the following bonds are made:

2 C=O bonds = 2 × 803

4 O–H bonds = 4 × 463

Therefore:

total of the bond energies for all bonds broken = (4 × 412) + (2 × 496) = 2640 kJ

total of the bond energies for all bonds made = (2 × 803) + (4 × 463) = 3458 kJ

$$\text{energy change} = \begin{matrix}\text{total of bond energies} \\ \text{of all bonds broken}\end{matrix} - \begin{matrix}\text{total of bond energies} \\ \text{of all bonds made}\end{matrix}$$

standard enthalpy of combustion of methane = 2640 − 3458 = −818 kJ

The reaction is exothermic because the energy change is negative. You would expect a combustion reaction to be exothermic as it releases heat to the surroundings.

Exam tip

Don't be tempted to multiply the value by 2 for a double covalent bond such as O=O or C=O because the bond energy value is the energy to break both the covalent bonds in O=O and C=O. For example, 496 kJ breaks both the covalent bonds in O=O.

H

Example 2

Ethanol undergoes complete combustion when it reacts with oxygen, forming carbon dioxide and water. The table below gives the bond energy values involved in this reaction:

Bond	Bond energy (kJ)
C–H	412
C–C	348
C–O	360
O–H	463
O=O	496
C=O	803

The equation for the reaction is given below, with the structures of the substances showing the bonds:

$$C_2H_5OH(l) + 3O_2(g) \rightarrow 2CO_2(g) + 3H_2O(l)$$

$$H{-}\underset{\underset{H}{|}}{\overset{\overset{H}{|}}{C}}{-}\underset{\underset{H}{|}}{\overset{\overset{H}{|}}{C}}{-}O{-}H + 3\,O{=}O \longrightarrow 2\,O{=}C{=}O + 3\,H{-}O{-}H$$

Calculate the energy change in this chemical reaction using the bond energies.

The following bonds are broken:

1 C–C bond = 348

5 C–H bonds = 5 × 412

1 C–O bond = 360

1 O–H bond = 463

3 O=O bonds = 3 × 496

The following bonds are made:

4 C=O bonds = 4 × 803

6 O–H bonds = 6 × 463

Therefore:

total of the bond energies for all bonds broken
= 348 + (5 × 412) + 360 + 463 + (3 × 496) = 4719 kJ

total of the bond energies for all bonds made = (4 × 803) + (6 × 463)
= 5990 kJ

energy change = total of bond energies of all bonds broken − total of bond energies of all bonds made

standard enthalpy of combustion of ethanol = 4719 − 5990 = −1271 kJ

Again the reaction is exothermic because the energy change is negative.

H

Example 3

Methane and steam (H_2O) react to form carbon dioxide and hydrogen according to the balanced symbol equation:

$$CH_4 + 2H_2O \rightarrow CO_2 + 4H_2$$

H–C(H)(H)–H + H–O–H, H–O–H → O=C=O + H–H, H–H, H–H, H–H

The bond energy values involved are given in the table below. Calculate the energy change for this reaction.

Bond	Bond energy (kJ)
C–H	412
O–H	463
C=O	803
H–H	436

The following bonds are broken:

4 C–H bonds = 4 × 412

4 O–H bonds = 4 × 463

The following bonds are made:

2 C=O bonds = 2 × 803

4 H–H bonds = 4 × 436

Therefore:

total of the bond energies for all bonds broken = (4 × 412) + (4 × 463) = 3500 kJ

total of the bond energies for all bonds made = (2 × 803) + (4 × 436) = 3350 kJ

$$\text{energy change} = \begin{array}{c}\text{total of bond energies}\\ \text{of all bonds broken}\end{array} - \begin{array}{c}\text{total of bond energies}\\ \text{of all bonds made}\end{array}$$

standard enthalpy of combustion of methane = 3500 − 3350 = +150 kJ

This reaction is endothermic because the energy change is positive.

Now test yourself

TESTED

4 Which of the following is endothermic: bond breaking or bond making?

5 For the reaction $H_2 + F_2 \rightarrow 2HF$, the bond energies for H–H, F–F and H–F are 436, 158 and 565 kJ respectively. Calculate the energy change.

6 For the combustion of ethene, $C_2H_4(g) + 3O_2(g) \rightarrow 2CO_2(g) + 2H_2O(l)$, calculate the energy change. Bond energies: C=C is 611 kJ, C–H is 412 kJ, O=O is 496 kJ, C=O is 803 kJ and O–H is 463 kJ.

H

Exam practice

1 What do these terms mean:
(a) exothermic
(b) endothermic? [2 marks]

2 What is meant by activation energy? [1 mark]

3 The energy change for the reversible reaction $2SO_2(g) + O_2(g) \rightleftharpoons 2SO_3(g)$ is –192 kJ.
(a) Explain whether the forward reaction is exothermic or endothermic. [1 mark]
(b) State the energy change for the reverse reaction. [1 mark]

4 The reaction between hydrogen and chlorine, forming hydrogen chloride, is given in the balanced symbol equation below:

$$H_2(g) + Cl_2(g) \rightarrow 2HCl(g)$$

Explain, in terms of bonds, why this reaction is exothermic. [3 marks]

5 A reaction profile is shown below:

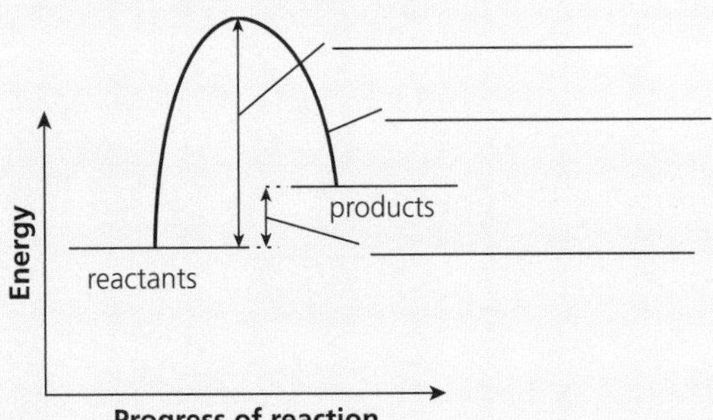

(a) Complete the three missing labels on the diagram. [3 marks]
(b) Explain whether this reaction is exothermic or endothermic. [1 mark]

6 For the following reactions:
A electrolysis of lithium chloride
B combustion of butane
C thermal decomposition of calcium carbonate
D neutralisation of sodium hydroxide solution with hydrochloric acid
Which reactions are exothermic and which are endothermic? [2 marks]

7 Hydrogen reacts with oxygen to form water according to the following balanced symbol equation:

$$2H_2(g) \quad + \quad O_2(g) \quad \rightarrow \quad 2H_2O(l)$$

H—H + O=O ⟶ H–O–H
H—H H–O–H

The bond energy values are given in the table below:

Bond	Bond energy (kJ)
H–H	436
O=O	496
O–H	463

Calculate the energy change for the reaction. [3 marks]

8 Hydrogen bromide undergoes thermal decomposition to form hydrogen and bromine.

$$2HBr(g) \rightarrow H_2(g) + Br_2(g)$$

(a) Explain, in terms of bonds, why this reaction is endothermic. [3 marks]
(b) The bond energies are: H–Br bond, 362 kJ; H–H bond, 436 kJ; and Br–Br bond,190 kJ. Calculate the energy change. [3 marks]
(c) Explain how the energy change supports the fact that the reaction is endothermic. [1 mark]

➜

9 The reaction between hydrogen and nitrogen, forming ammonia, is given in the balanced symbol equation below:

$$N_2(g) + 3H_2(g) \rightleftharpoons 2NH_3(g)$$

$$N{\equiv}N + 3\,H{-}H \rightleftharpoons 2\,H{-}\underset{\displaystyle H}{\underset{|}{N}}{-}H$$

The bond energies involved in this reaction are given in the table below:

Bond	Bond energy (kJ)
N≡N	916
H–H	436
N–H	386

(a) Calculate the energy change for this reaction. [3 marks]

(b) Explain whether the forward reaction is exothermic or endothermic. [1 mark]

10 Ethane reacts with oxygen to form carbon dioxide and water in a combustion reaction:

$$2C_2H_6(g) \quad + \quad 7O_2(g) \quad \rightarrow \quad 4CO_2(g) \quad + \quad 6H_2O(l)$$

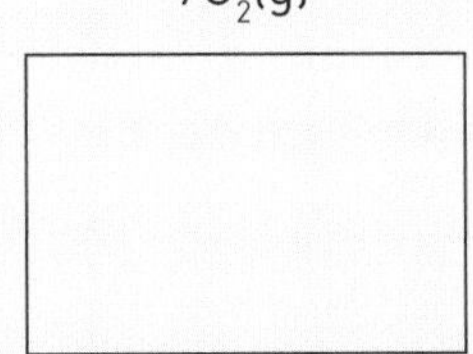
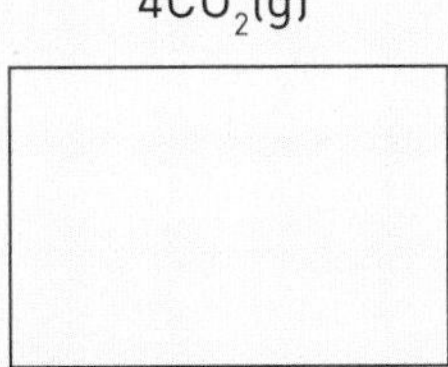
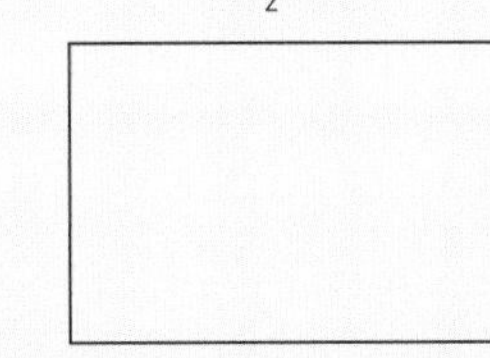

(a) In the boxes above list the total number of bonds for each of the reactants and products. [4 marks]

(b) The bond energies for the reactants and products are given in the table below. Calculate the energy change in the reaction. [3 marks]

Bond	Bond energy (kJ)
C–H	412
C–C	348
O=O	496
C=O	803
O–H	463

Answers online

ONLINE

29 Gas chemistry

Preparation and properties of gases

Gases in the atmosphere

REVISED

Atmosphere is the term used to describe the collection of gases that surrounds the Earth.

- The 'air' is the atmosphere near the surface.
- The composition of air has remained reasonably constant for about 200 million years, with 78% nitrogen, 21% oxygen and a small percentage of other gases, such as the noble gases (about 1% argon), water vapour and carbon dioxide (usually quoted to be around 0.03–0.04%).

Much of the non-metal chemistry section in the specification involves gases. For each gas it is important to know:

- the methods of preparation
- the physical properties
- the chemical properties
- the chemical test.

Exam tip

Make sure that you do not give chemical properties when asked for physical properties. Physical properties relate to the physical nature of the chemical, including its colour, odour (gases here are usually odourless or pungent), physical state (solid, liquid or gas) or melting point and boiling point (high or low), solubility in water (soluble or insoluble or low solubility in water) and density (less dense or denser than air for gases). Chemical properties relate to chemical reactions that they undergo.

Preparation of the gases

REVISED

You need to know:

- the reagents
- the apparatus used to prepare these gases:
  - hydrogen
  - oxygen
  - carbon dioxide
- the way in which a gas is collected.

Collection of gases

REVISED

Gas collection depends on the solubility of the gas in water and if it is soluble, collection depends on its density compared with air.

Collecting insoluble gases

Figure 29.1 shows how gases that are **insoluble** in water, or have a low solubility in water, can be collected. This is called **collection over water**.

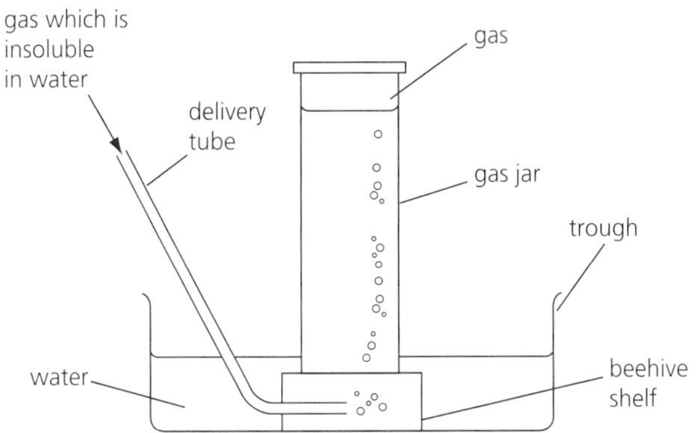

Figure 29.1 Collecting a gas over water

Collecting soluble gases

Gases that are soluble in water are collected by **displacement** of air and so the **density** of the gas compared with air must be considered:

- Soluble gases that are denser than air are collected by **downward delivery**.
- Soluble gases that are less dense than air must be collected by **upward delivery** (Figure 29.2).

(a) downward delivery

gas which is soluble in water and **denser** than air, for example sulfur dioxide, SO_2; hydrogen chloride, HCl

(b) upward delivery

gas which is soluble in water and **less dense** than air, for example ammonia, NH_3

Figure 29.2 Collecting a gas by (a) downward delivery and (b) upward delivery

Chemistry of specific gases

Nitrogen

REVISED

Physical properties

- Nitrogen is a colourless, odourless gas that is insoluble in water.
- It is a diatomic gas, N_2.
- It is an unreactive gas. The lack of reactivity of nitrogen is because of the triple covalent bond between the nitrogen atoms in N_2 molecules. The strong triple covalent bond requires substantial energy to break before the nitrogen atoms can react.

Uses of nitrogen

Nitrogen has two main uses:

- Liquid nitrogen is used as a coolant.
- In food packaging nitrogen creates an inert atmosphere to keep food fresh.

H Ammonia

REVISED

Physical properties

Ammonia, NH_3, is a colourless, pungent gas that is soluble in water and less dense than air.

Test for ammonia

Method: Dip a glass rod in concentrated hydrochloric acid and put this in a sample of the gas.

Test result: If ammonia is present, a white 'smoke' of ammonium chloride is observed.

Exam tip

Ammonia is often confused with 'ammonium'. Ammonia is a compound (NH_3) but 'ammonium' is an ion with the formula NH_4^+. Remember that any chemical with 'ammonium' as part of its name must contain NH_4 — ammonia is simply NH_3.

Ammonia gas reacts with hydrogen chloride gas, forming ammonium chloride, which appears as a white smoke or solid:

$$NH_3 + HCl \rightarrow \underset{\text{white smoke}}{NH_4Cl}$$

Uses of ammonia

Ammonia is used to produce fertilisers by reacting it with acids such as hydrochloric acid and nitric acid:

$$NH_3 + HCl \rightarrow NH_4Cl$$

$$NH_3 + HNO_3 \rightarrow NH_4NO_3$$

The ammonium salts (ammonium chloride, NH_4Cl, and ammonium nitrate, NH_4NO_3) are very soluble in water, so they can dissolve in soil water and be absorbed by plants.

Now test yourself

TESTED

1. Which gas makes up approximately 21% of the atmosphere?
2. Explain why nitrogen is unreactive.
3. Name the white smoke produced during the test for ammonia gas.

Hydrogen

REVISED

Physical properties of hydrogen

Hydrogen is a colourless, odourless gas that is insoluble in water and less dense than air. Hydrogen is diatomic, H_2.

Preparation of hydrogen

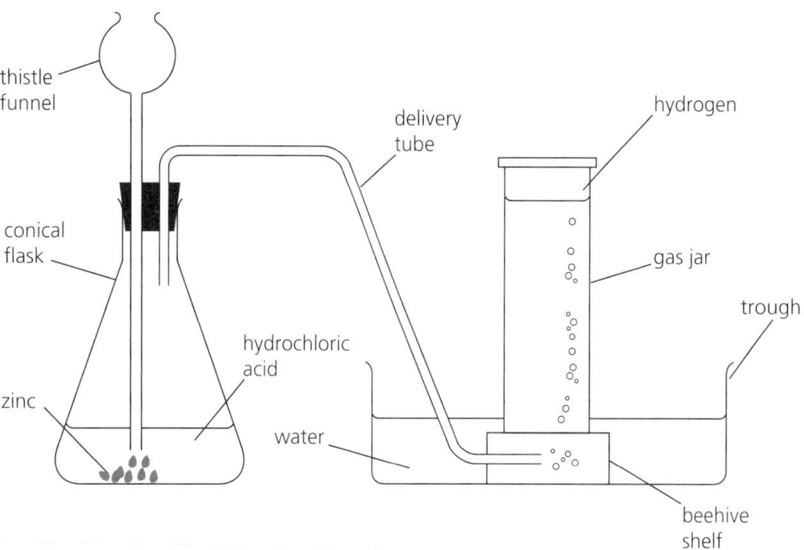

Figure 29.3 Preparing and collecting hydrogen gas

Hydrogen is prepared using zinc (or magnesium) and dilute hydrochloric acid:

$$Zn + 2HCl \rightarrow ZnCl_2 + H_2$$

$$Mg + 2HCl \rightarrow MgCl_2 + H_2$$

- Hydrogen is collected over water because it is insoluble in water (Figure 29.3).
- The reaction of dilute hydrochloric acid with magnesium is more vigorous and is not recommended for controlled preparation of hydrogen.

Exam tip

There are several important points to note when drawing the general gas preparation apparatus:
- The thistle funnel must be below the level of the solution in the conical flask.
- The delivery tube must not cut through the side of the trough.
- The water level in the trough must be *above* the beehive shelf.
- Water must be present in the gas jar.

Test for hydrogen

Method: Apply a lit splint.

Test result: It burns with a squeaky pop.

$$2H_2 + O_2 \rightarrow 2H_2O$$

Uses of hydrogen

These include:
- meteorological (weather) balloons
- hardening oils to form solid fats in margarines and butter-like spreads
- as a clean fuel.

Hydrogen as a clean fuel

Hydrogen is described as a clean fuel because the only product of combustion (water) is non-polluting.
- Hydrogen can be produced from the electrolysis of water — but this requires electricity. The electricity can be generated from combustion of fossil fuels or renewable sources of energy such as tidal, solar, wave or hydroelectric. The use of renewable sources of energy increases the potential of hydrogen as a clean fuel.
- Hydrogen can be used to power vehicles and would be supplied to consumers in liquid form. Keeping hydrogen as a liquid requires energy and specialised storage.
- Hydrogen is very flammable and so high-level safety controls would have to be in place to protect users from the risk of explosion.
- Technology is advancing to improve the supply, storage and use of hydrogen.

Now test yourself

TESTED ☐

4. State three physical properties of hydrogen.
5. Write a balanced symbol equation for the production of hydrogen from zinc and hydrochloric acid.
6. Describe the test for hydrogen gas.

Oxygen

REVISED

Physical properties of oxygen

Oxygen is a colourless, odourless gas that is only slightly soluble in water and is slightly denser than air. Oxygen is diatomic, O_2.

Laboratory preparation of oxygen

Oxygen is made using hydrogen peroxide and manganese(IV) oxide, using the same apparatus as in Figure 29.3. The zinc is replaced by manganese(IV) oxide and the hydrochloric acid by hydrogen peroxide. Manganese (IV) oxide is also called manganese dioxide — it is a catalyst (page 186) used to speed up the decomposition of hydrogen peroxide:

$$2H_2O_2 \rightarrow 2H_2O + O_2$$

Test for oxygen

Method: Apply a glowing splint.

Test result: The glowing splint relights.

Uses of oxygen

- In medicine
- In welding

Chemical properties of oxygen

Reaction with carbon

Observations: Black carbon burns with an orange, sooty flame (and sparks), forming a colourless gas — carbon dioxide:

$$C + O_2 \rightarrow CO_2$$

If there is a limited supply of oxygen, the combustion reaction produces carbon monoxide:

$$2C + O_2 \rightarrow 2CO$$

Carbon monoxide is toxic (page 198).

Reaction with sulfur

Observations: Yellow, solid sulfur melts to a red liquid and burns with a blue flame, giving a colourless, pungent gas — sulfur dioxide:

$$S + O_2 \rightarrow SO_2$$

Reaction with magnesium

Observations: Grey, solid magnesium burns with a bright, white light, releasing heat and producing a white solid — magnesium oxide:

$$2Mg + O_2 \rightarrow 2MgO$$

Reaction with iron

Observations: Grey, solid iron filings burn with orange sparks, producing a black solid — Fe_3O_4:

$3Fe + 2O_2 \rightarrow Fe_3O_4$

Reaction with copper

Observations: Red-brown solid glows red (there may be a blue-green flame) and forms a black solid — copper(II) oxide.

$2Cu + O_2 \rightarrow 2CuO$

Nature of the oxides

The oxides formed above are either basic or acidic.

Basic oxides (bases) are generally oxides of metals that react with acid to produce a salt and water. Examples are MgO, Fe_3O_4 and CuO. For example:

$MgO + 2HCl \rightarrow MgCl_2 + H_2O$

$CuO + H_2SO_4 \rightarrow CuSO_4 + H_2O$

MgO and CuO are basic oxides, but they are insoluble in water, so they do not produce an alkaline solution when added to water.

Acidic oxides are generally oxides of non-metals that will react with alkalis to form a salt and water. Examples are CO_2 and SO_2. For example:

$CO_2 + 2NaOH \rightarrow Na_2CO_3 + H_2O$

$SO_2 + 2KOH \rightarrow K_2SO_3 + H_2O$

K_2SO_3 is potassium sulfite. The sulfite ion is SO_3^{2-}.

Some acidic oxides like CO_2 and SO_2 react with water to form an acidic solution:

$CO_2 + H_2O \rightarrow H_2CO_3$
carbonic acid

$SO_2 + H_2O \rightarrow H_2SO_3$
sulfurous acid

Both of these acids are weak acids and would have a pH of around 3–6, depending on the concentration of the acid solution. Universal indicator would change to orange or yellow when added to these solutions.

Now test yourself

TESTED ☐

7 Name the solution and catalyst used in the preparation of oxygen gas.
8 What is observed when carbon burns in oxygen?
9 Is magnesium oxide an acidic or basic oxide?

Carbon dioxide

REVISED

Preparation of carbon dioxide

Carbon dioxide is prepared from calcium carbonate (marble chips) and hydrochloric acid using the same apparatus as in Figure 29.3 (page 234). The zinc is replaced by calcium carbonate (marble chips):

$$CaCO_3 + 2HCl \rightarrow CaCl_2 + CO_2 + H_2O$$

Carbon dioxide is collected over water because it has a low solubility in water.

Physical properties of carbon dioxide

Carbon dioxide is a colourless, odourless gas with a low solubility in water. It is denser than air.

Exam tip

Carbon dioxide does not support combustion, but this is a chemical property rather than a physical property because it involves combustion, which is a chemical reaction.

Test for carbon dioxide

Method: Bubble the gas through limewater.

Test result: The colourless solution becomes milky.

If carbon dioxide is bubbled through limewater until it is in excess, the colourless solution becomes milky (a white precipitate is formed) and then the precipitate dissolves to form a colourless solution:

$$CO_2 + \underset{\text{limewater}}{Ca(OH)_2} \rightarrow \underset{\text{white ppt}}{CaCO_3} + H_2O$$

$$\underset{\text{white ppt}}{CaCO_3} + CO_2 + H_2O \rightarrow \underset{\text{colourless solution}}{Ca(HCO_3)_2}$$

Chemical properties of carbon dioxide

Reaction with water

Carbon dioxide reacts with water to form the weak acid, carbonic acid, H_2CO_3:

$$CO_2 + H_2O \rightarrow H_2CO_3$$

The acid cannot be isolated from the solution and so is often simply written as $CO_2(aq)$. Carbonic acid causes the acidity in fizzy drinks.

Reaction with limewater

As discussed above, carbon dioxide reacts with an alkaline solution of calcium hydroxide, $Ca(OH)_2$, to form a white precipitate of calcium carbonate, $CaCO_3$. Addition of further carbon dioxide results in the precipitate dissolving to form a colourless solution of calcium hydrogencarbonate. The equations are given above.

Now test yourself

TESTED

10 Write a balanced symbol equation for the production of carbon dioxide from calcium carbonate and hydrochloric acid.
11 State two properties of carbon dioxide.
12 Name the chemical that causes the solution to change to 'milky' when carbon dioxide is bubbled into limewater.

Exam tip

Carbon dioxide is an acidic gas and will react with any alkali such as calcium hydroxide and sodium hydroxide. It forms the metal carbonate, so carbon dioxide reacts with sodium hydroxide to form sodium carbonate and water:

$$2NaOH + CO_2 \rightarrow Na_2CO_3 + H_2O$$

Uses of carbon dioxide

Carbon dioxide is used in fire extinguishers because it does not support combustion — it is also denser than air and so covers the burning fuel.

Carbon dioxide is used in making carbonated drinks because it has a low solubility in water. When a bottle of a carbonated drink is opened, the gas is released with a fizz. It also gives the drink an acidic taste due to the carbonic acid present.

Prescribed practical C6

Investigate the preparation, properties, tests and reactions of the gases hydrogen, oxygen and carbon dioxide

- Be able to carry out and describe practically the preparation of the three gases (H_2, O_2 and CO_2).
- Be able to carry out the tests for the gases and test their properties or recognise their properties from given information.
- Be able to carry out and describe the reactions of the gases, including observations.
- Be able to write balanced symbol equations for the reactions.

Exam practice

1 What is the most abundant gas in the atmosphere? [1 mark]

2 In each case, name two chemicals needed to prepare these gases:
(a) oxygen
(b) carbon dioxide
(c) hydrogen [6 marks]

3 State how you would carry out the test for the following gases and what you would observe if the gas was present:
(a) hydrogen [2 marks]
(b) carbon dioxide [3 marks]
(c) oxygen [2 marks]
H (d) ammonia [3 marks]

4 Explain why hydrogen is described as a clean fuel. [2 marks]

5 State two uses of carbon dioxide. [2 marks]

6 Write a balanced symbol equation for the production of oxygen gas from hydrogen peroxide solution. Include state symbols. [4 marks]

7 State two uses of oxygen. [2 marks]

8 What is the common name and chemical name for the solution used to test for carbon dioxide? [1 mark]

9 Why is nitrogen unreactive? [2 marks]

10 Write a balanced symbol equation for carbon dioxide reacting with water. [2 marks]

11 Sulfur reacts with oxygen when it is heated in air:
(a) Write a balanced symbol equation for sulfur reacting with oxygen when it is heated in air. [2 marks]
(b) What is observed when sulfur is heated in air? [3 marks]

12 Write a balanced symbol equation for calcium carbonate reacting with hydrochloric acid to produce carbon dioxide. [3 marks]

13 Name the catalyst used to decompose hydrogen peroxide. [1 mark]

14 State two physical properties of oxygen gas [2 marks]

15 In the following diagram for gas preparation:

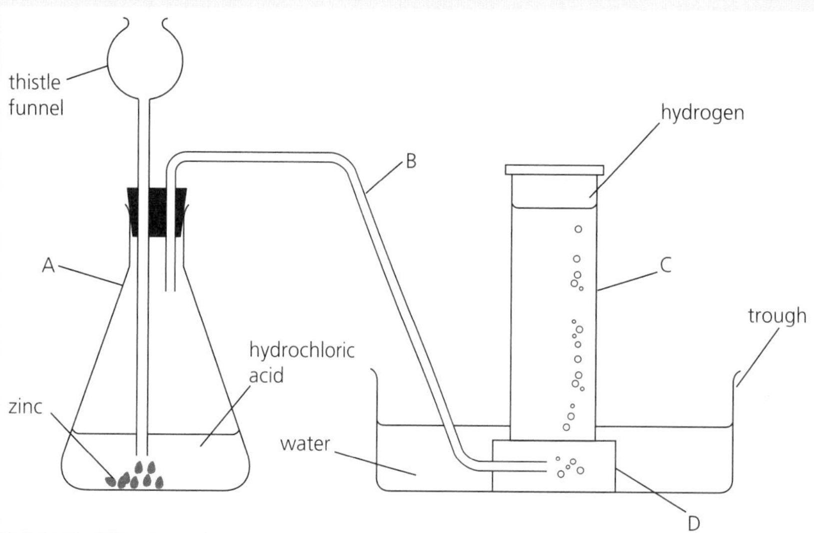

What labels should be placed at A, B, C and D? [4 marks]

Answers online

ONLINE

30 Motion

Motion in a straight line

REVISED

The **distance** between two points is how far they are apart.

Displacement measures their distance apart *and* specifies the direction.

In the same way, **speed** is the rate at which distance travelled changes with time, but **velocity** is the rate of change of displacement with time.

$$\text{average speed} = \frac{\text{total distance travelled}}{\text{total time taken}} \qquad \text{average velocity} = \frac{\text{total displacement}}{\text{total time taken}}$$

displacement is the distance between two points in a specified direction

speed is the rate at which distance changes with time

velocity is the rate of change of displacement with time

Exam tip

These formulae must be memorised.

Suppose Jo walks around the three sides of the sports pitch shown in Figure 30.1 in a time of 250 seconds.

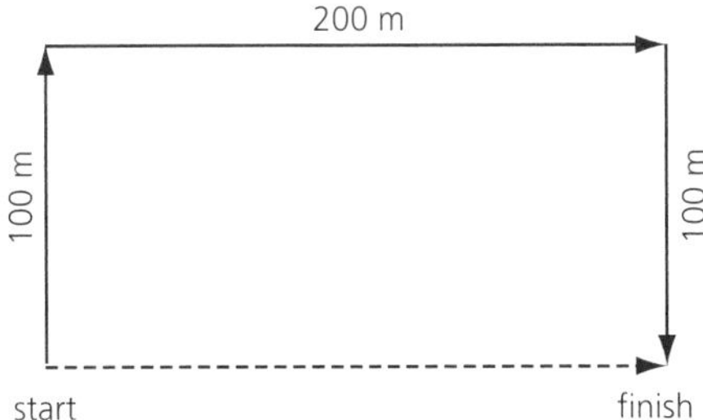

Figure 30.1

She has travelled a **distance** of **400 metres**. But her final **displacement** is **200 m to the right** of the starting position.

Her **average speed** is given by:

$$\text{average speed} = \frac{\text{total distance travelled}}{\text{total time taken}} = \frac{400}{250}$$

$$= 1.6\,\text{m/s}$$

average speed is the total distance travelled divided by the total time taken

The rate of change of speed with time is defined by the equation:

$$\text{rate of change of speed} = \frac{\text{final speed} - \text{initial speed}}{\text{time taken}}$$

Rate of change of speed is measured in m/s^2.

Rate of change of speed has no direction since speed has no direction.

If the rate of change of speed is constant, then average speed can be calculated from the formula:

$$\text{average speed} = \frac{\text{initial speed} + \text{final speed}}{2}$$

Exam tip

These formulae must be memorised.

H Her **average velocity** is given by:

$$\text{average velocity} = \frac{\text{total displacement}}{\text{total time taken}} = \frac{200}{250}$$

$$= 0.8\,\text{m/s to the right}$$

If the starting position and finishing position are the same, then the total displacement will be zero and hence the average velocity must be zero also.

average velocity is the total displacement divided by the total time taken

H **Acceleration** is defined as the rate of change of velocity with time. The definition can be written as an equation:

$$a = \frac{v - u}{t} = \frac{\text{change in velocity}}{\text{time taken}} = \frac{\Delta v}{t}$$

or

$$v = u + at$$

where a is the acceleration in m/s^2
u is the initial (starting) velocity in m/s
v is the final velocity in m/s
t is the time taken in seconds
Δv = change in velocity in m/s

acceleration is the rate of change of velocity with time

If the acceleration is constant, the average velocity can be calculated using the formula:

$$\text{average velocity} = \frac{\text{initial velocity} + \text{final velocity}}{2}$$

A positive acceleration means the velocity is increasing, while a negative acceleration means a decreasing velocity or retardation.

Exam tip

These formulae must be memorised.

Table 30.1 represents an acceleration of 2 m/s^2. Every second the velocity increases by 2 m/s.

Table 30.2 represents an acceleration of −2 m/s^2. Every second the velocity decreases by 2 m/s.

Table 30.1

Velocity/m/s	12	14	16	18	20
Time/s	0	1	2	3	4

Table 30.2

Velocity/m/s	19	17	15	13	11
Time/s	0	1	2	3	4

One of the prescribed practicals (P1) that you are required to carry out involves using simple apparatus, including trolleys, ball bearings, metre rulers, stop clocks and ramps, to investigate experimentally how the average speed of an object moving down a runway depends on the slope of the runway measured as the height of one end of the runway.

A possible approach is given below.

Prescribed practical P1

Average speed

Method

1 See Figure 30.2. Set up a ramp against a small pile of wooden blocks (or books).
2 With a ruler draw two pencil lines on the ramp, one at the top and the other at the bottom, 1.0 m apart. This is the distance, x.
3 Measure the height of the ramp, h. This is the independent variable. The dependent variable is the average speed of a trolley moving down the ramp.
4 Allow a trolley to roll down the ramp, starting from rest at the upper pencil line and finishing as its wheels reach the lower pencil line.
5 For each height, h, ranging from about 1 cm to about 5 cm, time this motion a total of *three* times using a stop clock and record the results in a table such as shown on page 243.
6 Calculate the average time, t.

→

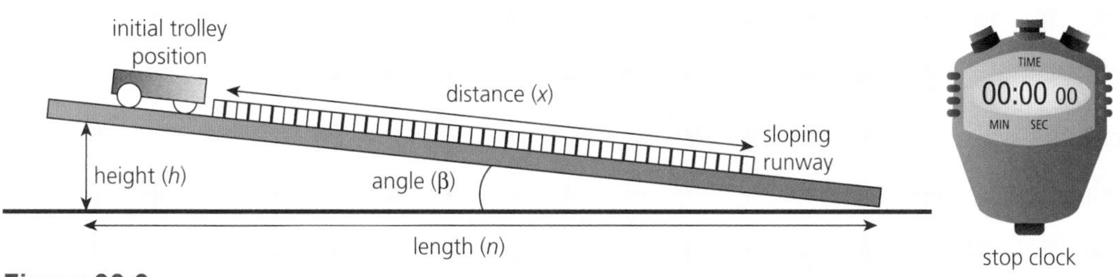

Figure 30.2

Results

The average speed is approximately equal to x/t.

Height (h)/cm	1	2	3	4	5
Time (t_1)/s	8.2	5.9	4.6	4.1	3.5
Time (t_2)/s	8.2	5.8	4.7	4.1	3.7
Time (t_3)/s	8.1	5.8	4.8	4.2	3.9
Average time (t)/s	8.2	5.8	4.7	4.1	3.7
Average speed (v)/cm/s	12.2	17.2	21.3	24.4	27.0

The data in the table are typical of what might be obtained.

7 Plot the graph of average speed (y-axis) against height, h (x-axis). The line of best fit is a curve through the origin of decreasing gradient. The graph shows that the average speed is **not** proportional to h, but it increases with h in some (unknown) non-linear way.

Many possible approaches to this experiment are possible. For example, the timing may be done with light gates and a data-logger, and the data analysis and graph plotting might be done using suitable computer software.

Ⓗ Vectors and scalars

REVISED

A **scalar** is a physical quantity that has magnitude, but not direction. Examples are distance, speed, rate of change of speed and time.

A **vector** is a physical quantity that has both magnitude and direction. Examples are displacement, velocity, acceleration and force.

For every physical quantity you encounter in your GCSE course, you need to be able to state its unit and whether it is a scalar or a vector.

scalar is a physical quantity that has magnitude but not direction

vector is a physical quantity that has both magnitude and direction

Now test yourself

TESTED

1 An athlete jogs five times around a rectangular track measuring 105 m by 150 m in a time of 850 s. Calculate:
  (a) the total distance travelled
  (b) the athlete's average speed.

2 A sports car can increase its speed from 3 m/s to 27 m/s in 8 s. Calculate:
  (a) its rate of increase in speed, assuming it is constant
  (b) its average speed
  (c) the distance travelled during the 8 s its speed was increasing.

3 A marble takes 3.7 s to roll down a runway of length 100 cm starting from rest.
  (a) Calculate:
    (i) the marble's average speed
    (ii) the marble's maximum speed
    (iii) the rate at which the speed of the marble was increasing.
  (b) What assumption did you make in parts (a)(ii) and (iii)?

4 A car decelerates uniformly from 28 m/s to rest in a time of 7.0 s. Calculate:
(a) its retardation
(b) the distance travelled while it was slowing down.

5 (a) Which of the following are vectors?
speed, distance, work, energy, power, time
(b) Explain why the sum of a 12 kg mass and an 8 kg mass is always 20 kg, but the sum of a 12 N force and an 8 N force is not always 20 N.

Motion graphs

REVISED

You need to be able to interpret distance–time and velocity–time graphs to solve questions of a mathematical nature. Here are some essential ideas which you must remember:

- The **gradient** of a **distance–time graph** represents an object's **speed**.
- The **gradient** of a **speed–time graph** represents an object's **rate of change of speed**.
- The **area** between a **speed–time graph** and the time axis represents the **distance moved**.
- The **gradient** of a **displacement–time graph** represents an object's **velocity**.
- The **gradient** of a **velocity–time graph** represents an object's **acceleration**.
- The **area** between a **velocity–time graph** and the time axis represents the **displacement**.

Examples

1 A speed–time graph for a tractor is shown in Figure 30.3. Calculate:
(a) the distance travelled between times t = 0 s and t = 30 s
(b) the rate of change of speed between times t = 15 s and t = 20 s.

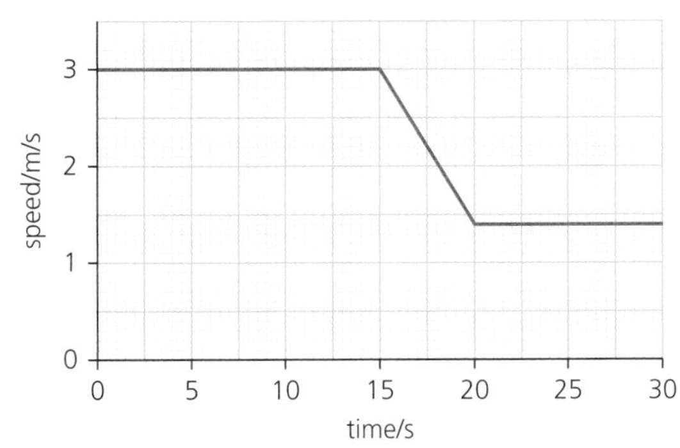

Figure 30.3

Answer

(a) distance = area between graph and the time axis. Divide graph into two rectangles and a triangle as shown in Figure 30.4.

distance =	(1.4 × 30)	+	(1.6 × 15)	+	(½ × 5 × 1.6)
	lower rectangle		upper rectangle		triangle
=	42	+	24	+	4

= 70 m

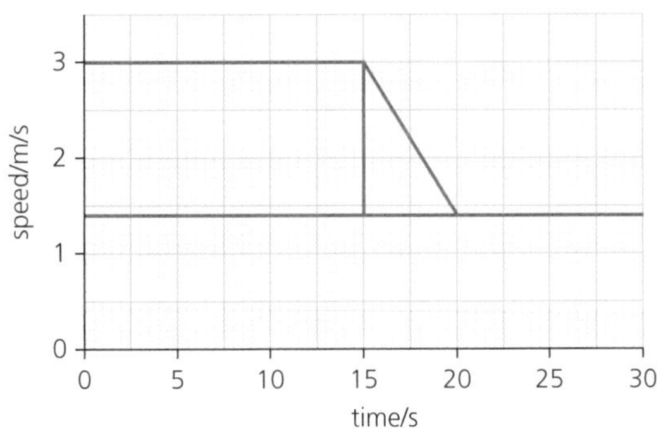

Figure 30.4

(b) rate of change of speed = gradient of speed–time graph

$$= \frac{1.4 - 3.0}{20 - 15} = \frac{-1.6}{5} = -0.32\,\text{m/s}^2$$

2 A pupil runs a race and the velocity–time graph for the race is shown in Figure 30.5 .Calculate:
(a) the length of the race
(b) the acceleration during the first 4 seconds.

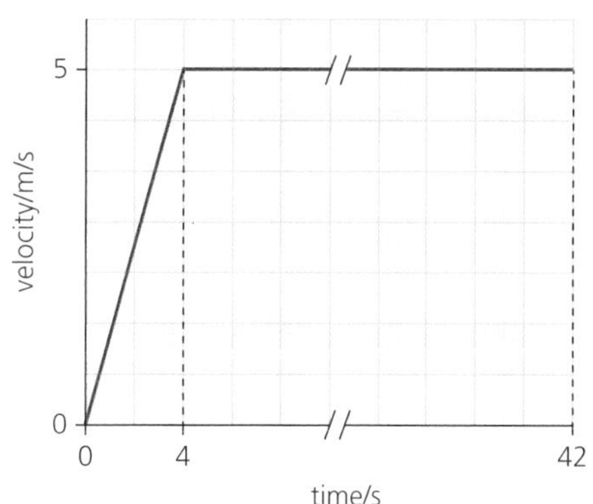

Figure 30.5

Answer

(a) displacement = area between v–t graph and the time axis

$$\begin{aligned}\text{displacement} &= \text{area of triangle} + \text{area of rectangle}\\ &= \tfrac{1}{2}(4 \times 5) \quad + \quad (38 \times 5)\\ &= 10 \quad + \quad 190\\ &= 200\,\text{m}\end{aligned}$$

(b) acceleration = gradient of velocity–time graph

$$\text{acceleration} = \frac{\text{change in velocity}}{\text{change in time}} = \frac{5-0}{4-0} = 1.25\,\text{m/s}^2$$

3 Jim runs home from school each day. The graph in Figure 30.6 shows part of his journey.
(a) How far from school is Jim after 15 seconds?
(b) What is Jim's steady speed during the first 15 seconds of his motion?
(c) Describe the motion during the last 10 seconds of the journey.
(d) Calculate Jim's average speed for the entire 25 seconds of the journey.

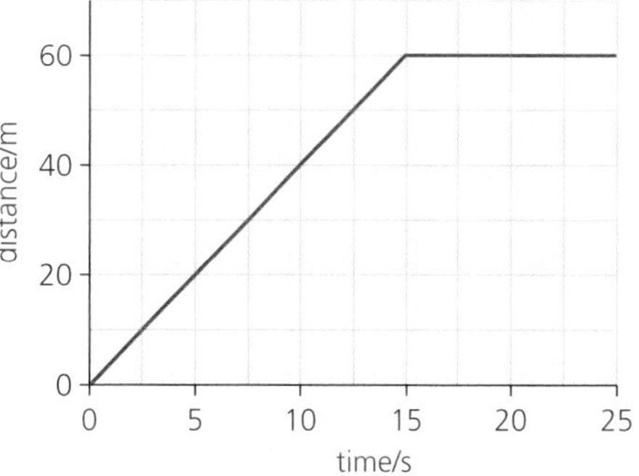

Figure 30.6

Answer

(a) At t = 15 s, distance to school is 60 m.

(b) $\text{speed} = \frac{\text{distance}}{\text{time}} = \frac{60}{15} = 4\,\text{m/s}$

(c) Jim is stopped for the last 10 s.

(d) $\text{average speed} = \frac{\text{total distance}}{\text{total time}} = \frac{60}{25} = 2.4\,\text{m/s}$

You should also remember that the average velocity along the straight part of a velocity–time graph is the average of the velocities at the start and end of the straight line. This is tested in question 4 in the exam practice at the end of this chapter.

Exam practice

1 A soldier marches 15 metres due east, then turns and marches 30 metres due west. He turns again and marches 30 metres due east. The total time taken is 25 seconds. Calculate:
(a) his final displacement from his starting position
(b) the total distance marched
(c) his average speed over the 25 seconds
(d) his average velocity.

→

H 2 A graph of velocity against time for a golf ball is shown in Figure 30.7.
(a) Use the graph to find the deceleration of the ball.
(b) How far does the ball travel in 6 seconds?

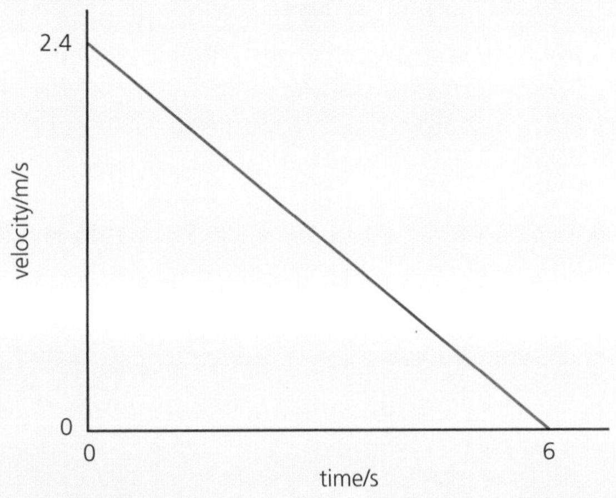

Figure 30.7

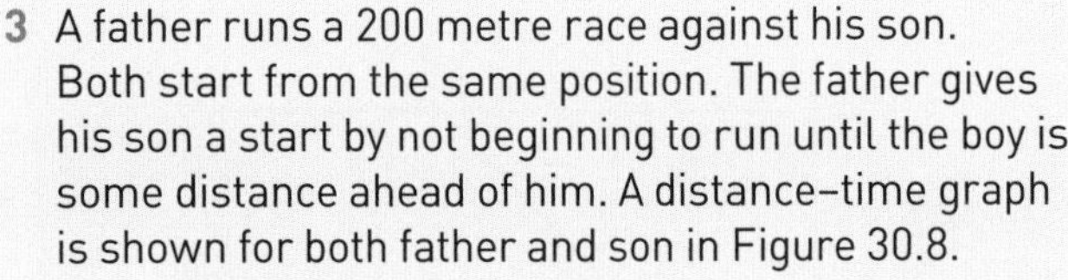

3 A father runs a 200 metre race against his son. Both start from the same position. The father gives his son a start by not beginning to run until the boy is some distance ahead of him. A distance–time graph is shown for both father and son in Figure 30.8.
(a) How far ahead was the son when the father began to run?
(b) How far from the start did the father overtake his son?
(c) When the father finished the race, how far behind was the son?

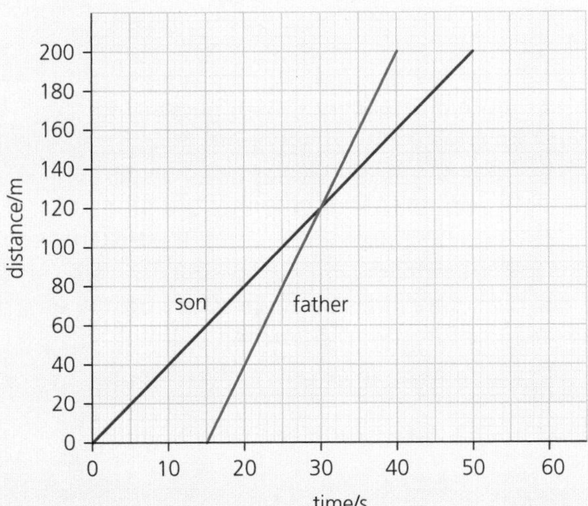

Figure 30.8

H 4 Car A and car B are in a race and their velocity–time graphs are shown in Figure 30.9.
(a) How far has car A travelled after 12 seconds?
(b) How far apart are the two cars after 12 seconds?
(c) Calculate the average velocity of car B.

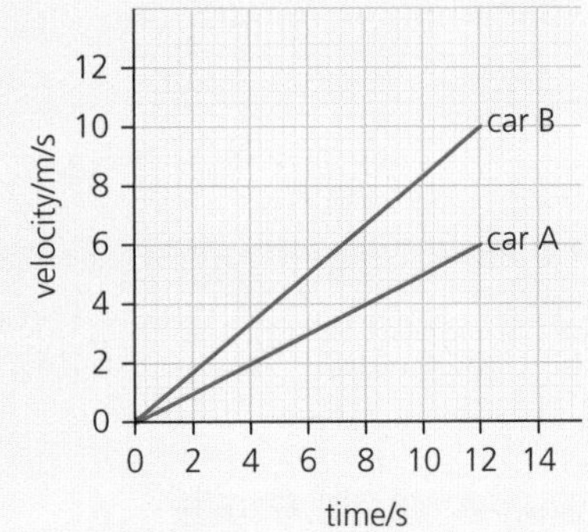

Figure 30.9

5 Maureen cycles to school each day. The graph in Figure 30.10 illustrates her journey. Use the graph to calculate her speed for the last part, BC, of the journey.

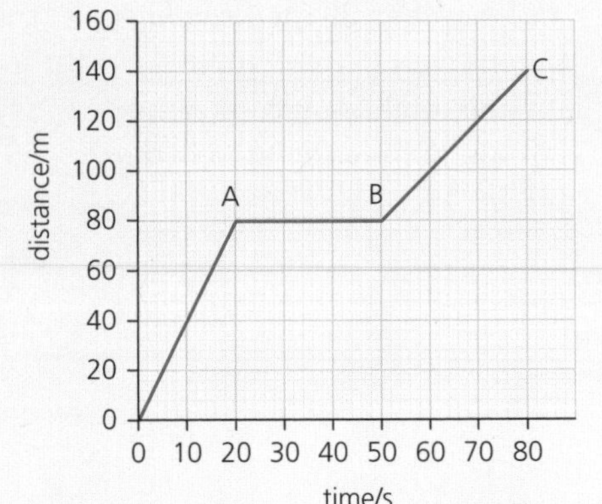

Figure 30.10

Answers online

31 Forces

Balanced and unbalanced forces

REVISED

Force is a vector — it has both size and direction. The size of the force is measured in **newtons** (N). In diagrams a force is represented by an arrow.

Equal forces acting in opposite directions are balanced. **Balanced forces** do not change the velocity of an object. This is summed up in **Newton's first law**:

> A body stays at rest, or if moving it continues to move with uniform velocity, unless an unbalanced force makes it behave differently.

balanced forces occur when there is no resultant force on an object

Exam tip

This law must be memorised.

Practical

Newton's first law

To demonstrate Newton's first law we use a linear air track and blower (to minimise friction), a glider and interrupt card, two light gates, a data logger and a computer (see Figure 31.1).

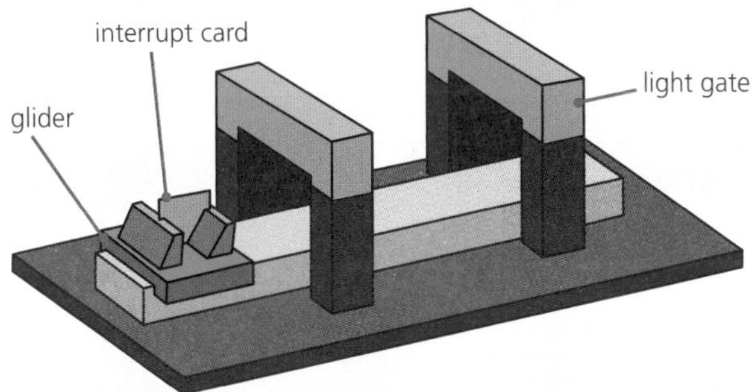

Figure 31.1

Method

1. Set the linear air track on a flat bench and adjust the feet on the air track to make sure that it is level.
2. Measure the length of the interrupt card and enter this in the data logger.
3. Connect up the light gates so that they measure the velocity of the glider at two points.
4. Give the glider a gentle push so that it passes through both light gates.
5. Confirm by looking at the results that the velocity does not change between the two positions of the light gate and so the glider is obeying Newton's first law of motion.
6. Repeat for other velocities and positions of the light gates.

Unbalanced forces will change the velocity of an object. Since velocity involves both speed and direction, unbalanced forces can make an object speed up, slow down or change direction. This is summed up in **Newton's second law**:

> The acceleration of a body is directly proportional to the force applied to it and inversely proportional to the object's **mass**.

Newton's second law can be written as an equation:

force = mass × acceleration

$$F = m \times a$$

Friction is a force which always opposes motion.

unbalanced forces occur when there is a resultant force causing an acceleration

mass is measured in kg and shows the amount of matter in an object

Exam tip

The equation $F = ma$ and the definition of friction must be memorised.

Practicals

Relationship between force, mass and acceleration

Apparatus

- runway
- trolley
- string
- double interrupt mask
- light gate and data logger
- pulley masses
- balance

Experiment 1: Keeping the mass being accelerated constant

1 Prepare a table for results as shown on page 249.
2 See Figure 31.2. To compensate for friction, tilt the runway until the trolley moves with a constant speed after it is given a gentle push.
3 Screw the clamped pulley to the end of the bench.
4 Attach a length of string to connect the end of the trolley to a slotted mass carrier and pass the string over the clamped pulley.
5 Position the light gate in such a way that the mask on top of the trolley passes through it without hitting anything and passes through it before the masses on the end of the string hit the ground.
6 Use the light gate to measure the acceleration of the trolley for various masses on the mass carrier from 100 g to 600 g, repeating each measurement two times and taking an average. Remember, each 100 g mass is equivalent to 1 N.
7 Find the weight of the masses and record this as the value for the resultant force *F*.
8 Plot a graph of acceleration (*y*-axis) versus force (*x*-axis).

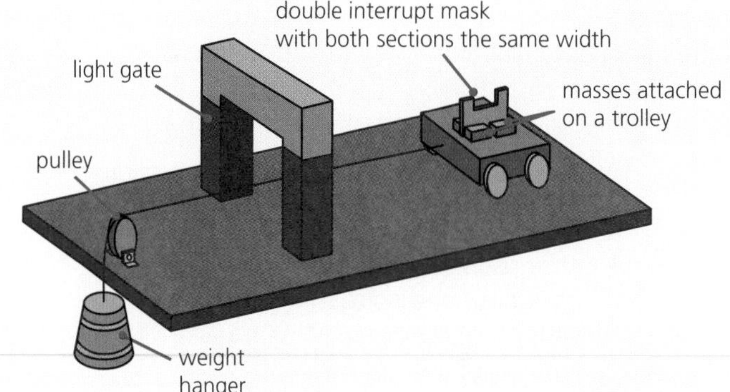

Figure 31.2 Diagram of apparatus used to show that $F = ma$

→

Resultant force/N	1.0	2.0	3.0	4.0	5.0	6.0
Acceleration/m/s²	0.80	1.59	2.40	3.21	4.10	4.80
Acceleration/m/s²	0.80	1.61	2.40	3.19	3.90	4.80
Mean acceleration/m/s²	0.80	1.60	2.40	3.20	4.00	4.80

A graph of mean acceleration (vertical axis) against resultant force (horizontal axis) gives a straight line through the origin (Figure 31.3), showing that the acceleration is directly proportional to the resultant force, when the mass being accelerated is constant.

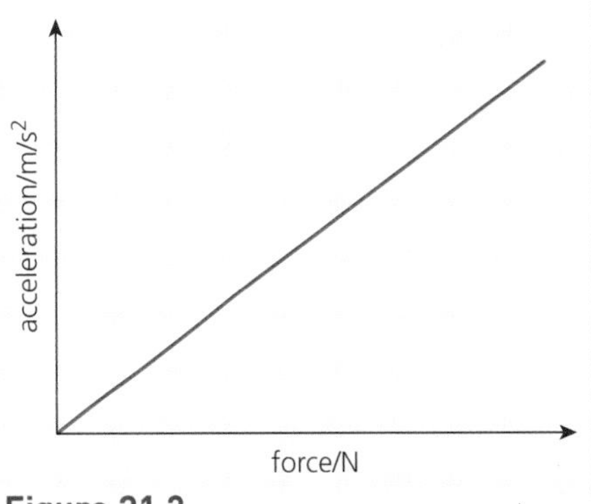

Figure 31.3

Experiment 2: Keeping the accelerating force constant

1. Prepare a table for results as shown below.
2. See Figure 31.2. To compensate for friction, tilt the runway until the trolley moves with a constant speed after it is given a gentle push.
3. Screw the clamped pulley to the end of the bench.
4. Attach a length of string to connect the end of the trolley to a slotted mass carrier and pass the string over the clamped pulley.
5. Position the light gate in such a way that the mask on top of the trolley passes through it without hitting the light gate and passes through it before the masses on the end of the string hit the ground.
6. Choose a suitable value for the driving force provided by the falling weights, e.g. 500 g (5 N).
7. Use the light gate to measure the acceleration of the trolley for various masses of trolley by either adding slotted masses to the trolley or stacking trolleys on top of each other.
8. Repeat each measurement and calculate the mean acceleration.
9. Plot graphs of (a) acceleration (*y*-axis) versus mass of trolley (*x*-axis) and (b) acceleration against 1/mass.

Mass/kg	1.0	2.0	3.0	4.0	5.0	6.0
Acceleration/m/s²	2.40	1.20	0.79	0.62	0.48	0.39
Acceleration/m/s²	2.40	1.20	0.81	0.58	0.48	0.41
Mean acceleration/m/s²	2.40	1.20	0.80	0.60	0.48	0.40

A graph of mean acceleration (vertical axis) against 1/mass (horizontal axis) yields a straight line through the origin (Figure 31.4) showing that the acceleration is **inversely proportional** to mass (or directly proportional to 1/mass) when the accelerating force is constant.

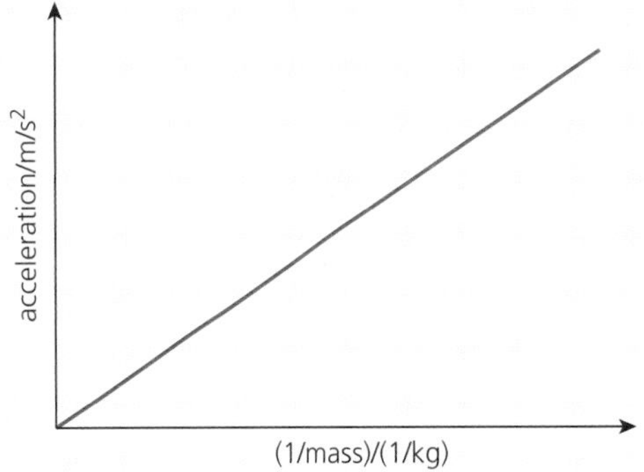

Figure 31.4

Examples

1 Calculate the force needed to give a car of mass 900 kg an acceleration of $0.5\,m/s^2$.

Answer

$$F = m \times a$$
$$= 900 \times 0.5$$
$$= 450\,N$$

2 A forward thrust of 400 N exerted by a tractor enables it to travel along a lane at constant velocity. The tractor has a mass of 1200 kg. Calculate the thrust required to accelerate the tractor at $1.5\,m/s^2$.

Answer

The phrase 'at constant velocity' is a clue to use Newton's first law. If the thrust exerted by the engine is 400 N, there must be an equal and opposite force of 400 N due to friction on the tractor. To calculate the force to accelerate the tractor we should draw a force diagram (Figure 31.5).

Figure 31.5

The unbalanced force is thrust minus friction, $F - 400$ N.

$$\text{unbalanced force} = \text{mass} \times \text{acceleration}$$
$$(F - 400) = 1200 \times 1.5$$
$$F - 400 = 1800$$
$$F = 2200\,N$$

Now test yourself

TESTED

1 A car accelerates at $3.0\,m/s^2$ along a road. The mass of the car is 1200 kg and all the resistive forces add up to 400 N. Calculate the forward thrust exerted by the car's engine.

2 Calculate the force of friction on a car of mass 1100 kg if it accelerates at $2\,m/s^2$ when the engine force is 3000 N.

H 3 The upward drag force on a parachutist of mass 60 kg is 480 N. Calculate her acceleration.

4 A forward thrust of 400 N exerted by a speedboat engine enables the speedboat to go through the water at a constant speed (Figure 31.6). The speedboat has a mass of 500 kg. Calculate the thrust required to accelerate the speedboat at $2\,m/s^2$.

5 A cyclist has a mass of 65 kg. When she provides a thrust of 100 N to her bicycle, she accelerates at $1.0\,m/s^2$. When the thrust is 140 N, the acceleration is $1.5\,m/s^2$. The friction force is constant. Calculate the mass of the bicycle and the size of the friction force.

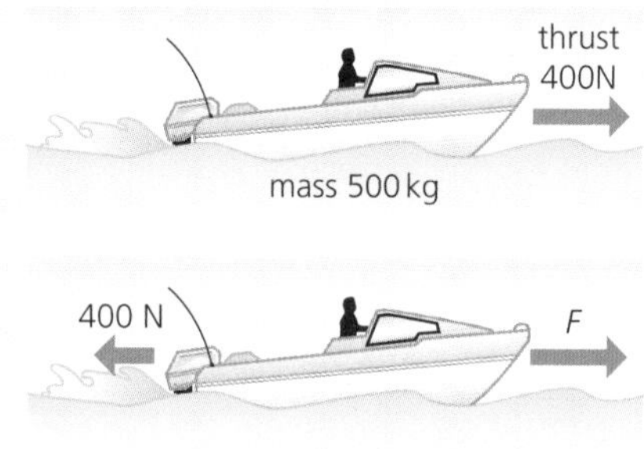

Figure 31.6

Mass and weight

REVISED

Mass is defined as the **amount of matter in a body**. Mass is measured in kilograms (kg). It is another example of a scalar quantity.

Weight is the **force of gravity on an object**. Since objects close to the Earth all experience the same acceleration, the **acceleration of free fall**, we can apply Newton's second law. In our case, the force is exerted by the Earth:

weight = force of gravity = mass × acceleration due to gravity

$$W = m \times g$$

weight is the force of gravity on an object

acceleration of free fall is the acceleration of an object towards the surface of the Earth when the only force on it is the gravitational force

Near the surface of the Earth, there is a gravitational force of **10 N** *on each* **1 kg** of mass. We say that the Earth's **gravitational field strength, *g***, is 10 N/kg.

Exam tip

This equation must be memorised.

The Moon is smaller than the Earth and pulls objects towards it less strongly. On the Moon's surface the value of g is 1.6 N/kg.

In deep space, far away from the planets, there are no gravitational pulls, so g is zero, and therefore everything is weightless.

gravitational field strength, *g* is the gravitational force on an object of mass 1 kg close to the surface of a planet

The size of g also gives the gravitational acceleration, because from Newton's second law:

$$\text{acceleration} = \frac{\text{force}}{\text{mass}}$$

or

$$g = \frac{\text{weight}}{\text{mass}}$$

So an alternative unit for g is m/s^2.

Now test yourself

TESTED

6 Describe three differences between mass and weight.
7 A parachutist of weight 620 N falls vertically through the air at a constant speed of 5 m/s.
 (a) State the resultant force on the parachutist.
 (b) State the size and direction of the frictional force on the parachutist.
8 Explain why there must be a resultant force on an object moving in a circular track at a steady speed.
9 Deep in space (where friction can be taken as zero), an astronaut throws a hammer. It leaves her hand with a speed of 3 m/s. Describe its subsequent motion. On what law of physics does your description depend?

Free fall

REVISED

Galileo showed that two lead balls of different diameter hit the ground at the same instant when dropped from the top of the leaning tower of Pisa.

All bodies in the absence of air resistance fall at the same rate of 10 m/s^2 near the surface of the Earth. It is a common misconception to think that a more massive object falls faster than a less massive one. It is true that there is a greater force on the more massive object but the acceleration, which is the ratio of force to mass, will be the same for both bodies:

$$a = \frac{F}{m} \text{ or in this case } g = \frac{W}{m}$$

H This means that if there is no air resistance, as in Figure 31.7, the speed of a falling object will increase by 10 m/s every second, i.e. its acceleration is 10 m/s^2.

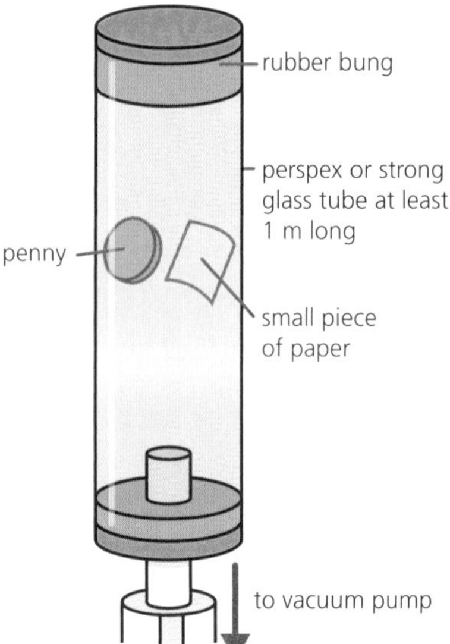

Figure 31.7

Now test yourself

TESTED

10 Julie said, 'My weight is 35 kg.' What is wrong with this statement and what do you think her weight really is?

11 A ball bearing is gently dropped into a tall cylinder of oil, which resists its motion. Describe what will happen to the ball bearing.

12 An astronaut standing on the surface of the Moon releases a hammer and a feather from the same height. What will happen and why?

13 Why does a parachute slow down a falling parachutist?

H Vertical motion under gravity

REVISED

When a body is thrown vertically upwards, its motion is opposed by the force of gravity. The velocity of the body will *decrease* by 10 m/s in each and every subsequent second until its vertical velocity is zero. The body has experienced negative acceleration, more often referred to as **deceleration** or **retardation**.

Example

A ball is thrown vertically upwards with an initial velocity of 50 m/s. How long will the ball take to reach the top of its motion?

Answer

Initial velocity $u = 50$ m/s

Final velocity $v = 0$ m/s

Time of vertical motion = t

Acceleration = $a = -10$ m/s^2

$$a = \frac{v - u}{t}$$

$$-10 = \frac{0 - 50}{t}$$

$$t = \frac{-50}{-10}$$

$$t = 5\text{ s}$$

Hooke's law states that the extension is directly proportional to the applied load, up to a limit known as the limit of proportionality

natural length is the length of a spring when no stretching or compressing forces are applied

extension is the difference between the stretched length of a spring and its natural length

Hooke's law

REVISED

Hooke's law describes the behaviour of metal wires and springs when they are stretched or compressed. The difference between its extended length and its **natural length** is called its **extension**.

Hooke's law states that the extension is directly proportional to the applied load, up to a limit known as the limit of proportionality

natural length is the length of a spring when no stretching or compressing forces are applied

extension is the difference between the stretched length of a spring and its natural length

Exam tip

This law should be memorised.

Prescribed practical P2

Hooke's law

Apparatus

Figure 31.8 shows the apparatus.

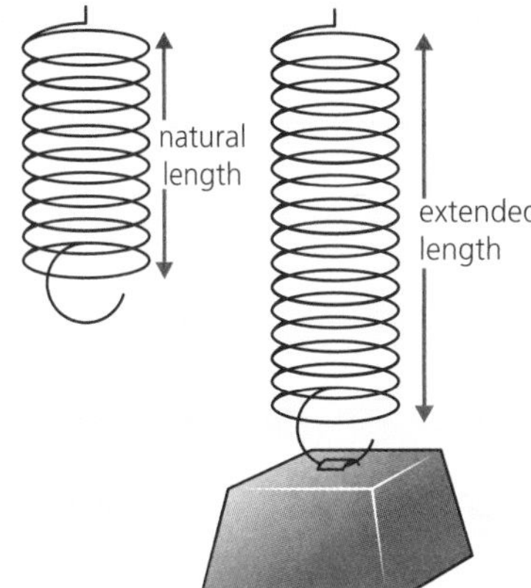

Figure 31.8

Method

1. Measure the natural length of the spring with a metre ruler.
2. Add a 100 gram (weight = 1.0 N) mass hanger.
3. Measure the extended length of the spring.
4. Calculate and record the extension.
5. Add an additional 100 gram mass.
6. Repeat the measurements and record the results in a table, as shown below.
7. Draw a graph of load/N on the *y*-axis against extension/cm on the *x*-axis.

Results

Load/N	0.0	1.0	2.0	3.0	4.0	5.0
Total length/cm	4.5	6.1	7.7	9.3	10.9	12.5
Extension/cm	0.0	1.6	3.2	4.8	6.4	8.0

Figure 31.9 shows the graph produced. It should show a straight line through the origin. This is part of the region (AB) in which Hooke's law is obeyed. If you were to continue to add slotted masses and take measurements, the graph would go beyond the limit of proportionality and eventually cease to be a straight line and curve as shown.

In the straight line region the spring is **elastic**. This means it will return to its original length when the load is removed. In the curved region the spring is **plastic**. In the plastic region it will no longer return to its original length when the load is removed.

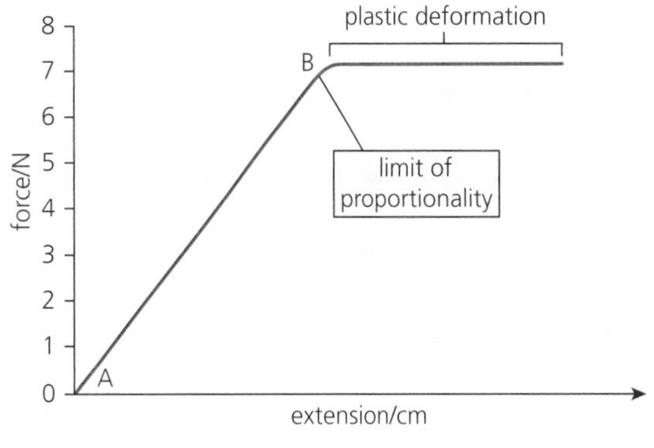

Figure 31.9

elastic materials return to their original length when the stretching force is removed

plastic materials do not return to their original length when the stretching force is removed

Hooke's law equation

Hooke's law can be written as an equation:

$$F = ke$$

where F is the applied force in N
e is the extension in cm (or mm)
k is the spring constant in N/cm or N/mm and is equal to the gradient of the force–extension graph

(Note that k is sometimes called the stiffness constant or Hooke's law constant.)

Example

A spring has a natural length of 8 cm. When loaded with a 5 N weight the total length of the spring is 23 cm.
(a) What weight would stretch the spring to a total length of 14 cm?
(b) What is the total length of the spring when the load is 7 N?

Answer

(a) Extension with a 5 N weight = 23 − 8 = 15 cm

$$\text{So, } k = \frac{F}{e} = \frac{5\,\text{N}}{15\,\text{cm}} = 0.3333\,\text{N/cm}$$

When total length = 14 cm, extension, e = 14 − 8 = 6 cm

$F = ke = 0.3333 \times 6 = 2\,\text{N}$

(b) $e = \frac{F}{k} = \frac{7}{0.3333} = 21\,\text{cm}$

total length = natural length + extension = 8 + 21 = 29 cm

Now test yourself

TESTED

14 State Hooke's law.
15 Explain what is meant by elastic and plastic deformation.
16 Explain what is meant by the limit of proportionality.

Pressure

REVISED

Pressure is defined by the equation:

$$\text{pressure} = \frac{\text{force}}{\text{area}}$$

or

$$P = \frac{F}{A}$$

where force, F, is measured in N,
area, A, is measured in mm^2, cm^2 or m^2, and
pressure, P, is measured in N/mm^2, N/cm^2 or N/m^2

Note that the unit N/m^2 was renamed the **pascal (Pa)** after the French scientist of that name.

Table 31.1 lists some common situations where knowledge of physics helps you to understand what is happening.

pressure is the force per unit area on a surface

pascal is the SI unit of pressure and is the same as $1\,\text{N/m}^2$

Table 31.1

Situation	Comment
A wise chef sharpens a carving knife before cutting a joint of meat.	A sharp knife has a tiny area of contact with the meat, so that even a small force can produce a very large pressure. This makes cutting the joint easier.
The blade of an ice skate is sharp.	The sharp blade causes a large pressure on the ice, producing a fine layer of water between blade and ice. The water reduces friction and makes skating effortless.
A high stiletto heel can cause considerable damage to a wooden floor.	Much of the woman's weight is borne by the tiny area of the heel. This causes huge pressure on the floor. If the floor is quite soft (like wood), this causes damage.
In some places houses must be built on rafts made of concrete.	If the surrounding ground is soft there is a danger that the weight of the house will cause it to sink. The large area of the concrete spreads the weight of the house, reduces the pressure and prevents sinking.

Example

Figure 31.10 shows three stacked building blocks. Each block has a weight of 225 N.

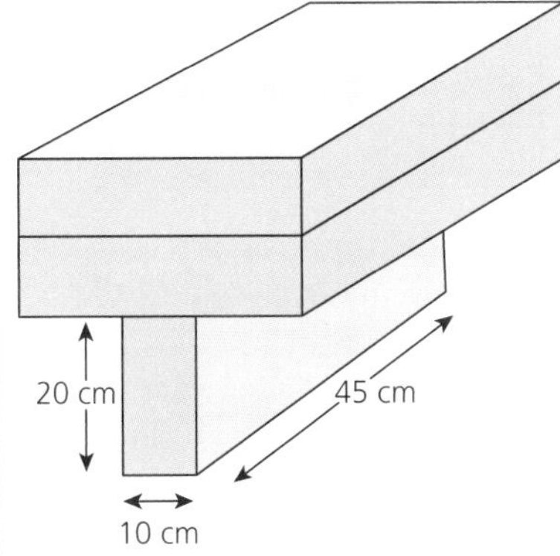

Figure 31.10

(a) With the help of the measurements in Figure 31.10, calculate the pressure exerted on the ground. Give your answer in pascals.

(b) It is possible to stack the blocks on top of one another so that the pressure exerted on the ground is less than the value you have calculated. Suggest how the blocks should be arranged to give this smaller pressure.

Answer

(a) total weight = 3 × 225 = 675 N

area in contact with ground = 0.1 m × 0.45 m

= 0.045 m^2

$$\text{pressure} = \frac{F}{A} = \frac{675}{0.045} = 15\,000\,\text{Pa}$$

(b) Have the 20 cm × 45 cm surface in contact with the ground and the other two blocks on top of it.

Now test yourself

TESTED

17 The tip of a drawing pin has an area of $1.0 \times 10^{-8}\,m^2$. Find the pressure exerted if the force applied to it is 10 N.
18 A car has a weight of 9000 N. The tyre pressure is $18\,N/cm^2$. Calculate the area of each tyre in contact with the ground.
19 A large concrete cube is of side 0.9 m. If it exerts a pressure of 61 000 Pa, calculate the cube's weight.

Centre of gravity

REVISED

The **centre of gravity** of an object is the point through which the entire weight of that object may be thought to act.

Only for regularly shaped objects is the centre of gravity, *G*, at the centre of the object. Centre of gravity is often wrongly thought of as the point where the object would balance.

The centre of gravity of a rectangular lamina, for example, is at its geometric centre (the point where its diagonals cross).

The idea of centre of gravity is closely linked with our notion of **stability** and **equilibrium**.

A ball on flat ground is in **neutral equilibrium**. When gently pushed the ball rolls, keeping its centre of gravity at the same height above the point of contact with the ground (Figure 31.11).

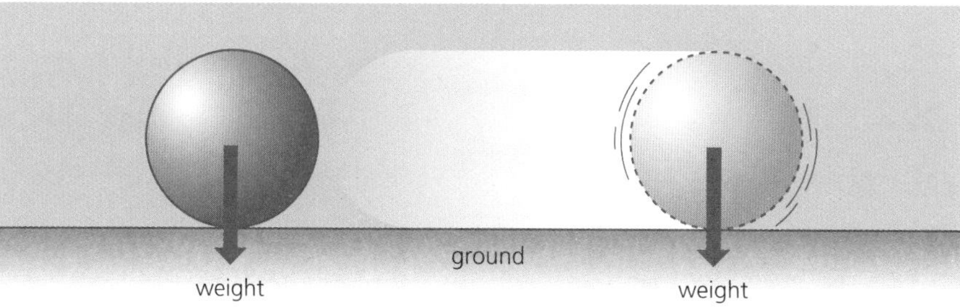

Figure 31.11

A tall radio mast is in **unstable equilibrium**. A small push from the wind will cause it to topple, bringing its centre of gravity closer to the ground. To prevent the mass toppling, it is stabilised with strong cables.

A car on the road is in **stable equilibrium**. If the car is tilted, the centre of gravity rises. Provided the weight, acting from the centre of gravity, passes through the wheel base, the car will not topple over.

Modern agricultural tractors are designed with a big wheel base to allow farmers to plough steeply sloping ground safely.

centre of gravity is the point through which the entire weight of a body appears to act

Exam tip

This definition must be memorised.

equilibrium is a state in which opposing forces or moments are balanced

neutral equilibrium is a state in which a slight displacement causes the object neither to move very far from its original position nor return to it (e.g. a ball sitting on a bench)

unstable equilibrium is a state in which a slight displacement causes the object to move a long way from its original position (e.g. a pencil balanced on its end

stable equilibrium is a state in which a slight displacement causes the object to return to its original position (e.g. a traffic cone standing on its wide base)

Exam tip

Remember that for maximum stability:

- the centre of gravity should be as low as possible
- the area of the base should be as large as possible.

Moments and levers

REVISED

Opening a door, pushing a wheelbarrow, cutting with scissors and using a nutcracker are all examples of the application of **moments** or turning forces. So what is a moment?

The moment of a force about a pivot is the product of the force and the perpendicular distance to the pivot (Figure 31.12). If the distance is in cm, then the moment is in N cm; if the distance is in metres then the moment is in N m. Moments have a direction — they are either **clockwise** or **anticlockwise**.

moment = force × distance to pivot

N cm   N   cm

The moment in Figure 31.12 is clockwise.

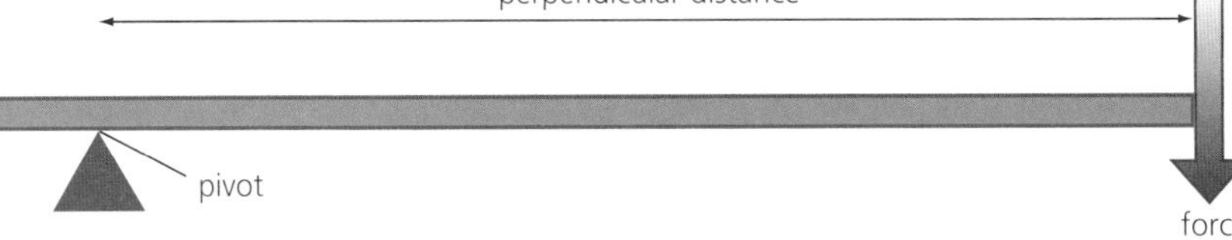

Figure 31.12

The behaviour of levers is summed up in a law known as the **principle of moments**. This law states:

> When a lever is balanced, the sum of the clockwise moments about any point equals the sum of the anticlockwise moments about the same point.

the **moment** about a point is the product of a force and its perpendicular distance from the point

the **principle of moments** states that when a lever is balanced, the sum of the clockwise moments about any point equals the sum of the anticlockwise moments about the same point

Exam tip

This law must be memorised.

Prescribed practical P3

Principle of moments

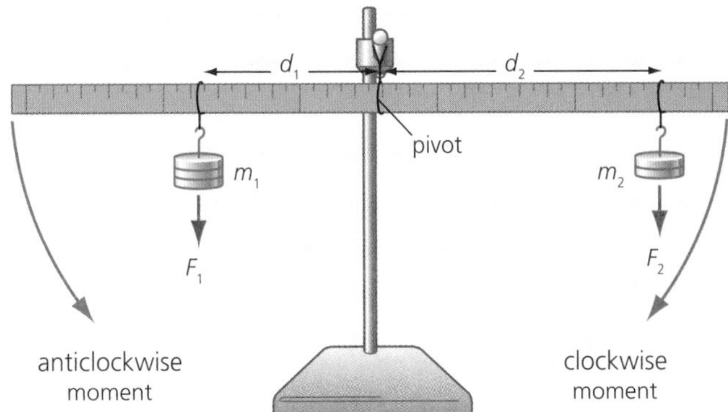

Figure 31.13

Method

1. Suspend and balance a metre ruler at the 50 cm mark using twine (see Figure 31.13).
2. Adjust the position of the twine so that the rule does not rotate.
3. Hang unequal masses, m_1 and m_2 (100 g slotted masses), from either side of the metre ruler as shown in Figure 31.13.
4. Adjust the position of the masses until the metre ruler is balanced (in equilibrium) again.
5. Gravity exerts forces F_1 and F_2 on the masses m_1 and m_2. Remember that a 100 g slotted mass is equivalent to 1 N.

→

6 Record the results in a table like the one below and repeat for other loads and distances.

m_1/g	F_1/N	d_1/cm	$F_1 \times d_1$/N cm	m_2/g	F_2/g	d_2/cm	$F_2 \times d_2$/N cm

- Force F_1 is trying to turn the metre ruler anticlockwise,and its moment is $F_1 \times d_1$.
- Force F_2 is trying to turn the metre ruler clockwise — its moment is $F_2 \times d_2$.
- When the metre ruler is balanced (i.e. in equilibrium), the results should show that the anticlockwise moment $F_1 \times d_1$ equals the clockwise moment $F_2 \times d_2$.

Another important consequence of the fact that the metre ruler is in equilibrium is that the forces acting on the ruler in any direction *must* balance. The **upward forces** must balance the **downward forces**. This idea is useful when doing problems.

Exam tip

At GCSE level you will never be asked to solve problems involving more than three forces.

Example

A boy, weighing 300 N, sits 0.9 m away from the pivot of a see-saw, as shown in Figure 31.14.

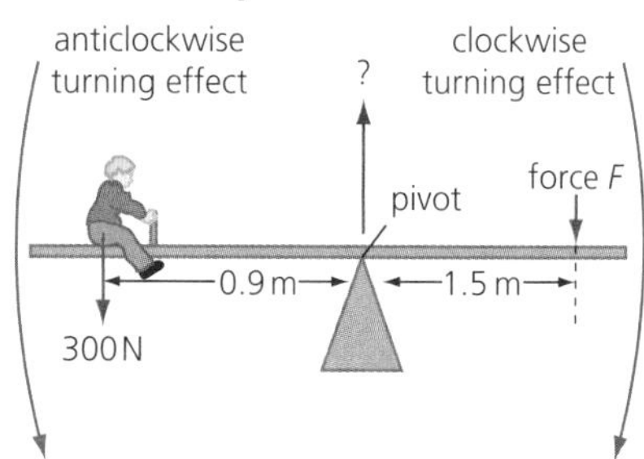

Figure 31.14

(a) What force 1.5 m from the pivot is needed to balance the see-saw?
(b) Find the size of the upward force exerted by the pivot.

Answer

(a) The force F exerts a clockwise turning effect about the pivot while the boy's weight exerts an anticlockwise turning effect.

By the principle of moments:

clockwise moment = anticlockwise moment

$F \times$ distance from pivot = 300 N × distance from pivot

$F \times 1.5\,\text{m} = 300\,\text{N} \times 0.9\,\text{m}$

$$F = \frac{300 \times 0.9}{1.5} = 180\,\text{N}$$

(b) Upward force = downward force = 300 + 180 = 480N.

Typical mistake

Examiners often see students giving the direction of a moment as 'up' or 'down'. This is quite wrong. Moments can only be clockwise or anticlockwise.

To summarise:

- balanced forces — if the object is stationary it will remain stationary; if it is moving it will carry on moving at the same speed and in the same direction
- unbalanced forces — an unbalanced force on an object causes the object to accelerate, according to $F = ma$.

Exam practice

H 1 (a) State the mathematical equation linking force, mass and acceleration. [1]
(b) An aircraft of mass 2000 kg lands at an airport. As it travels along the runway, friction forces on the aircraft add up to 8000 N. The speed of the aircraft reduces to 5 m/s in 9 s.
(i) Calculate the aircraft's deceleration. [1]
(ii) Calculate the speed of the aircraft when its wheels first touch the runway. [2]

2 When the resultant force on a vehicle is 3000 N, its acceleration is 4 m/s^2. Calculate the acceleration of the vehicle when the engine force is 3600 N and the friction force is 1725 N. [2]

3 International Cycle Union rules require racing bicycles to have a mass of 6.8 kg or more. A cyclist of mass 53 kg can produce an acceleration of 2 m/s^2 when the resultant force on the bicycle is 120 N. Is the mass of the bicycle within the limit set by the ICU? [2]

4 (a) Calculate the weight of an object of mass 70 kg on Earth. [2]
(b) The same object is taken to the Moon, where $g = 1.6\,m/s^2$. Calculate:
(i) its mass and
(ii) its weight on the Moon. [3]
(c) On another planet, a mass of 12 kg weighs 105.6 N. Calculate the value of g on this planet. [2]
(d) Comment on the units for g. [2]

5 A bullet, fired vertically upwards from a pistol, rises to a maximum height of 1875 m above a planet in a time of 25 seconds.
(a) Calculate the average velocity of the bullet during this time. [3]
(b) Using your answer to part (a), or otherwise, calculate the maximum velocity of the bullet. [3]
The bullet then takes another 25 seconds to fall back to the planet's surface.
(c) What is the average velocity of the bullet over the entire distance covered? [1]
(d) Give a reason for your answer to part (c). [1]
(e) Determine the acceleration due to gravity on the planet. [3]

6 The friction force opposing the motion of a locomotive of mass 25 000 kg is 100 000 N.
(a) What forward force must the locomotive provide if it is to travel along a straight, horizontal track at a steady speed of 1.5 m/s? [1]
(b) What is the acceleration of the locomotive if the forward force increases to 175 000 N and the friction force is unchanged? [3]

7 An elephant's mass is 5500 kg. The total area of its feet in contact with the ground is 0.14 m^2. Calculate the pressure exerted on the ground. [5]

8 The side of a large lorry measures 15 m × 2 m (Figure 31.15). On a windy day the pressure on one side is 5 kPa. Calculate the force on that side of the lorry. [5]

Figure 31.15

9 A wheelbarrow and its load together weigh 600 N. The distance between the pivot and the wheelbarrow's centre of gravity is 75 cm. The distance between the handles and the pivot is 225 cm (Figure 31.16).

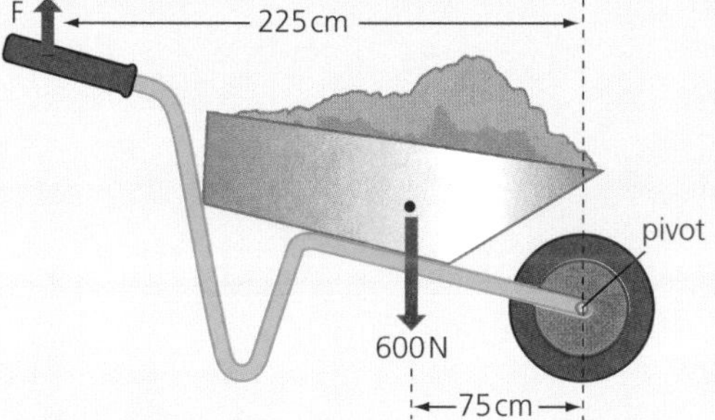

Figure 31.16

(a) Calculate the size of the smallest force, F, needed to lift the wheelbarrow at the handles. [4]
(b) Calculate the force acting through the pivot. [2]

→

10 The centre of gravity of an 80 cm snooker cue is 15 cm from its thick end (Figure 31.17).

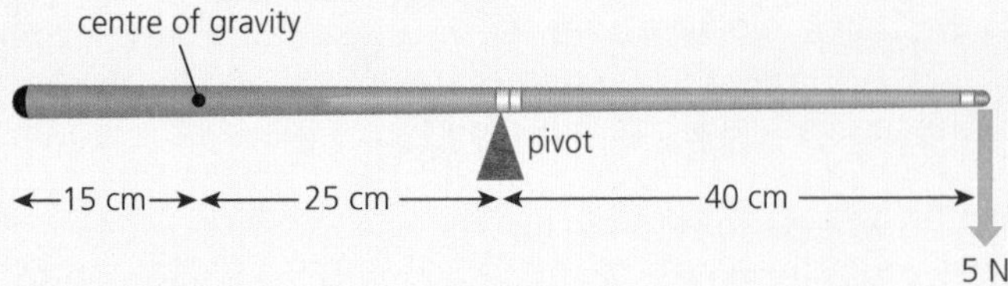

Figure 31.17

The cue balances on a pivot 40 cm from its thick end when a force of 5 N is applied to the thin end.

(a) Calculate the moment of the 5 N force about the pivot and state the direction in which it acts. [4]

(b) Calculate the weight of the snooker cue. [4]

11 A uniform plank of weight 150 N and length 4 m rests on a pivot at one end. It is balanced by an upward force, F, which acts at 3 m from the pivot. Calculate the size of force F and the size and direction of the force acting through the pivot.

Answers online

ONLINE

32 Density and kinetic theory

Density

REVISED

The **density** of a material is defined as the mass per unit volume.

It is calculated using the formula:

$$\text{density} = \frac{\text{mass}}{\text{volume}}$$

The unit of density is kilogram per cubic metre (kg/m^3). Sometimes you will also see the unit gram per cubic centimetre (g/cm^3).

density is the mass of an object divided by its volume

Example

Taking the density of mercury as $14\,g/cm^3$, find:

(a) the mass of $7\,cm^3$ of mercury

(b) the volume of 42 g of mercury.

Answer

(a) $\text{density} = \frac{\text{mass}}{\text{volume}}$

$14 = \frac{\text{mass}}{7}$

$\text{mass} = 14 \times 7 = 98\,g$

(b) $\text{density} = \frac{\text{mass}}{\text{volume}}$

$14 = \frac{42}{\text{volume}}$

$\text{volume} = \frac{42}{14} = 3\,cm^3$

Measuring density

Common practical

Mass and volume

This practical requires you to investigate the relationship between the mass and volume of liquids and regular solids.

To determine the density of a substance you need to measure its mass and its volume.

1 Liquids

Apparatus

- digital balance
- measuring cylinder
- liquid

Method

1. Find the mass of a dry, empty, graduated cylinder using a digital balance (see Figure 32.1).
2. Pour the liquid under test into the cylinder and measure the volume.
3. Find the mass of the cylinder and liquid (see Figure 32.1).
4. Subtract the mass of the empty cylinder from the combined mass of cylinder and liquid.

→

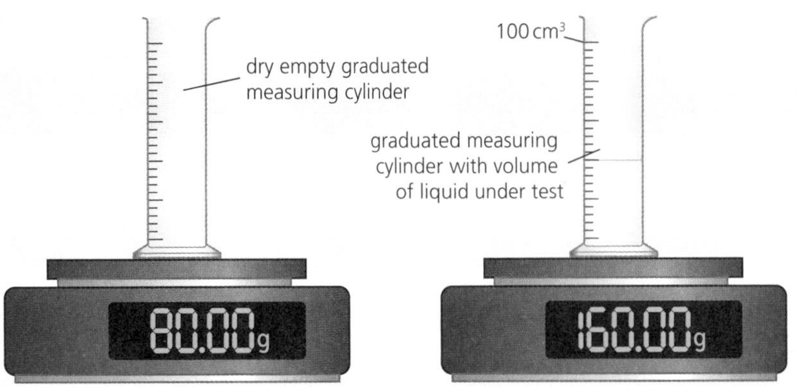

Figure 32.1 Finding the density of a liquid

5 Find the density of the liquid by dividing the mass of the liquid by the volume of the liquid.
6 For reliability, repeat the experiment and average the calculated densities.

2 Regularly shaped object

Apparatus

- digital balance
- ruler

Method

1 Find the mass of the object using a digital balance.
2 Find its dimensions with a ruler (or digital callipers) and use the appropriate formula to determine its volume.
For example:
volume of a rectangular block = length × breadth × height
volume of a cylinder = π × radius^2 × height
3 Find the density by dividing the mass by the volume.

3 Irregularly shaped object

If the shape of the object is too irregular for the volume to be determined using formulae, then a displacement method is used, as shown in Figure 32.2.

As before, the mass is found using a digital balance and the density calculated as outlined previously.

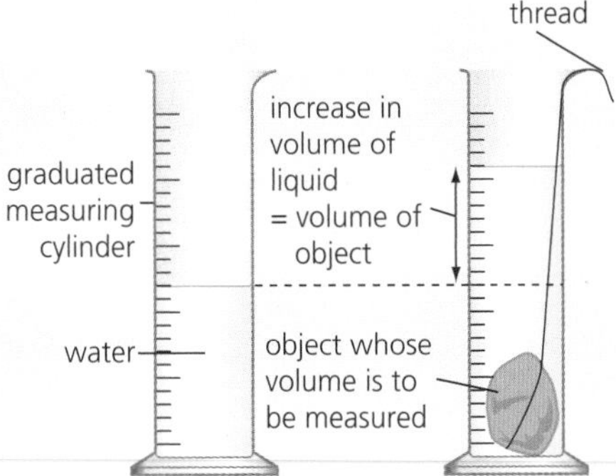

Figure 32.2 Measuring the volume of a small irregularly shaped object

Graphical treatment of density

The gradient of a mass–volume graph for different materials represents the density of a particular material (Figure 32.3).

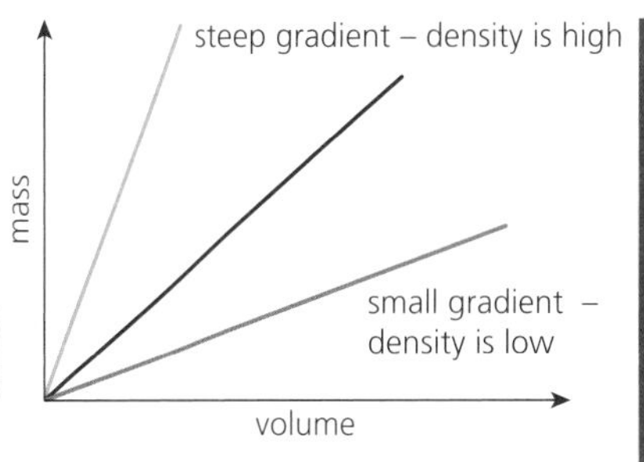

Figure 32.3

Now test yourself

TESTED

1 A substance of mass 75.6 g has a volume of 6.0 cm^3. Calculate its density.
2 Aluminium has a density of 2.7 g/cm^3.
  (a) What is the mass of 20 cm^3 of aluminium?
  (b) What is the volume of 54 g of aluminium?
3 Calculate the mass of air in a room of dimensions 10 m by 5 m by 3 m, if air has a density of 1.26 kg/m^3.
4 A stone of mass 60 g is lowered into a measuring cylinder causing the liquid level to rise from 15 cm^3 to 35 cm^3. Calculate the density of the stone in g/cm^3.
5 The capacity of a petrol tank in a car is 0.08 m^3. Calculate the mass of petrol in a full tank if the density of petrol is 800 kg/m^3.
6 The mass of an evacuated steel container, of internal volume 1000 cm^3, is 350 g. The mass of the steel container when full of air is 351.2 g. Calculate the density of air.

kinetic theory explains the properties of solids, liquids and gases according to the arrangement and motion of their molecules

Kinetic theory

REVISED

Table 32.1 Explaining the variation in density of solids, liquids and gases using kinetic theory

Solids	Liquids	Gases
• Molecules are packed close together. • The molecules vibrate about fixed positions. • Solids have a fixed shape and volume. • Solids have a high density because their molecules are so tightly packed. • There are strong forces of attraction between the molecules.	• The molecules are **close together** but not as close as they are in solids. • The molecules can move around in any direction and are **not fixed in position**. • Liquids have a fixed volume but take on the shape of the container. • Liquids have a medium density. • The forces of attraction between are still **quite strong** but, again, not as strong as in solids.	• There are large distances between the molecules. • The molecules move around very quickly in all directions. • Gases completely fill their container. • Gases have a very low density because their molecules are so far apart. • The forces of attraction between them are negligible.

Exam practice

1 (a) Explain what is meant by density. [2]
(b) Describe briefly how you could use a measuring cylinder half filled with water to find the volume of a bracelet. In your description state what measurements you would make and what calculation you would carry out. [4]
(c) A certain bracelet has a volume of 2.4 cm^3 and a mass of 46 g. Calculate its density. [3]
(d) The bracelet is made from a metal which is almost 100% pure. Use your answer to part (c) and the table below to find out what the metal is. [1]

Metal	Copper	Gold	Lead	Platinum
Density/g/cm^3	8.9	19.3	11.3	21.5

2 (a) A rectangular concrete slab of mass 520 kg is 1.8 m long, 1.2 m wide and 0.1 m deep. Use these data to calculate the volume of this concrete slab. [4]
(b) For bridge construction the concrete slabs must have a density of at least 2350 kg/m^3. Is this particular slab dense enough to be used for bridge construction? Show clearly how you get your answer. [4]

3 The density of aluminium is 2.7 g/cm^3.
(a) Calculate the number of cm^3 in 1 m^3. [1]
(b) Calculate the mass in grams of 1 m^3 of aluminium. [1]
(c) Calculate the density of aluminium in kg/m^3. [2]

4 A student places an empty measuring cylinder on an electronic balance and adds different volumes of liquid. Each time she measures the volume of the liquid she also records the reading on the electronic balance. She plots her results as a graph as shown in Figure 32.4.
(a) What is the mass of the empty measuring cylinder? [1]
(b) The table shows four different liquids and their densities.

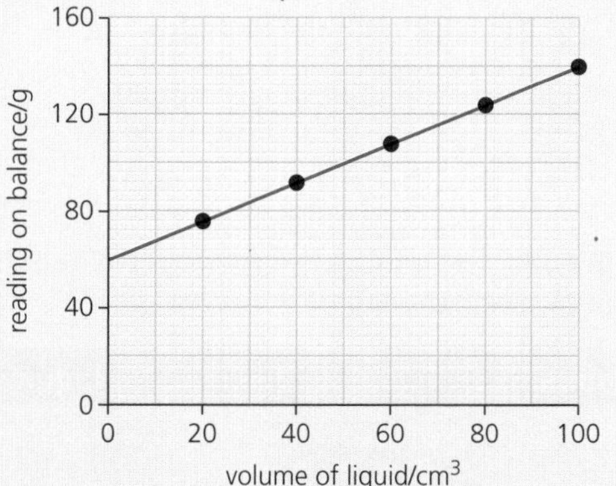

Figure 32.4

Liquid	Density/g/cm^3
Petrol	0.7
Castor oil	0.9
Water	1.0
Ethanol	0.8

Using the data from Figure 32.4 and your answer to part (a) identify the liquid the student used. [4]
(c) The student repeats the same procedure using the same measuring cylinder but using a liquid of lower density than the first liquid. Copy the grid in Figure 32.4 and on it draw the straight line she would expect to obtain. [2]
(d) Ice has a density of 0.9 g/cm^3 and water has density of 1.0 g/cm^3.
What does this tell you about the spacing of the molecules in the two states? [1]

Answers online

ONLINE

33 Energy

Energy forms

REVISED

It is important to understand the difference between **energy forms** and **energy resources**. Energy forms are the different ways in which energy can appear, such as heat, light, sound, nuclear, **kinetic**, **gravitational potential** and chemical energy. Energy resources are the different ways of supplying a particular energy form. Table 33.1 summarises some of the main energy forms.

Table 33.1

Energy form	Definition	Examples of resources
Chemical	The energy stored within a substance, which is released on burning	Coal, oil, natural gas, peat (turf), wood, food
Gravitational potential	The energy a body contains as a result of its height above the ground	Stored energy in the dam (reservoir) of a hydroelectric power station
Kinetic	The energy of a moving object	Wind, waves, tides
Nuclear	The energy stored in the nucleus of an atom	Uranium, plutonium

Other common energy forms are electrical energy, magnetic energy and strain potential energy — the energy a body has when it has been stretched or squeezed out of shape and will return to its original shape when the force is removed, such as a wind-up toy.

One of the fundamental laws of physics is the **principle of conservation of energy**. This states that:

Energy can neither be created nor destroyed, but it can change its form.

In other words, energy can be changed from one form into another, but the total amount of energy does not change.

energy forms are the different ways in which energy can appear, such as heat, light, sound etc.

energy resources are the different ways of supplying a particular energy form such as coal, oil, wind etc.

kinetic energy is the energy possessed by a mass due to its speed

gravitational potential energy is the energy possessed by a mass due to its height above the ground

the **principle of conservation of energy** states that energy can change from one form to another but can never be created or destroyed

Energy resources

REVISED

Energy resources can be classified as **renewable** or **non-renewable**. Renewable resources are those that are replaced by nature in less than a human lifetime. Non-renewable resources are those that are used faster than they can be replaced by nature. We will eventually run out of non-renewable energy resources. Table 33.2 lists the common renewable energy resources and Table 33.3 lists common non-renewable resources.

renewable energy is energy that is being replaced by nature in less than a human lifetime, so we will never run out of it

non-renewable energy is energy that will eventually run out

Table 33.2 Renewable resources

Resource	Comment
Solar cells	Solar cells convert sunlight (**solar energy**) directly into electricity; they are joined together into arrays.
Hydroelectric power stations	Water behind a high dam contains **gravitational potential energy**. The water is allowed to fall from the dam through a pipe and it gains **kinetic energy** as it falls. The fast-flowing water falls on a **turbine**, which then drives a **generator**. The output from the generator is **electrical energy**.
Tidal barrages	This is created when a **dam** is built across a **river estuary**. As the tide rises and falls every 12 hours, water will flow through a gate in the dam. The moving water drives a turbine, which is made to turn a generator to produce electrical energy.
Wave machines	Waves are produced largely by the action of the wind on the surface of water. The **wave machine** floats on the surface of the water and the up-and-down motion of the water is converted to rotary motion of a turbine–generator unit to produce electricity.
Wind turbines	As the wind blows, the large blade turns and this drives a turbine. The turbine drives a generator, which produces electricity. Large numbers of turbines are often grouped together to form a **wind farm**.
Geothermal power stations	These use heat from the hot rocks deep inside the Earth. Cold water is passed down a pipe to the rocks. The rocks heat the water and the hot water is then pumped to the surface. The steam generated is used to drive a turbine–generator to produce electricity.
Biomass	The timber from fast-growing trees is harvested. The wood is dried and turned into woodchips, which are then burned in power stations to produce electricity or sold for solid fuel heating. There are many other forms of biomass. Oil from **oil-bearing seeds** can be converted into **biodiesel** for road transport. Grass can be fermented in a **bio-digester** and turned into **biogas** for heating.

Table 33.3 Non-renewable resources

Resource	Comment
Fossil fuels	These are oil, natural gas, coal, turf (peat) and lignite. They are burned in power stations to produce steam, which drives a turbine, turns a generator and so produces electricity.
Nuclear power	Large nuclei (uranium or plutonium) in a nuclear reactor are made to split into lighter nuclei (**by nuclear fission**) with the release of large amounts of kinetic energy (of sub-atomic particles). This energy is used to produce steam, which drives a turbine, turns a generator and so produces electricity.

All energy resources have unique **advantages** and **disadvantages**. The main features of the most common resources are listed in Table 33.4.

Table 33.4 Advantages and disadvantages of the main energy resources

Energy resource	Advantages	Disadvantages	Other comments
Fossil fuels — coal, oil, natural gas, lignite, turf	Relatively cheap to start up. Moderately expensive to run. Large world reserves of coal (much less for other fossil fuels).	All fossil fuels are non-renewable. All fossil fuels release carbon dioxide on burning and so contribute to global warming. Burning coal and oil also releases sulfur dioxide gas, which causes acid rain.	Coal releases the most carbon dioxide and natural gas the least per unit of electricity produced. Removing sulfur or sulfur dioxide is very expensive and adds greatly to the cost of electricity production.
Nuclear fuels — mainly uranium	Do not produce carbon dioxide. Do not emit gases that cause acid rain.	The waste products will remain dangerously radioactive for tens of thousands of years. As yet, no one has found an acceptable method to store these materials cheaply, safely and securely for such a long time. Nuclear fission fuels are non-renewable. An accident could release dangerous radioactive material, which would contaminate a wide area, leaving it unusable for decades.	Nuclear fuel is relatively cheap on world markets. Nuclear power station construction costs are much higher than fossil fuel stations, because of the need to take expensive safety precautions. Decommissioning nuclear power stations is particularly long and expensive, requiring specialist equipment and personnel.
Wind farms	A renewable energy resource. Low running costs. Conserve fossil fuels.	Wind farms are: ● unreliable ● unsightly ● very noisy ● hazardous to birds	Wind farms take up much more ground per unit of electricity produced than conventional power stations.
Waves	A renewable energy resource. Low running costs. Conserve fossil fuels.	Wave generators at sea are: ● unreliable ● unsightly ● hazardous to shipping	Many turbines are needed to produce a substantial amount of electricity.
Tides	A renewable energy resource. Low running costs. Conserve fossil fuels.	Tidal barrages are built across river estuaries and can cause: ● navigation problems for shipping ● destruction of habitats for wading birds and the mud-living organisms on which they feed	Tides (unlike wind and waves) are predictable, but they vary from day to day and month to month. This makes them unsuitable for producing a constant daily amount of electrical energy.

Now test yourself

TESTED

1 Name four fossil fuels.
2 Name six different energy forms.
3 Explain what is meant by:
   (a) a renewable energy resource
   (b) a non-renewable energy resource.
4 Name four renewable energy resources and four non-renewable energy resources.
5 Why is it misleading to say that electricity is a 'clean' fuel?
6 State the principle of conservation of energy.

Energy flow diagrams

REVISED

It is common for examiners to ask about the main energy flow through various devices. The kind of question that has appeared in recent examination questions is shown in the example.

Example

Fill in the spaces below to show the main types of energy change each device is designed to bring about. (Here the answers have all been provided to show you what the examiners expect.)

Main input energy	Device	Required output energy
Sound	Microphone	Electricity
Electricity	Room heater	Heat
Chemical	Burning match	Heat

Typical mistake

A common mistake is to write that renewable resources are those that can be used over and over again. This is quite wrong. Once a unit of energy has been used, that particular unit of energy can never be used again. So *learn the correct definitions* of renewable and non-renewable resources.

The Sun

REVISED

Almost all energy resources rely on the energy of the Sun. Fossil fuel resources come from the dead remains of plants and animals laid down many millions of years ago. The plants obtained their energy from the Sun by **photosynthesis**. Herbivores ate the plants, while carnivores ate the herbivores. Under the Earth's surface, these remains slowly fossilised into coal, peat, gas and oil. Other processes also rely on the Sun's energy. Hydroelectric energy depends on the water cycle, which begins when ocean water evaporates as a result of absorbing radiant energy from the Sun. Wind and waves rely on the Earth's weather, which is largely controlled by the Sun.

Only geothermal and nuclear energy do not depend directly on the energy emitted by the Sun.

Now test yourself

TESTED

7 (a) Copy and complete the boxes below to show the useful energy change that takes place in a wind turbine.

[ ]	→	[ ]
input		useful output

(b) The wind is a renewable energy resource. What does this mean?

(c) Give two other examples of renewable energy resources.

8 Figure 33.1 represents a typical hydroelectric power station. Copy and complete the boxes below to show the energy changes taking place in a hydroelectric power station.

[ ]	→	[ ]	→	[ ]
(energy stored in the upper lake)		(energy in the moving water)		(output energy from the power station)

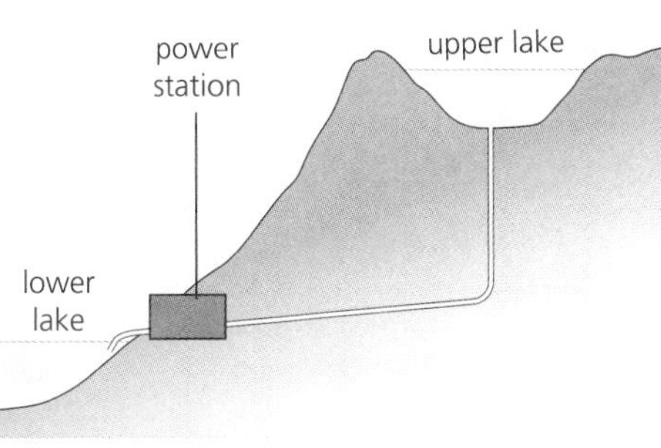

Figure 33.1

9 For each of the devices or situations shown below, construct a flow diagram to show the main energy change that is taking place. The first has been done for you.

Device/situation	Input energy form		Useful output energy form
Microphone	**Sound** energy	→	**Electrical** energy
Loudspeaker	________energy	→	________energy
Electric smoothing iron	________energy	→	________energy
Coal burning in an open fire	________energy	→	________energy
Weight falling towards the ground	________energy	→	________energy
Candle flame	________energy	→	________energy and ________energy
Battery-powered electric drill	________energy → ________ energy	→	________energy

Work

REVISED

Work is only done when a **force** causes **movement**. The amount of work done is given by the formula:

work done = force × distance moved in the direction of the force

or

$W = F \times d$

where work, W, is measured in newton metres (N m) or joules (J), force, F, is measured in newtons (N), distance, d, is measured in metres (m).

work is the product of the force and the distance moved in the direction of the force

Exam tip

This formula must be memorised.

Examples

1 How much work is done when a packing case is dragged 60 cm across the floor at a steady speed against a frictional force of 45 N? How much energy is needed?

Answer

There are two points to watch here. First, the frictional force and the direction of motion are in opposite directions. But the case moves at a steady speed, so the forward force must also be equal to the frictional force (45 N). Second, the distance moved is given in cm, so be sure to convert it to metres.

So work done = $F \times d = 45 \times 0.6 = 27$ J

2 A crane does 1560 J of useful work when it lifts a load vertically by 40 cm. Find the weight of the load.

Answer

Since the load is being lifted, the minimum upward force is the weight of the load (Figure 33.2).

So, $W = F \times d$

Always write the equation first, *then* make the substitutions.

$1560 = F \times 0.4$ (convert 40 cm to 0.4 m)

$$F = \frac{1560}{0.4} = 3900\,\text{N}$$

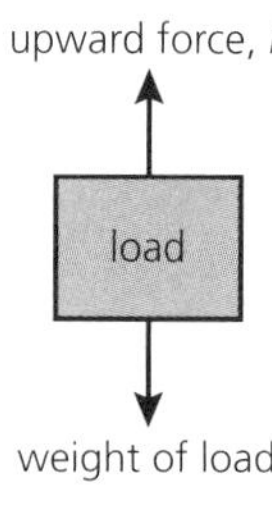

Figure 33.2

So, weight of load = 3900 N

3 How much work is done by an electric motor pulling a 130 N load 6.5 m up the slope shown in Figure 33.3 if the constant tension in the string is 60 N?

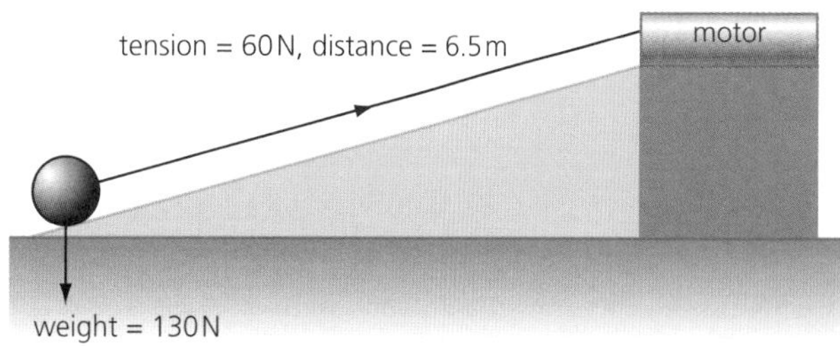

Figure 33.3

Answer

Since the tension force and distance moved are both parallel to the slope, they are both used to find the work done. Neither the weight of the load, nor the vertical distance between the weight and the motor are used in this question.

$W = F \times d$ moved in the direction of the force

$= 60 \times 6.5 = 390$ J

Work and energy

REVISED

Energy is the *ability* to do work. So if a machine has 100J of stored energy, this means it can do 100J of work. Similarly, work is sometimes thought of as the amount of energy transferred. Note that both work and energy are measured in joules.

energy is what is required to do work

Example

A battery stores 15kJ of energy. If the battery is used to drive an electric motor, what is the *maximum* height to which it could raise a 750N load, if it was lifted vertically?

Answer

The battery stores 15kJ or 15000J, so it can do a maximum of 15000J of work.

$$W = F \times d$$

$$15\,000 = 750 \times d$$

$$d = \frac{15\,000}{750} = 20\,\text{m}$$

Power

REVISED

Power is the *rate* of doing work. This means that the power of a machine is the work it can do in a second. The formula for calculating power is therefore:

$$\text{power} = \frac{\text{work done}}{\text{time taken}}$$

or

$$P = \frac{W}{t}$$

where power, P, is measured in joules per second (J/s) or watts (W)
work, W, is measured in joules (J)
time, t, is measured in seconds (s)

power is the rate of doing work

Exam tip

This formula must be memorised.

Examples

1 An electric motor is used to raise a load of 210N. The load rises vertically 3m in a time of 6s. Find the work done and the power of the motor.

Answer

Note: This is a two-part question. First, calculate the work done using $W = F \times d$, then find the power using $P = \frac{W}{t}$.

$$\text{work done} = \text{force} \times \text{distance} = 210 \times 3 = 630\,\text{J}$$

$$\text{power} = \frac{\text{work done}}{\text{time taken}} = \frac{630}{6} = 105\,\text{W}$$

→

2 A crane has a power of 1200W. How much work could it do in 1 hour?

Answer

In power calculations, the unit of time is the second. So first convert 1 hour to seconds:

$$1 \text{ hour} = 60 \text{ minutes} = 60 \times 60 \text{ seconds} = 3600 \text{ seconds}$$

$$\text{power} = \frac{\text{work done}}{\text{time taken}}$$

$$1200 = \frac{\text{work done}}{3600}$$

$$\text{work done} = 1200 \times 3600 = 4\,320\,000\,\text{J or } 4.32\,\text{MJ}$$

Measuring power

REVISED

Prescribed practical P4

Measuring personal power

An experiment to measure personal power is prescribed by the specification. This means you must do it as an experiment in class. In recent years, a description of the experiment has been asked as a 6-mark QWC question.

Figure 33.4

In this experiment be sure to remember:

- what is measured
- the apparatus used to make the measurements
- the calculations required to find the output power.

To measure your personal power, you need to find out how long it takes you to do a given amount of work. First find your weight in newtons. The easy way to do this is to find your mass in kilograms using bathroom scales, and then use the fact that 1 kg has a weight of 10 N. Then you need to find the height of a staircase.

This can be done by measuring the average height of a riser (stair) using a metre ruler and multiplying by the number of risers in the staircase. Finally, you need to have someone who will time you as you run up the stairs using a stopwatch (Figure 33.4).

For reliability, the experiment should be repeated and the average power of the pupil determined.

➜

Measurements

Some typical measurements that might be obtained are:

- mass of student in kg: 45
- weight of student in N: 450
- height of risers in cm: 14.0, 13.8, 13.8, 14.0, 13.9
- average riser height in cm: 13.9
- number of risers: 30
- staircase height: 13.9 × 30 = 417 cm = 4.17 m
- time to run upstairs in s: 5.2, 5.1, 4.9, 5.0, 4.8
- average time taken in s: 5.0

Calculations

$$\text{work} = \text{force} \times \text{distance} = 450 \times 4.17 = 1876.5\,\text{J}$$

$$\text{power} = \frac{\text{work}}{\text{time}} = \frac{1876.5}{5.0} = 375\,\text{W (approx.)}$$

An alternative method involves a student of known mass, *m*, being timed to do, say, 100 'step-ups' on to a platform such as the first step of a flight of stairs. If the time taken is *t* and the height of the platform is *h* metres, then the power, *P*, of the student is given by:

$$P = 100 \times \frac{mgh}{t}$$

The apparatus used to obtain the measurements are the same as described in the stairs experiment above.

Efficiency

REVISED

Efficiency is defined by the formula:

$$\text{efficiency} = \frac{\text{useful energy output}}{\text{total energy input}}$$

Alternative definitions are possible involving work and power. Learn one and stick to it.

Exam tip

This formula must be memorised.

efficiency is the ratio of the useful output work to the total energy input

Examples

1 An electric kettle is rated 2500 W. It uses 2500 J of eletrical energy every second. The kettle takes 160 seconds to boil some water and during this time 360 000 J of heat energy pass into the water. Find the kettle's efficiency.

Answer

$$\text{useful energy output (passed into water)} = 360\,000\,\text{J}$$

$$\text{total energy input} = 2500 \times 160 = 400\,000\,\text{J}$$

$$\text{efficiency} = \frac{\text{useful energy output}}{\text{total energy input}} = \frac{360\,000}{400\,000} = 0.9$$

2 A motor lifts a load of 80 N to a height of 90 cm. In the process, 88 J of energy are wasted as heat and sound. Find the motor's efficiency.

Answer

$$\text{useful energy output} = \text{work done by motor} = \text{force} \times \text{distance}$$
$$= 80 \times 0.9 = 72\,\text{J}$$

$$\text{total energy input} = \text{total energy output} = \text{useful energy} + \text{wasted energy}$$
$$= 72 + 88 = 160\,\text{J}$$

$$\text{efficiency} = \frac{\text{useful energy output}}{\text{total energy input}} = \frac{72}{160} = 0.45$$

Typical mistake

A common mistake is to have the input on the top line and the output on the bottom. As efficiency is a ratio, it has no units. And since *some* energy is wasted in *every* physical process, the efficiency of a machine is *always less than 1.*

Exam tip

Avoid the use of percentages in efficiency calculations. For example, giving the efficiency of the motor in the last example as 45% will get full marks. But leaving out the % sign and writing only 45 will cost you at least 1 mark.

Gravitational potential energy

REVISED

You are expected to remember that 1 kg has a weight of 10 N on Earth.

This is just another way of saying that the gravitational field strength, g, on Earth is 10 N/kg. So the weight of an object, W, is given by:

$$W = mg$$

When any object with mass is lifted, work is done on it against the force of **gravity**. The greater the mass of the object and the higher it is lifted, the more work has to be done. The work that is done is only possible because some energy has been used. This energy is stored in the object as **gravitational potential energy (GPE)**.

GPE = work done in raising a load (m) against the force of gravity (g) through a height (h)

$$\text{GPE} = mgh$$

where m is the mass in kg
g is the gravitational field strength in N/kg
h is the vertical height in m.

Exam tip

This formula must be memorised.

Examples

1 Find the gravitational potential energy of a mass of 600 g when raised to a height of 250 cm. Take g = 10 N/kg.

Answer

First note that 600 g is 0.6 kg and 250 cm is 2.5 m.

$$\text{GPE} = mgh = 0.6 \times 10 \times 2.5 = 15\,\text{J}$$

2 How much heat and sound energy is produced when a mass of 1.2 kg falls to the ground from a height of 5 m? Take g = 10 N/kg.

Answer

Heat and sound energy produced = original GPE = mgh = 1.2 × 10 × 5
= 60 J

3 A book of mass 500 g has a gravitational potential energy of 3.2 J when at a height of 4 m above the surface of the Moon. Find the gravitational field strength on the Moon.

Answer

$$\text{GPE} = mgh = 3.2$$

$$3.2 = 0.5 \times g \times 4 = 2g$$

$$g = \frac{3.2}{2} = 1.6\,\text{N/kg}$$

Kinetic energy

REVISED

The **kinetic energy (KE)** of an object is the energy it has because it is moving. It can be shown that an object's kinetic energy is given by the formula:

$$\text{KE} = \tfrac{1}{2} mv^2$$

where m is the mass in kg
v is the speed of the object in m/s

Examples

1 A car of mass 800 kg is travelling at 15 m/s. Find its kinetic energy.

Answer

$$KE = \frac{1}{2} mv^2 = \frac{1}{2} \times 800 \times 15^2 = 0.5 \times 800 \times 225 = 90\,000\,\text{J}$$

2 A bullet has a mass of 20 g and is travelling at 300 m/s. Find its kinetic energy.

Answer

We must first change the bullet's mass from g to kg by dividing by 1000.

$$KE = \frac{1}{2} mv^2 = \frac{1}{2} \times \left(\frac{20}{1000}\right) \times 300^2 = 0.5 \times 0.02 \times 90\,000 = 900\,\text{J}$$

3 The input power of a small hydroelectric power station is 1 MW. If 18 000 000 kg of water flows past the turbines every hour, find the average speed of the water.

Answer

1 hour = 60 × 60 seconds = 3600 seconds
mass of water flowing every second = 18 000 000/3600 = 5000 kg/s

Since a 1 MW power station produces 1 000 000 J of electrical energy per second, the minimum KE of the water passing every second is 1 000 000 J.

So KE = ½ mv^2

$$1\,000\,000 = \frac{1}{2} \times 5000 \times v^2$$

$$v^2 = \frac{1\,000\,000}{2500}$$

$$v^2 = 400$$

$$v = \sqrt{(400)} = 20\,\text{m/s}$$

Exam tip

If you find it hard to rearrange the kinetic energy formula to find speed, you might remember

$$v = \sqrt{\frac{2 \times KE}{m}}$$

Now test yourself

TESTED

10 Explain why a waiter holding a heavy tray at rest for 5 minutes is doing no work.
11 Write down the equations for work, power and efficiency.
12 Explain the difference between gravitational potential energy and kinetic energy and write down the equation for each.
13 Explain why no machine can have an efficiency greater than 1.
14 Explain why efficiency has no unit.
15 According to your specification it takes approximately 1 J to lift an apple vertically 1 m. Use this information to calculate the typical mass of an apple.
16 (a) How much work is done by a tractor when it lifts a load of 8000 N to a height of 1.8 m?
(b) The output power of the tractor is 5.2 kW. How long does it takes to do 26 000 J of work?
(c) The efficiency of the tractor is 0.26 (or 26%). If the output power of the tractor is 5.2 kW, calculate the input power.
17 Stephen weighs 550 N. How much work does he do in climbing up to a diving board which is 3.0 m high?
18 A basketball player throws a ball vertically up into the air. Place a tick (✓) in the appropriate column to show what happens to each quantity as the ball rises. Ignore the effects of friction.

Quantity	Increases	Decreases	Remains constant
Speed of ball			
Potential energy of ball			
Total energy of ball			
Kinetic energy of ball			

Exam practice

1 Competitors in the World's Strongest Man competition must throw a cement block of mass 100 kg over a wall 5.5 m high. How much work is done if the block just clears the top of the wall? [4]

2 A gardener pushes a lawn mower 200 m with a force of 60 N. The work takes her 4 minutes.
(a) How much work does she do altogether? [4]
(b) Calculate her average power. [4]

3 The electrical energy used by a boiler is 1050 kJ. The useful output energy is 840 kJ.
(a) Calculate the efficiency of the boiler. [3]
(b) Suggest what might have become of the energy wasted by the boiler. [1]

4 A car engine has an efficiency of 0.28. How much input chemical energy must be supplied if the total output useful energy is 140 000 kJ? [4]

5 The power of the electric motor in a lift is 3600 W. How much electrical energy is converted into other energy forms in 3 minutes if the lift has been rising continuously? [4]

6 A barrel of weight 1000 N is pushed up a ramp. The barrel rises vertically 40 cm when it is pushed 1 m along the ramp.
(a) Calculate how much useful work is done when the barrel is pushed 1 m along the ramp. [4]
(b) To push the barrel 1 m along the ramp requires 1200 J of energy. Calculate the efficiency of the ramp. [3]

7 A communications satellite of mass 120 kg orbits the Earth at a speed of 3000 m/s. Calculate its kinetic energy. [3]

8 On planet X an object of mass 2 kg is raised 10 m above the surface. At that height the object has a gravitational potential energy of 176 J. Details of three planets are given in the table. Which one of these three planets is most likely to be planet X? [4]

Planet's name	Mercury	Venus	Jupiter
Gravitational field strength, *g*, in N/kg	3.7	8.8	26.4

9 A ball of mass 2 kg falls from rest at a height of 5 m above the ground. Copy the table below and complete it to show the gravitational potential energy, the kinetic energy, speed and the total energy of the falling ball at different heights above the surface. [7]

Height above ground/m	Gravitational potential energy/J	Kinetic energy/J	Total energy/J	Speed/m/s
5.0		0	100	0
4.0				4.47
	64			
1.8		64		
0.0	0			

10 A bouncing ball of mass 200 g leaves the ground with a kinetic energy of 10 J.
(a) If the ball rises vertically, calculate the maximum height it is likely to reach. [3]
(b) In practice, the ball rarely reaches the maximum height. Explain why this is so. [1]

Answers online

ONLINE

34 Atomic and nuclear physics

The structure of atoms

REVISED

All matter is made up of **atoms**, but what are atoms made of? Experiments carried out by Thomson and Rutherford in the early part of the twentieth century led physicists to believe that atoms themselves had a structure.

an **atom** is the smallest part of an element that can exist on its own

The relative masses and charges of the particles that make up the atom are given in Table 34.1.

Table 34.1

Particle	Location	Relative mass	Relative charge
Proton	Within the nucleus	1	+1
Neutron	Within the nucleus	1	0
Electron	Orbiting the nucleus	1/1840	–1

Atomic number and mass number

The number of protons in a nucleus is called the **atomic number** and is given the symbol Z. The atomic number also tells you the number of electrons in the neutral atom.

Typical mistake

The atomic number is given the symbol Z (not A). The mass number is given the symbol A (not M).

The **mass number** (or nucleon number) is the sum of the number of protons and the number of neutrons. Mass number is given the symbol A.

- Atomic number, Z = number of protons
- Mass number, A = number of protons + number of neutrons = number of nucleons

Every nucleus can therefore be written in the form: ${}^{A}_{Z}X$ where X is the chemical symbol, A is the mass number and Z is the atomic number.

For example, the element uranium has the chemical symbol U. All uranium nuclei have 92 protons in the nucleus. One form of uranium, called uranium-235, has a mass number of 235. This means it has 92 protons and 143 neutrons (235 – 92 = 143). A uranium nucleus is given the symbol:

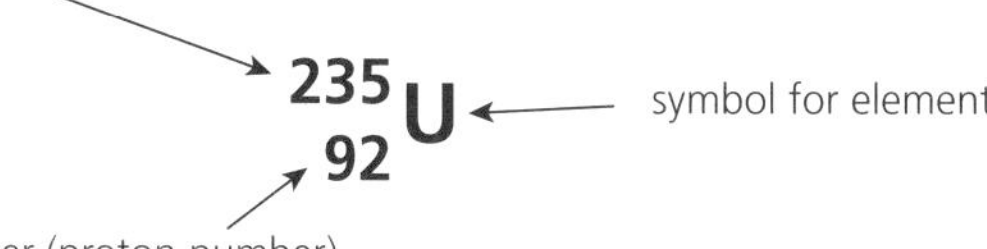

It is important to realise that this is the symbol for the **nucleus** of the atom. Orbiting electrons are completely ignored.

Isotopes

Not all the atoms of the same element have the same mass, but they all have the same number of protons. Physicists call atoms with the same number of protons but a different number of neutrons, **isotopes**.

Isotopes are atoms of the same element that have the same atomic number but different mass number. The main isotopes of helium, for example, are: ${}^{3}_{2}He$ and ${}^{4}_{2}He$.

Now test yourself

TESTED

1. An atom contains *electrons*, *protons* and *neutrons*. Which of these particles:
   (a) are outside the nucleus
   (b) are uncharged
   (c) have a negative charge
   (d) are nucleons
   (e) are the lightest?
2. The element sodium has the chemical symbol Na. In a particular sodium isotope there are 12 neutrons. In a neutral sodium atom there are 11 orbiting electrons. Write down the symbol for the nucleus of this isotope.
3. In what way are the nuclei of isotopes the same? In what way are they different?

Nuclear radiation

REVISED

French scientist Henri Becquerel discovered that certain rocks containing uranium gave out strange radiation that could penetrate paper and fog photographic film. He called the effect **radioactivity**. Three separate types of radiation, called **alpha** (α), **beta** (β) and **gamma** (γ) radiation were identified.

The atoms that emit these radiations are said to be **radioactive**. The particles and waves are referred to as **nuclear radiation**. The disintegration is called **radioactive decay**.

radioactivity is the process that occurs when alpha particles, beta particles or gamma waves are emitted from an unstable nucleus

alpha particles are the helium nuclei containing two protons and two neutrons which are emitted by radioactive nuclei

beta particles are fast-moving electrons emitted by radioactive nuclei

gamma radiation is a high-energy electromagnetic wave emitted by radioactive nuclei

Ionising radiation

Ions are charged atoms (or molecules). Atoms become ions when they lose (or gain) electrons (Figure 34.1).

Nuclear radiation can become dangerous by removing electrons from atoms in its path, so it has an ionising effect.

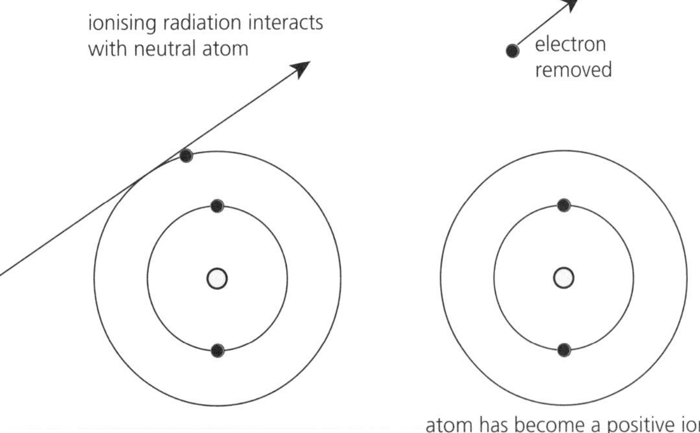

Figure 34.1

When this happens with molecules in living cells, the genetic material of a cell is damaged and the cell may become cancerous. Other forms of ionising radiation include ultraviolet and X-rays.

Alpha radiation is not as dangerous if the radioactive source is *outside* the body, because it cannot pass through the skin and is unlikely to reach cells inside the body. Beta and gamma radiation can penetrate the skin and cause damage to cells. Alpha radiation will damage cells if the radioactive source has been breathed in or swallowed.

A summary of the properties of radiation is given in Table 34.2.

Table 34.2 summary of the nature and properties of nuclear radiations

Type of radiation	Alpha particle (α)	Beta particle (β)	Gamma rays (γ)
Nature	Each particle is two protons plus two neutrons; it is a nucleus of helium-4	A fast electron	Very high-energy electromagnetic waves
Source	A radioactive nucleus	A radioactive nucleus	A radioactive nucleus
Relative charge (compared with charge on proton)	+2	−1	0
Mass	High compared to beta	Low	0
Speed	Up to 0.1 × speed of light	Around 10% of the speed of light	Speed of light
Ionising effect	Strong	Weak	Very weak
Penetrating effect	Very low penetrating power, stopped by a few cm of air or thin tissue paper	Penetrating, but stopped by a few millimetres of aluminium or other metal	Very penetrating: never completely stopped, although lead and thick concrete will reduce intensity

Now test yourself

TESTED

4 Name a radioactive isotope that occurs naturally in living things.
5 Which of the three types of radiation (alpha, beta or gamma):
   (a) is a form of electromagnetic radiation
   (b) carries positive charge
   (c) is made up of electrons
   (d) travels at the speed of light
   (e) is the most ionising
   (f) can penetrate a thick sheet of lead
   (g) is stopped by skin or thick paper?

Dangers of radiation

REVISED

Most radioactive **background activity** comes from natural sources such as cosmic rays from space, rocks and soil, some of which contain radioactive elements such as radon gas. Living things and plants absorb radioactive materials from the soil, which are then passed along the food chain.

Human behaviour also adds to the background activity that we are exposed to through medical X-rays, radioactive waste from nuclear power plants and the radioactive fallout from nuclear weapons testing.

Radioactive material is found naturally all around us and inside our bodies. Traces of radioactive elements, for example potassium, can be found in our food. Certain rocks contain uranium, all the isotopes of which are radioactive, and this decays giving radon, a radioactive gas. There is also radiation reaching Earth from outer space, referred to as cosmic rays.

All these natural sources are known together as **background radiation.**

Protection when handling radioactive material

We can minimise the risk to those using radioactive materials by:

- wearing protective clothing
- keeping the source as far away as possible by using tongs
- working quickly to keep exposure to the source to as short a time as possible
- keeping radioactive materials in lead-lined containers.

Students under the age of 16 are forbidden to handle radioactive sources.

Nuclear disintegration equations

REVISED

Symbol equations can be written to represent alpha and beta decay. The alpha particle can be written as ${}^{4}_{2}\alpha$ or ${}^{4}_{2}He$ and the beta particle as ${}^{0}_{-1}\beta$ or ${}^{0}_{-1}e$.

For example, the symbol equation for the alpha decay of uranium-238 is:

$${}^{238}_{92}U \rightarrow {}^{234}_{90}Th + {}^{4}_{2}He \text{ (or } \alpha)$$

The symbol equation for the beta decay of carbon-14 is:

$${}^{14}_{6}C \rightarrow {}^{14}_{7}N + {}^{0}_{-1}e \text{ (or } \beta)$$

When writing symbol equations it is important to remember the following:

- The sum of the mass numbers (at the top) on the left-hand side of the equation must equal the sum of the mass numbers on the right-hand side.
- The sum of the atomic numbers (at the bottom) on the left-hand side of the equation must equal the sum of the atomic numbers on the right-hand side.

Exam tip

For all particles, the top number shows the mass and the bottom number gives the charge.

Example

Radium-226 decays to polonium-222. Radium (Ra) has atomic number 86 and polonium (Po) has atomic number 84. Which type of decay occurs?

Answer

$${}^{226}_{86}Ra \rightarrow {}^{222}_{84}Po + {}^{a}_{b}X$$

Balancing mass numbers: $226 = 222 + a$

$a = 4$

Balancing atomic numbers: $86 = 84 + b$

$b = 2$

The particle with a mass number of 4 and an atomic number of 2 is helium and so X is an alpha particle. **Alpha decay** is shown by the equation:

$${}^{A}_{Z}X \rightarrow {}^{A-4}_{Z-2}Y + {}^{4}_{2}He \text{ (or } {}^{4}_{2}\alpha)$$

If the mass number of the parent nucleus does not change and the atomic number of the daughter nucleus increases by 1, then the reaction must be beta decay. **Beta decay** is exemplified by:

$$^{A}_{Z}X \rightarrow ^{A}_{Z+1}Y + ^{0}_{-1}e \text{ (or } ^{0}_{-1}\beta)$$

To take a specific case, radium-228 decays to actinium-228 by emitting a β-particle:

$$^{228}_{88}Ra \rightarrow ^{228}_{89}Ac + ^{0}_{-1}\beta$$

In **gamma decay**, the excited parent nucleus relaxes by emitting gamma ray(s). There is no change in the nature of the nucleus, so the mass number and the atomic number stay the same. The γ-radiation is usually emitted at the same time as the α- and β-particle emissions and represents the excess energy of the daughter nucleus as it settles down into a more stable condition.

An asterisk indicates that a nucleus is in an excited state with excess energy. The process of getting rid of that energy by gamma-ray emission is called relaxation.

$$^{A}_{Z}X^{*} \rightarrow ^{A}_{Z}X + \gamma$$

Now test yourself

TESTED

6 Copy and complete the table below.

Radiation	Atomic number (*Z*)	Mass number (*A*)
α-emission	Decreases by 2	
β-emission		Unchanged
γ-emission		

7 Complete the following equations for alpha decay and beta decay.

(a) $^{238}_{92}U \rightarrow ^{?}_{?}Th + ^{4}_{2}He$

(b) $^{14}_{6}C \rightarrow ^{?}_{?}N + ^{0}_{-1}e$

8 Work out the type of decay in each of the following examples:

(a) bismuth-213 to polonium-213

(b) radium-226 to radon-222

(c) francium-221 to actinium-217.

Radioactive decay

REVISED

Radioactive decay is **random** and **spontaneous**. Random means that we cannot predict when a particular nucleus will disintegrate. Spontaneous means that the rate of decay is unaffected by any physical changes such as temperature, pressure or chemical changes. However, some types of nuclei are more unstable (Figure 34.2) than others and decay at a faster rate.

half-life is the time taken for the activity of a radioactive material to fall by half

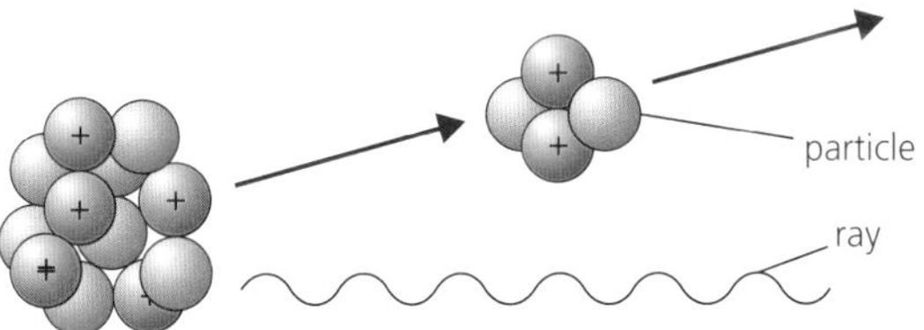

Figure 34.2 An unstable nucleus emitting a particle and a ray

Rate of decay and half-life

The **half-life** of a radioactive material is the time taken for its activity to fall to half its original value (Figure 34.3).

Calculations involving half-life are generally best solved using a table, as in Exam practice question 1 at the end of this chapter.

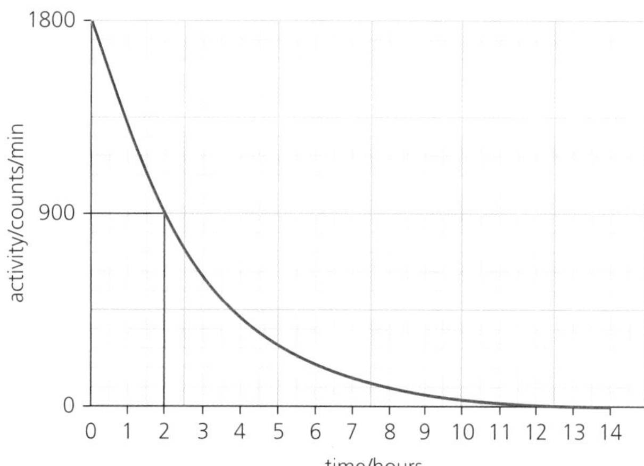

Figure 34.3 The radioactive decay curve for a substance with a half-life of 2 hours

Each isotope has a specific and constant half-life. Some half-lives are very short — a matter of seconds or even a fraction of a second — and others can be thousands of years. Table 34.3 gives the half-lives of some common radioactive isotopes.

Table 34.3

Isotope	Half-life
Uranium-238	4 500 000 000 years
Carbon-14	5730 years
Phosphorus-30	2.5 minutes
Oxygen-15	2.06 minutes
Barium-144	114 seconds
Polonium-216	0.145 seconds

The unit for radioactivity is the becquerel (Bq). 1 Bq = 1 disintegration per second.

Now test yourself

TESTED

9 Calculate the half-lives of the following samples:
  (a) A sample of iodine-123 whose activity falls from 1000 Bq to 250 Bq in 14.4 hours.
  (b) A sample of technetium-99 whose activity falls from 200 Bq to 25 Bq in 18 hours.
  (c) A sample of strontium-90 whose activity falls from 500 Bq to 62.5 Bq in 86.4 years.

10 Calculate how long it would take for the following to decay to an activity of 1 Bq.
  (a) A sample of cobalt-60 (half-life = 5.27 years) whose original activity is 64 Bq.
  (b) A sample of iodine-131 (half-life = 8 days) whose original activity is 128 Bq.
  (c) A sample of polonium-210 (half-life = 138 days) whose original activity is 32 Bq.

Uses of radiation

REVISED

In medicine

Gamma radiation from the cobalt-60 isotope can be used to treat tumours.

Different radioisotopes are used to monitor the function of organs by injecting a small amount into the bloodstream and detecting the emitted radiation. The tracers used in this case must have a short half-life.

Iodine-131 is used in investigations of the thyroid gland.

Surgical instruments and hospital dressings can be sterilised by exposure to gamma radiation. The source should have a very long half-life so that it does not need to be replaced on a regular basis.

In agriculture

Gamma radiation can be used to treat fresh food. By killing bacteria on the food, the radiation helps the food to have a longer shelf life. The use is controversial, however, as many people are worried about eating food exposed to radiation. Ideally the radioisotope used in a food processing plant should have a very long half-life so that it is a long time before it needs to be replaced.

The ease with which a plant absorbs a fertiliser can be found by putting a small amount of radioactive isotope in the fertiliser. If parts of the plants are then checked for radioactivity you can tell how much fertiliser has been taken up by the plant.

In industry

Beta radiation can be used to monitor the thickness of a sheet of paper or aluminium (Figure 34.4). An emitter is placed on one side of the sheet and a detector on the other. As the sheet moves past, the activity detected will be the same as long as the thickness remains unchanged.

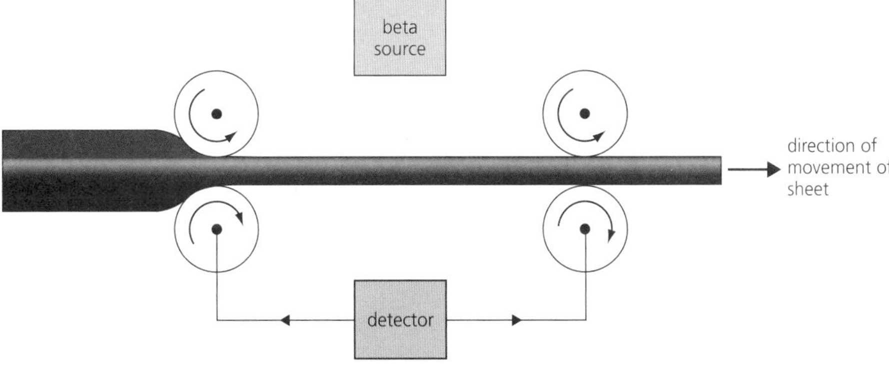

Figure 34.4 **A long half-life beta source is used to control the thickness of an aluminium sheet**

Radioactive tracers

A suitable radioactive isotope can be used to provide information about fluid movement and mixing to monitor, for example, leaks in underground pipes (Figure 34.5). The tracer is added to the fluid in the pipe and builds up in the ground if there is a leak. The radiation needs to penetrate many centimetres of soil to reach the detectors — this means that it must be a gamma emitter. Only gamma rays have sufficient penetrating power. But to avoid dangerous radioactive materials being in the ground for a long time the source should have a short half-life.

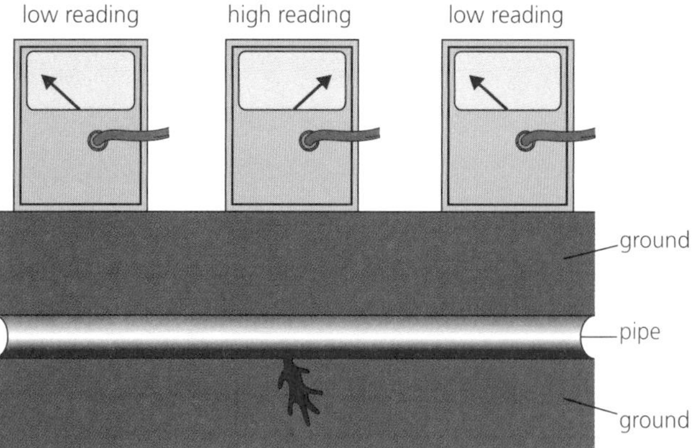

Figure 34.5 Radioactive tracers locating leaks in pipes

Practical work with radioactive materials

Students under the age of 16 are forbidden to handle radioactive sources.

The most common type of radiation detector is the Geiger–Müller tube (GM tube) connected to a counter or ratemeter. It is not necessary to know how a GM tube works, but it is important to know how it could be used to do practical work on radiation.

Background radiation

REVISED

Background activity is that which is detected when no known radioactive sources are present. Fortunately, the background count in Northern Ireland does not present a serious health risk.

The background count must always be *subtracted* from any other count when measuring the activity from a specific source.

Safety precautions when using closed radioactive sources in schools

- Always store the sources in a lead-lined box, under lock and key, when not required for experimental use.
- Always handle sources using tongs, holding the source at arm's length and pointing it away from any bystander.
- Wear protective clothing.
- Always work quickly and methodically with sources to minimise the time of exposure and hence the dose to the user.

Measuring the approximate range of radiation

Alpha

- Place a GM tube on a wooden cradle and connect it to a ratemeter.
- Hold an alpha source directly in front of the window of the tube and slowly increase the distance between the source and the tube. At about 3 cm (depending on the source used) the ratemeter reading falls dramatically to that of background radiation.
- Place a thin piece of paper in contact with the window of the GM tube. Bring the alpha source up to the paper so that the casing of the source touches it. The reading on the ratemeter is now equal to the background count showing that the alpha particles are unable to penetrate the paper.

Beta

- Place a 1 mm thick piece of aluminium in contact with the window of the GM tube.
- Bring the beta source up to the aluminium so that the casing of the source touches it.
- The reading on the ratemeter is observed to be significantly above the background count, showing that some beta particles have penetrated the aluminium.
- Repeat the process with 2 mm, 3 mm etc. thick pieces of aluminium. At about 5 mm there is a significant reduction in the count rate on the ratemeter, indicating the approximate range of beta particles in aluminium.

Gamma

If the beta particle experiment is repeated with a gamma source, there is practically no reduction in the count rate for a 5 mm thick piece of aluminium. If the aluminium sheets are replaced with lead, it will be found that even school sources will give gamma radiation that can easily penetrate several centimetres of lead.

Figure 34.6 shows a summary for range of penetration of the three types of radiation.

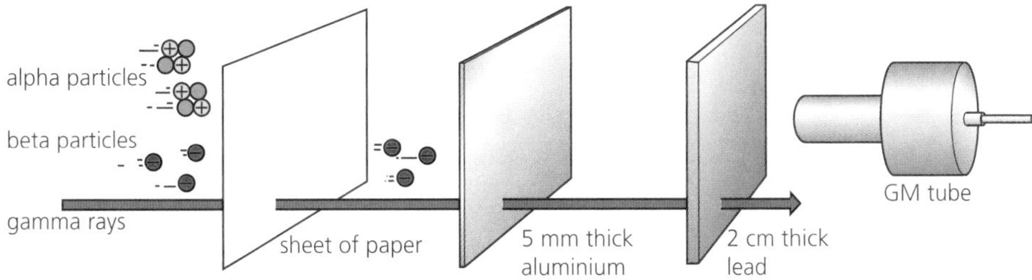

Figure 34.6 The penetrative range of the three types of radiation

Nuclear fission

REVISED

Some heavy nuclei, such as uranium, can be forced to split into two lighter nuclei. The process is called **nuclear fission**. It occurs when a uranium nucleus is struck by a slow neutron (Figure 34.7). The heavy nucleus splits and the fragments move apart at very high speed, carrying with them a vast amount of energy. At the same time, two or three fast neutrons are also emitted. These are the fission neutrons and they go on to produce further fission and so create a chain reaction.

In a nuclear power station, steps are taken to ensure that, on average, just one of the fission neutrons goes on to produce further fission. This is **controlled nuclear fission**. The heat produced in the reaction is used to turn water into steam and drive a turbine to generate electricity. In a nuclear bomb there is no attempt to control the fission process (Figure 34.8).

nuclear fission is the process by which a uranium (or plutonium) nucleus absorbs a slow neutron and then splits to produce two or more lighter nuclei and several neutrons, together with a vast quantity of energy

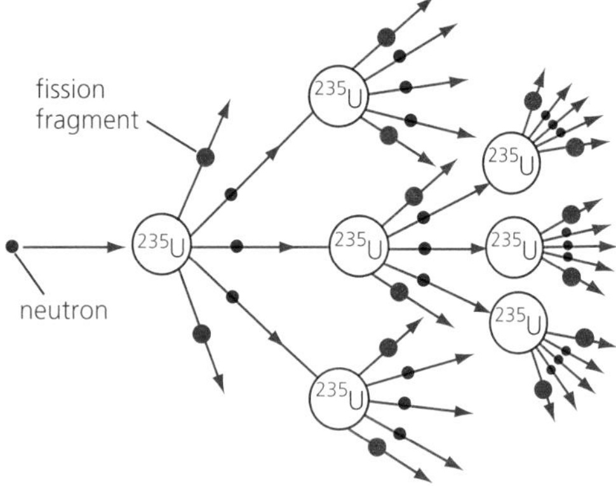

Figure 34.8 A chain reaction in uranium-235

A major disadvantage of all fission processes is that the fission fragments are almost always highly radioactive. This type of radioactive waste is *extremely dangerous* and expensive measures must be taken to store it until the level of activity is sufficiently small. In some cases this means the waste must be stored deep underground, in a vitrified (glass-like) state, for tens of thousands of years. One danger is that over time, the containers may leak and cause underground water pollution. Another danger is that earthquakes can rupture containers of radioactive waste buried underground, causing it to leak into the soil and water systems.

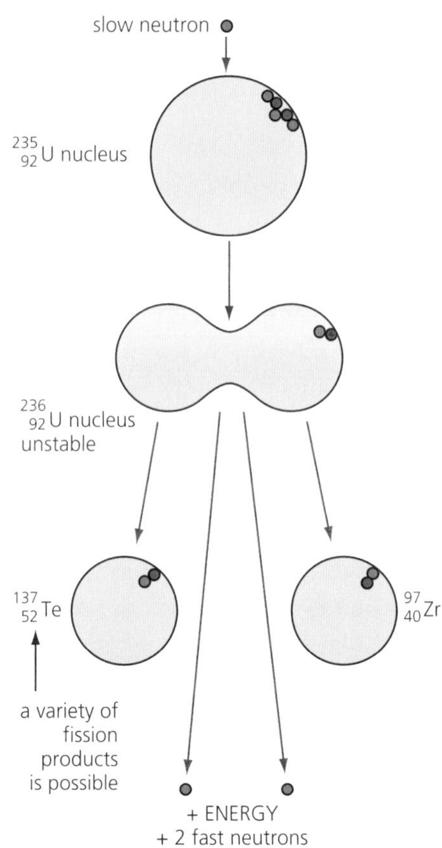

Figure 34.7 The fission of a uranium-235 nucleus

Issues relating to the use of nuclear energy to generate electricity

Nuclear energy is a controversial topic, which gives rise to strong political, social, environmental and ethical arguments on both sides of the debate.

Arguments in favour of nuclear energy

- It can produce vast amounts of energy/electricity.
- It produces very little carbon dioxide (CO_2) and hence does not contribute to global warming.
- It provides the 'base-load' for national electricity generation.
- It is a high-density source of energy.
- It provides employment opportunities for many people.
- Additionally, many nations are planning to build more nuclear reactors.

Arguments against nuclear energy

- The by-product of nuclear energy — nuclear waste and its disposal — has created one of the greatest problems of the twenty-first century.
- Many people are concerned about living close to nuclear power plants and the storage facilities used for radioactive waste. This gives rise to the acronym NIMBY (not in my back yard).
- This fear has been increased by
  - the disaster at Chernobyl in the Ukraine
  - the earthquake and tsunami in Fukushima, Japan in 2011, when several reactors were damaged leading to a meltdown and release of radiation.

Although nuclear fission does not release carbon dioxide, the mining, transport and purification of the uranium ore releases significant amounts of greenhouse gases into the atmosphere.

Nuclear fusion

REVISED

Nuclear fusion happens in stars like our Sun. At the centre of the Sun the temperature is about 15 000 000°C. At this temperature, the nuclei of all atoms are stripped of their orbiting electrons and they are moving at a tremendous speed. Being positively charged, the nuclei would normally repel each other, but if they are moving fast enough, they can join (or fuse) to form a new nucleus (Figure 34.9). This causes the release of a vast amount of energy.

nuclear fusion is the process in which light nuclei such as hydrogen combine together to produce a heavier nucleus such as helium and emit a vast quantity of energy.

The equation representing the fusion process is:

$${}^{2}_{1}H + {}^{3}_{1}H \rightarrow {}^{4}_{2}He + {}^{1}_{0}n + \text{energy}$$

Difficulties with fusion

The main problem is how to *contain* the reacting plasma at a *high enough temperature* and at *high particle densities* for a sufficiently *long time* for the reaction to take place.

A large advantage of fusion is that the isotopes of hydrogen — deuterium and tritium — are widely available as the constituents of sea water and so are nearly inexhaustible. Furthermore, fusion does not emit carbon dioxide or other greenhouse gases into the atmosphere since its major by-product is helium, an inert non-toxic gas.

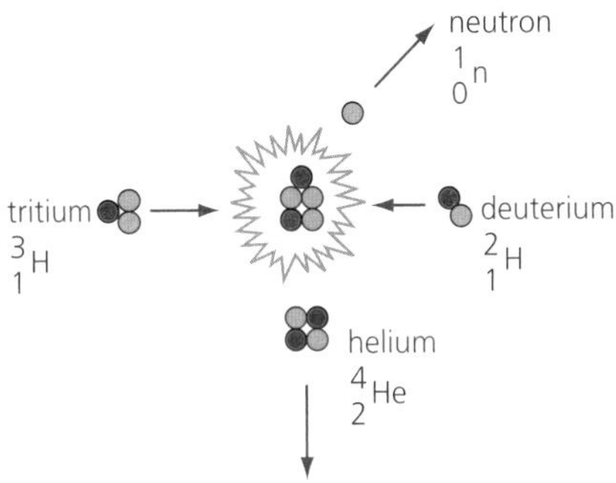

Figure 34.9 The fusion of deuterium and tritium

Fusion versus fission

Fusing nuclei together in a controlled way releases four million times more energy per kg than a chemical reaction such as the burning of coal, oil or gas; and fusing nuclei together in a controlled way releases four times as much as nuclear fission reactions per kg of fuel.

Typical mistake

Many students spell fission with one 's' and/or fusion with 'ss'. Make sure you can spell these terms correctly to be sure of scoring the mark.

Fusion: solution to world's energy crisis?

There are many difficulties to overcome before nuclear fusion provides electricity on a commercial scale and it may be another 50 years before that happens. Nuclear fusion reactors will be expensive to build, and the system used to contain the plasma will be equally expensive because of the very high temperatures needed for the nuclei to fuse.

What is ITER?

REVISED

ITER ('the way' in Latin) is one of the most ambitious energy projects in the world today. ITER stands for 'International Thermonuclear Experimental Reactor'.

ITER is the 'International Thermonuclear Experimental Reactor', an experimental nuclear fusion device being built in southern France and involving the cooperation of scientists and engineers from 35 nations.

In southern France, 35 nations are collaborating to build the world's largest tokamak, a magnetic fusion device that has been designed to prove the feasibility of fusion as a large-scale and carbon-free source of energy based on the same principle that powers our Sun and stars.

ITER will be the first fusion device to:

- produce net energy
- maintain fusion for long periods of time

ITER has been designed mainly to:

- produce 500 MW of fusion power
- demonstrate the integrated operation of technologies for a fusion power plant

Exam practice

1 What mass of nitrogen-13 would remain if 80 g were allowed to decay for 30 minutes? Nitrogen-13 has a half-life of 10 minutes.

2 How long would it take for 20 g of cobalt-60 to decay to 5 g? The half-life of cobalt-60 is 5.26 years.

3 Strontium-93 takes 32 minutes to decay to 6.25% of its original mass. Calculate the value of its half-life.

4 When a radioactive material of half-life 24 hours arrives in a hospital its activity is 1000 Bq. Calculate its activity 24 hours before and 72 hours after its arrival.

5 (a) The table below shows the particles that make up a neutral carbon atom. Copy and complete the table showing the mass, charge, number and location of the particles. Some information has already been added for you. [7]

Particle	Mass	Charge	Number	Location
Electron		–1		
Neutron	1		6	in the nucleus
Proton			6	

(b) Radon is a naturally occurring radioactive gas.

(i) Explain what is meant by radioactive. [2]

(ii) Explain the danger of breathing radon gas into the lungs. [2]

(iii) Explain, in terms of the particles that make up the nucleus, the meaning of *isotope*. [2]

6 $^{12}_{6}C$ and $^{14}_{6}C$ are both isotopes of carbon.

(a) (i) Write down one similarity about the nucleus of each isotope. [1]

(ii) Write down one difference in the nucleus of these isotopes. [1]

H (b) $^{14}_{6}C$ is radioactive. It decays to nitrogen by emitting a beta particle.

Copy and complete the equation below which describes the decay.

$^{14}_{6}C \rightarrow ^{?}_{?}N + ^{?}_{?}\beta$ [4]

(c) $^{14}_{6}C$ is present in all living materials and in all materials that have been alive. It decays with a half-life of 6000 years.

(i) Explain the meaning of the term *a half-life of 6000 years*. [2]

(ii) The activity of a sample of wood from a freshly cut tree is measured to be 80 disintegrations per second. Estimate the decrease in activity of the sample after 3 half-lives. [3]

7 A certain material has a half-life of 12 minutes. What proportion of that material would you expect still to be present an hour later? [3]

8 Four unknown nuclei are labelled W, X, Y and Z. Their full symbols are given below.

$^{30}_{15}W$ $^{30}_{16}X$ $^{32}_{17}Y$ $^{33}_{16}Z$

(a) Which, if any, of these nuclei are isotopes of the same element? [1]

(b) Explain your answer to part (a). [2]

9 (a) Describe the process of nuclear fusion. Your description should include:

- the particles involved
- what happens when nuclear fusion takes place
- where nuclear fusion occurs naturally. [6]

(b) A great deal of money is being invested on research into nuclear fusion.

(i) Suggest a reason why. [1]

(ii) Give two practical difficulties that must be overcome before fusion reactors become viable. [2]

(c) (i) What is ITER?

(ii) Explain the need for the ITER project. [3]

Answers online

ONLINE

35 Waves

Waves transfer **energy** from one point to another but they do not, in general, transfer matter. *All* waves are produced as a result of **vibrations** and can be classified as longitudinal or transverse.

Longitudinal waves

REVISED

A **longitudinal wave** is one in which the particles vibrate parallel to the direction in which the wave is travelling. The only types of longitudinal waves relevant to your GCSE course are:

- sound waves
- ultrasound waves
- slinky spring waves
- P-type earthquake waves.

a **longitudinal wave** is one in which the particles vibrate parallel to the direction in which the wave is travelling

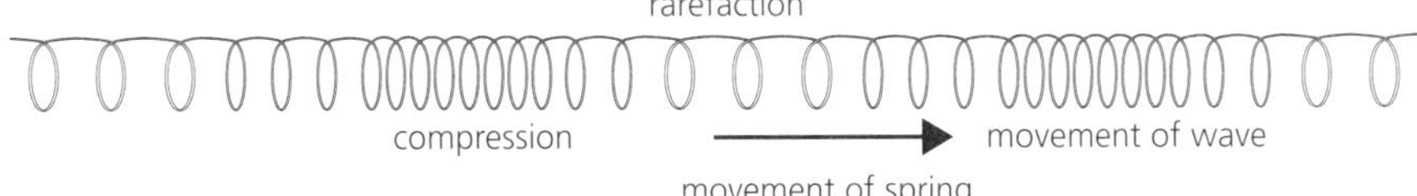

Figure 35.1 A longitudinal wave moving along a slinky spring

It is easy to demonstrate longitudinal waves by holding a slinky spring at one end and moving your hand backwards and forwards parallel to the axis of the stretched spring (Figure 35.1). **Compressions** are places where the coils (or particles) bunch together. **Rarefactions** are places where the coils (or particles) are furthest apart. All longitudinal waves are made up of compressions and rarefactions.

compressions are places where the coils (in a slinky) or particles in a longitudinal wave bunch together

rarefactions are places where the coils (in a slinky) or particles in a longitudinal wave are furthest apart

Transverse waves

REVISED

A **transverse wave** is one in which the vibrations are at 90° to the direction in which the wave is travelling. Most waves in nature are transverse — some examples are:

- water waves
- slinky spring waves (Figure 35.2)
- waves on strings and ropes
- electromagnetic waves.

a **transverse wave** is one in which the particles vibrate perpendicular to the direction in which the wave is travelling

A transverse wave pulse can be created by shaking one end of a rope. The pulse moves along the rope, but the final position of the rope is exactly the same as it was at the beginning. None of the material of the rope has moved permanently. But the wave pulse has carried energy from one point to another.

Water waves are clearly transverse. A cork floating on the surface of water bobs up and down as the waves pass.

Exam tip

Slinky springs can be used to demonstrate both longitudinal *and* transverse waves — so it is best not to quote it if you are asked to give an example of a longitudinal (or a transverse) wave.

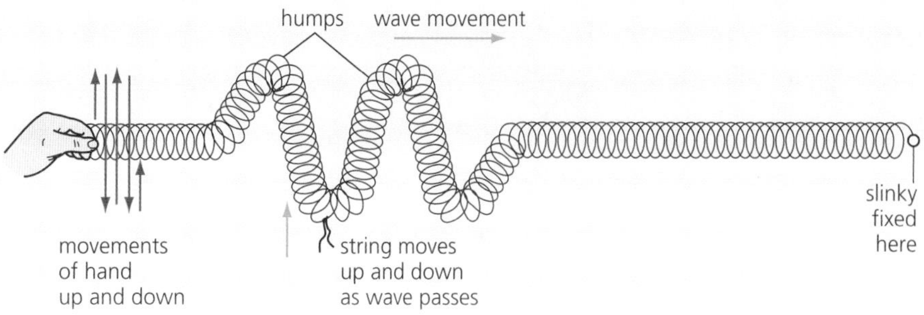

Figure 35.2 Transverse waves travelling through a slinky

Describing waves

REVISED

There are a number of important definitions relating to waves that must be learned.

The **frequency** of a wave is the number of complete waves passing a fixed point in a second. Frequency is given the symbol f, and is measured in units called hertz (abbreviation Hz).

The **wavelength** of a wave is the distance between two consecutive crests or troughs. Wavelength is given the symbol λ, and is measured in metres. λ is the Greek letter 'l' and is pronounced 'lamda'.

The **amplitude** of a wave is the greatest displacement of the wave from its undisturbed (equilibrium) position. Amplitude is measured in metres.

frequency is the number of complete waves passing a fixed point in a second

wavelength is the distance between two consecutive crests or troughs

amplitude is the greatest displacement of the wave from its undisturbed position

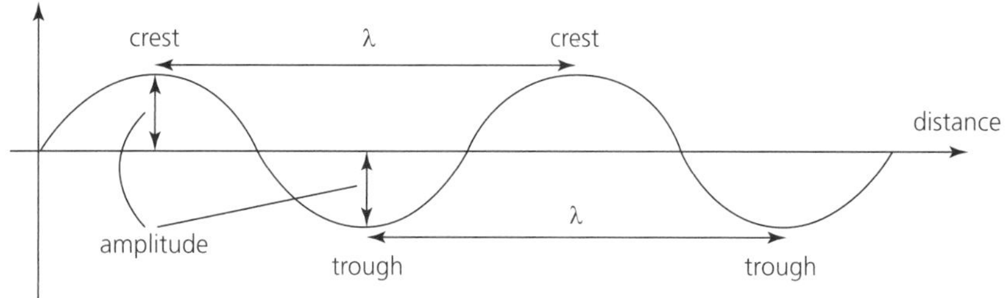

Figure 35.3 Displacement–distance graph to illustrate wavelength and amplitude

Wavelength and amplitude of longitudinal waves

For a **longitudinal** wave, the wavelength is the distance between the centre of one compression and the next. The amplitude of a longitudinal wave is the maximum distance a particle moves from the centre of this motion (Figure 35.4).

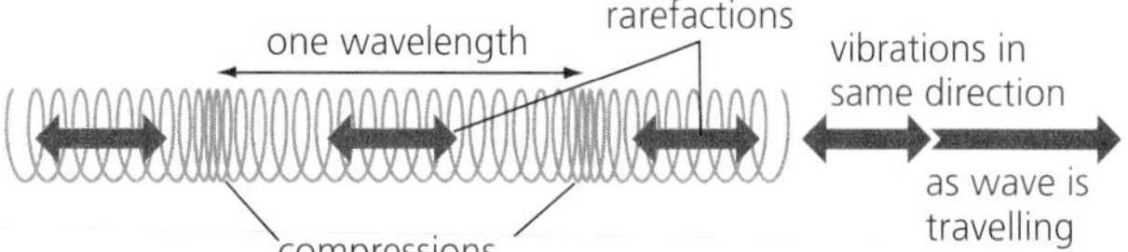

Figure 35.4 In longitudinal waves, the vibrations are along the same direction as the wave is travelling

Now test yourself

TESTED

1. Describe the difference between a transverse wave and a longitudinal wave.
2. Give two examples of transverse waves and two examples of longitudinal waves.
3. Define the terms *wavelength*, *frequency* and *amplitude* and state a unit in which each could be measured.
4. What evidence can you give that microwaves transmit energy?
5. How could you use a slinky spring to demonstrate:
   (a) a longitudinal wave
   (b) a transverse wave?
6. What happens to the compressions in a longitudinal wave on a slinky spring if the wavelength is increased?

The wave equation

REVISED

Imagine a wave with wavelength λ (metres) and frequency f (hertz). Then the speed of the wave, v, is given by:

wave speed = frequency × wavelength

$$v = f \times \lambda$$

Exam tip

This important equation must be learned for the GCSE examination.

Note that the units used in the wave equation must be consistent, as shown in Table 35.1.

Table 35.1

Frequency	Wavelength	Speed
always in Hz	cm	cm/s
	m	m/s
	km	km/s

Graphs and waves

REVISED

We can represent waves on graphs like those shown in Figure 35.5.

Note carefully: The upper graph is displacement against **time**. The lower graph is displacement against **distance**.

The vertical axis in both cases is displacement — so we can find the amplitude from either graph.

The red line in the upper graph shows the time, T, between the crests passing a fixed point. This time is known as the period.

Students often wrongly think T is a distance (the wavelength).

The blue line in the lower graph shows the distance between consecutive crests. This is the wavelength, λ.

The period, T, in the upper graph is 4 seconds, and the wavelength λ is 10 metres.

How could we find the speed from these data?

The graphs tell us that the wave travels 10 metres in 4 seconds.

$$\text{speed} = \frac{\text{distance}}{\text{time}} = \frac{10\,\text{m}}{4\,\text{s}} = 2.5\ \text{m/s}$$

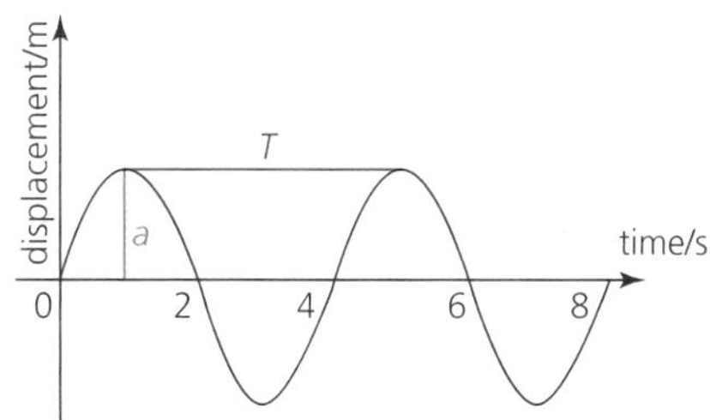

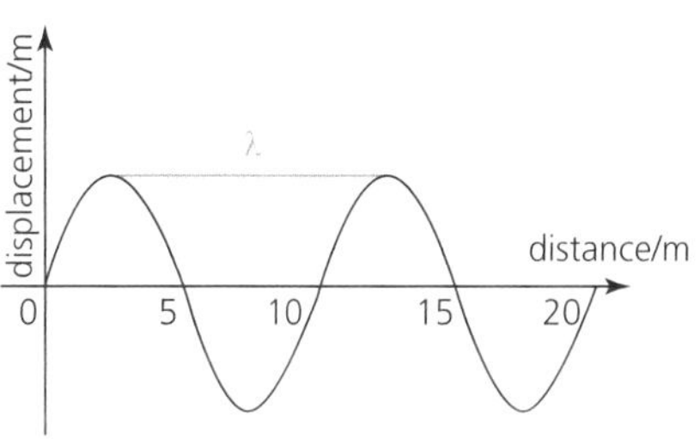

Figure 35.5 Graphs showing displacement against time and displacement against distance

frequency, $f = \frac{1}{T} = 0.25$ Hz

We can use the frequency to confirm the speed using the wave equation:

$v = f \times \lambda = 0.25\,\text{Hz} \times 10\,\text{m} = 2.5\,\text{m/s}$

Now test yourself

TESTED

7 Copy and complete the table below. Note carefully the unit in which you are to give your answers. The first one has been done for you as an example.

Wavelength	Frequency	Speed
5 m	100 Hz	500 m/s
12 m	50 Hz	______m/s
3 cm	60 kHz	______m/s
______m	4 Hz	20 cm/s
______m	5 kHz	2.5 km/s
16 mm	______Hz	80 cm/s
6×10^4 m	______Hz	3×10^8 m/s

8 The vertical distance between a crest and a trough is 24 cm and the horizontal distance between the first and the fifth wave crests is 40 cm. If 30 such waves pass a fixed point every minute, find the amplitude, frequency, wavelength and speed of the waves.

Exam tip

Make sure you know the following:

- the wave equation and the units in which speed, wavelength and frequency are measured
- how to find the frequency of a wave from a graph of displacement against time
- how to find the wavelength of a wave from a graph of displacement against distance.

Echoes

REVISED

Like all waves, sound and ultrasound can be made to reflect. When this happens the angle of incidence is always equal to the angle of reflection (Figure 35.6).

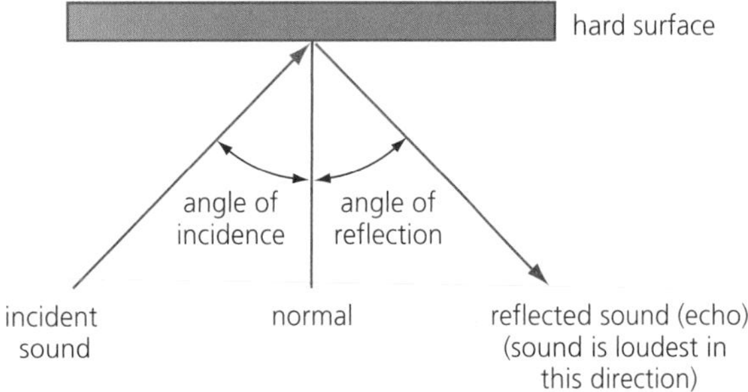

Figure 35.6 The reflection of sound waves from a surface

Audible sound ranges in frequency from 20 Hz to 20 000 Hz. Sound above 20 000 Hz is called ultrasound and cannot be heard by humans. It can, however, be detected by bats, dogs, dolphins and many other animals.

Reflected sound (and ultrasound) is called an echo. Humans have found clever ways to use ultrasound echoes:

- scanning metal castings for faults or cracks (e.g. in rail tracks)
- scanning a woman's womb to check on the development of her baby
- scanning soft tissues to diagnose cancers
- fish location by seagoing trawlers
- mapping the surface of the ocean floor in oceanography.

An application of ultrasound in medicine

In an ultrasound scan of an unborn baby, a probe is moved across the mother's abdomen. The probe sends out ultrasound waves and also detects the reflections. The other end of the probe is connected to a computer.

By examining the reflected waves from the womb, the computer builds up a picture of the foetus (unborn baby). Unlike X-rays, ultrasound is quite safe for this purpose.

Ultrasound can also be used to measure the diameter of the head of the baby as it develops in the womb (Figure 35.7). When the ultrasound reaches the baby's head at A, some ultrasound is reflected back to the detector and produces pulse A on the cathode ray oscilloscope (CRO). Some ultrasound passes through the head to point B, and is then reflected back to the detector. This reflection produces pulse B on the CRO.

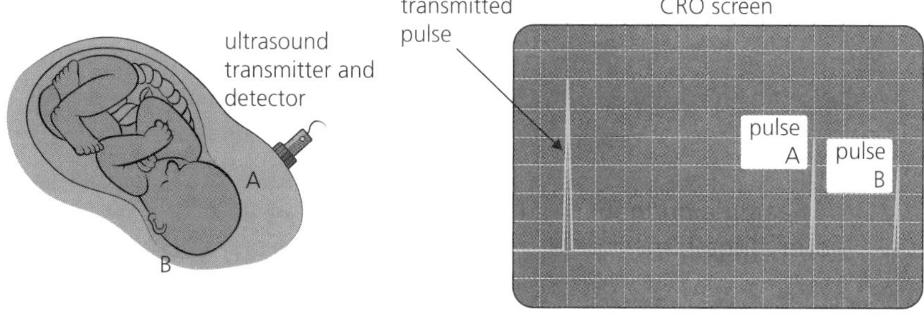

Figure 35.7 The diameter of a baby's head can be measured using ultrasound

In the diagram of the CRO screen, each horizontal division corresponds to a time of 40 microseconds ($40\,\mu s = 40 \times 10^{-6}\,s$).

The time interval between the arrival of pulse A and the arrival of pulse B at the detector corresponds to three divisions on the CRO. Since each division is $40\,\mu s$ this represents a total time of $120\,\mu s$. This additional $120\,\mu s$ is the time taken for ultrasound to travel from A to B and back to A. The time for ultrasound to travel from B to A is therefore half of that or $60\,\mu s$.

Now physicists know that ultrasound travels at a speed of 1500 m/s in a baby's head. So the width of the head can be found as follows:

width of head = speed × time = $1500\,m/s \times 60 \times 10^{-6}\,s = 0.09\,m = 9\,cm$

Scanning metal castings

Railway tracks do not last forever. They wear out. So it is important that we find out early if they are developing cracks or flaws below the surface.

We can do this with ultrasound scanners attached to specially fitted rail carriages (Figure 35.8). Ultrasound passes through the track. If there is a crack or other flaw it can be imaged (using the same science that allows us to obtain a picture of a baby in the womb).

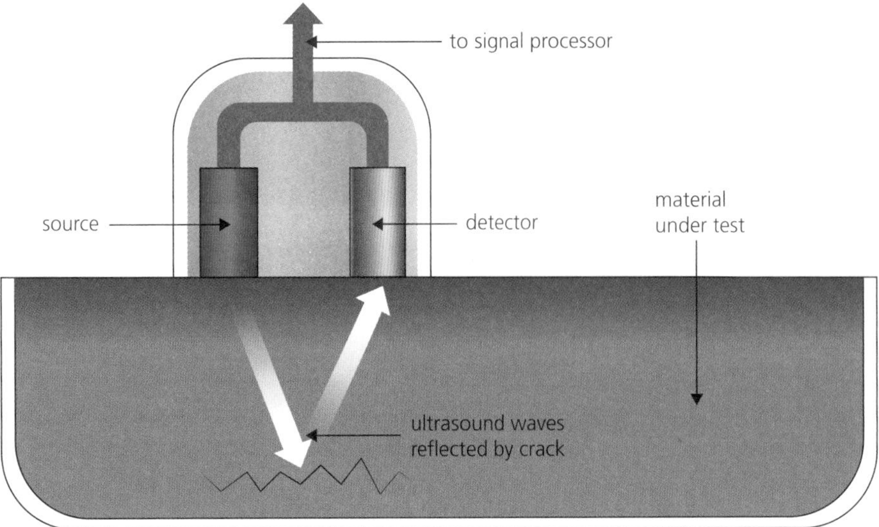

Figure 35.8 Detecting cracks in metals

Sonar

Sonar stands for **SO**und **N**avigation **A**nd **R**anging and was originally developed to detect submarines in the early twentieth century. However, the following example illustrates its use by fishermen to detect shoals of fish and to measure how far they are below the surface.

Example

A fishing trawler produces an ultrasound pulse and 0.4 s after it is transmitted an echo from a shoal of fish is detected. Assuming the speed of ultrasound in seawater is 1500 m/s:

(a) calculate the total distance travelled by the ultrasound
(b) calculate the distance from the trawler to the shoal of fish
(c) explain why a second echo is detected shortly after the first.

Answer

(a) total distance = speed × time = 1500 m/s × 0.4 s = 600 m
(b) distance from the trawler to the fish = ½ × total distance = ½ × 600 m = 300 m
(c) The first echo is from the fish. The second echo is from the sea bed, which is further away from the trawler than the fish. So the ultrasound from the sea bed takes longer to reach the trawler than the echo from the fish.

Radar

Radar stands for **RA**dio **D**etection **A**nd **R**anging and was originally developed during World War II to detect enemy aircraft and find their distance from the radar station.

Radar waves are in the microwave section of the electromagnetic spectrum. Think of a radar beam as a powerful beam of microwaves. They have wavelengths ranging from a few mm to just over a metre. Because radar waves are incredibly fast at 300 000 000 m/s, they are used extensively to track very fast objects that may be a large distance away. So, for example, they are used by air traffic controllers to track passenger airliners, by the military to track missiles and by the coastguard to detect ships.

Radar cannot be used under water. The water absorbs the radar (microwaves) within a metre or so. However, the physics of the application is very similar to that of sonar. If you are asked to solve a mathematical problem involving radar it is likely that numbers will be given in index form as shown in the example.

Example

Figure 35.9 shows a ground-based radar station A. It transmits a radar beam, which reflects off aircraft B. The radar echo is received at A 2.2×10^{-4} s after the original radar transmission. If the speed of radar waves is 3×10^{8} m/s, calculate the distance AB in km.

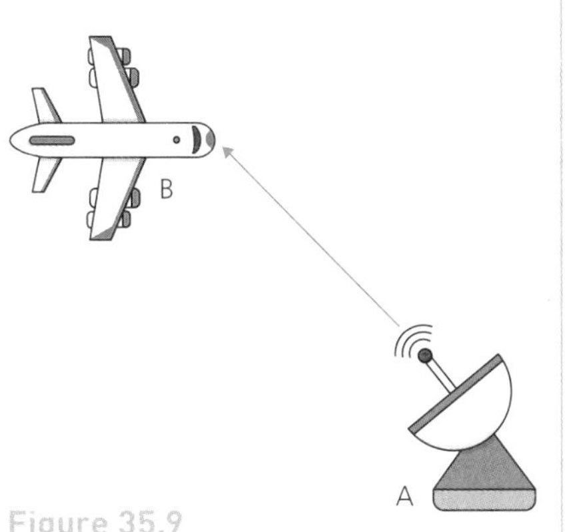

Figure 35.9

Answer

The time taken for radar to travel from A to B and back to A is 2.2×10^{-4} s

So, the time taken for radar to travel from A to B is $\frac{1}{2} \times 2.2 \times 10^{-4}$ s

$= 1.1 \times 10^{-4}$ s

Distance AB = speed × time = 3×10^{8} m/s × 1.1×10^{-4} s

$= 3.3 \times 10^{4}$ m = 33 km

Electromagnetic waves

REVISED

Electromagnetic waves are members of a family with common properties called the electromagnetic spectrum. They:

- can travel in a vacuum (unique property of electromagnetic waves)
- travel at exactly the same speed in a vacuum
- are transverse waves.

Electromagnetic waves also show properties common to all types of wave. They:

- carry energy
- can be reflected
- can be refracted.

There are seven members of the electromagnetic family. The properties of electromagnetic waves depend very much on their wavelength. In Table 35.2 they are arranged in order of increasing wavelength (or decreasing frequency). CCEA students need to be able to list these waves in order of increasing (and decreasing) wavelength. But you do not need to remember the wavelengths! Table 35.3 lists some dangers of electromagnetic waves.

Table 35.2 The electromagnetic spectrum

Electromagnetic wave	Typical wavelength
Gamma (γ) rays	0.01 nm*
X-rays	0.1 nm
Ultraviolet light	10 nm
Visible light	500 nm
Infrared light	0.01 mm
Microwaves	3 cm
Radio waves	1000 m

*1 nanometre (nm) = 1×10^{-9} m

Table 35.3 Dangers of electromagnetic waves

Electromagnetic wave	Dangers
Gamma (γ) rays	Damage cells and disrupt DNA, which may lead to cancer
X-rays	Damage cells and disrupt DNA, which may lead to cancer
Ultraviolet light	Certain wavelengths can damage skin cells, disrupting DNA and leading to skin cancer
Visible light	Intense visible light can damage the eyes (e.g. snow blindness)
Infrared light	Felt as heat and can cause burns
Microwaves	Cause internal heating of body tissues which, some say, can lead to eye cataracts
Radio waves	Large doses of radio waves are believed to cause cancer, leukaemia and other disorders and some people claim the very low-frequency radio waves from overhead power cables near their homes has affected their health

Now test yourself

TESTED

The following question shows that while you do not need to remember the wavelengths of the members of the electromagnetic spectrum, questions on their wavelengths can be asked.

9 Below are three members of the electromagnetic spectrum, not arranged in any particular order.

gamma (γ) rays radio waves ultraviolet light

Copy and complete the table below, writing the name of the missing electromagnetic wave opposite its typical wavelength.
(Hint: first identify the missing members of the spectrum and write them in order of increasing wavelength.)

Wave				
Typical wavelength/m	1×10^{-10}	6×10^{-7}	1×10^{-5}	1×10^{-3}

10 The emitter in Figure 35.10 sends out a pulse of sound. An echo from the object is detected after 2.5 ms. If sound travels at 340 m/s in the air, calculate the distance marked *d*.

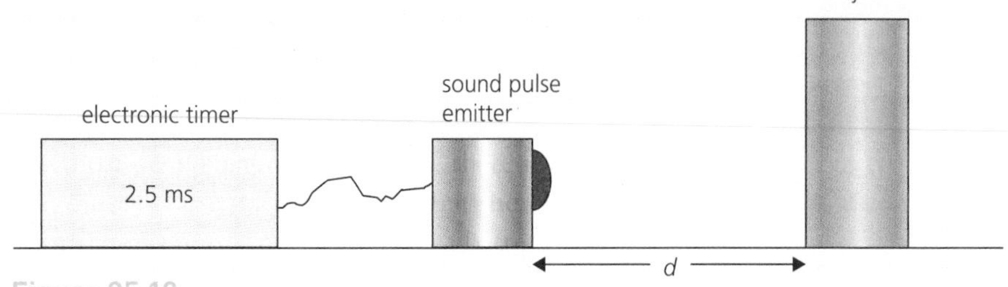

Figure 35.10

Exam practice

1 What physical property of a water wave never changes as a result of:
  (a) reflection
  (b) refraction? [2]

2 The graphs in Figure 35.11 relate to a water wave.

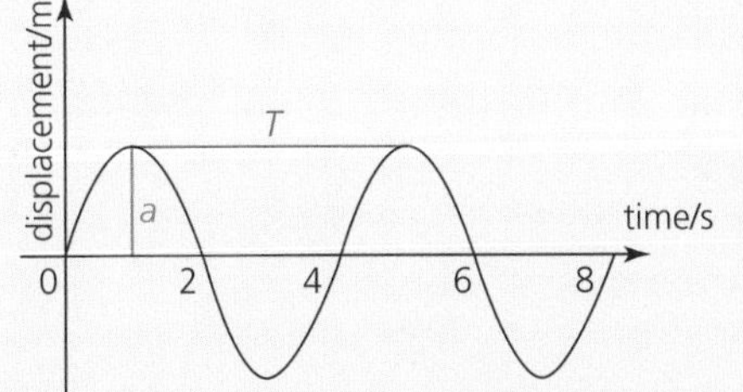

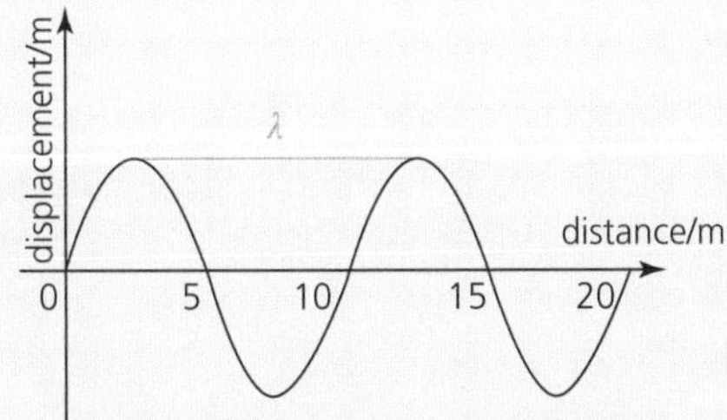

Figure 35.11

Find the frequency, wavelength and speed of the wave. [6]

3 (a) What is ultrasound? [1]
  (b) In what way is ultrasound different from the sound of human speech? [1]
  (c) State two differences between ultrasound waves and electromagnetic waves. [2]

The following question is provided to give you practice at the kind of question that could appear in Part B of Unit 3 (Part B of Unit 7 in Double Award Science).

4 When the frequency of sound is changed, the wavelength also changes. The table below shows the results of an experiment to measure the wavelength of sound at different frequencies. The unit for $1/\lambda$ is 1/m.

Wavelength, λ/m	0.7	1.0	1.5	2.5	4.0
Frequency, *f*/Hz	460	320	210	130	80
$1/\lambda$				0.40	0.25

  (a) Complete the table by entering the missing numbers in the third row. Two entries have already been done for you. [3]
  (b) Plot (on graph paper) the graph of *f*/Hz on the vertical axis against $1/\lambda$ on the horizontal axis and draw the straight line of best fit. [4]
  (c) Find the gradient of your line of best fit and state its unit. [3]
  (d) What is the physical significance of the gradient of the line of best fit? [1]
  (e) Use your graph to find the wavelength of sound of frequency 250 Hz. [2]

Answers online

ONLINE

36 Light

Reflection of light

REVISED

Figure 36.1 shows the reflection of light from a straight plane mirror.

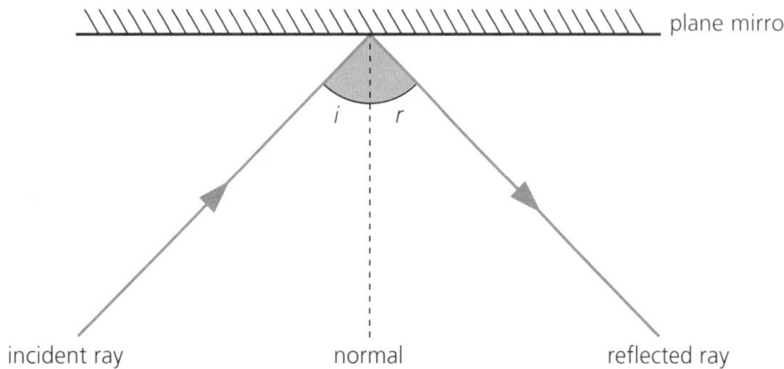

Figure 36.1 Reflection by a plane mirror

Experiments show that the **angle of incidence** is always *equal to* the **angle of reflection**. This is known as the **law of reflection**.

You should be able to describe the following experiment to demonstrate this law.

> **angle of incidence** (i) is the angle between the normal and the incident ray
>
> **angle of reflection** (r) is the angle between the normal and the reflected ray

Practical

The law of reflection

1. With a sharp pencil and a ruler, draw a straight line AOB on a sheet of white paper.
2. Use a protractor to draw a normal, N, at point O.
3. With the protractor, draw straight lines at various angles to the normal ranging from 15° to 75°.
4. Place a plane mirror on the paper so that its back rests on the line AOB (Figure 36.2).
5. Using a ray box, shine a ray of light along the line marked 15°.
6. Mark two crosses on the reflected ray on the paper.
7. Remove the mirror, and using a ruler join the crosses on the paper with a pencil and extend the line backwards to point O — this line shows the reflected ray.
8. Measure the angle of reflection with a protractor.
9. Record in a table the angles of incidence and reflection.
10. Repeat the experiment for different angles of incidence up to 75°.

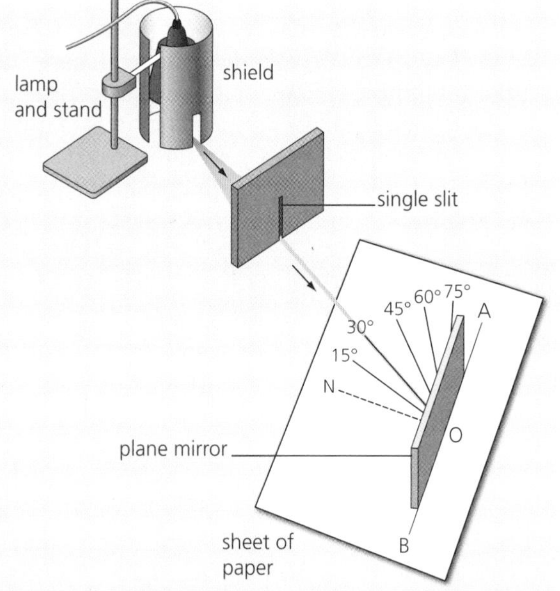

Figure 36.2 Experiment to demonstrate the law of reflection

Practical

Locating the image position in a plane mirror

1. Support a plane mirror vertically on a sheet of white paper, and with a pencil draw a straight line at the back to mark the position of the reflecting surface.
2. Use a ray box to direct two rays of light from point O towards points A and B on the mirror.
3. Mark the position of point O with a cross using a pencil (Figure 36.3).
4. Mark two crosses on each of the real reflected rays.
5. Remove both the ray box and the mirror.
6. Using a ruler, join the crosses with a pencil line so as to obtain the paths of the real rays from A and B.
7. Extend these lines behind the mirror (these are called virtual rays) — they meet at I, the point where the image was formed.
8. Measure the distance from the image I to the mirror line (IN) and the distance from the object O to the mirror line (ON) — they should be the same.
9. Repeat the experiment for different positions of the object O.
10. In each case, the object O and its image I should be the same perpendicular distance from the mirror.

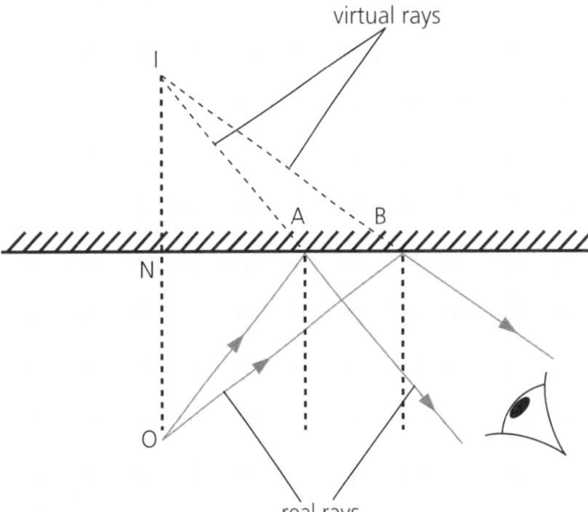

Figure 36.3 **The image in a plane mirror is the same distance behind as the object is in front**

Remember: the image in a plane mirror is:

- virtual (cannot be projected on to a screen)
- the same size as the object
- laterally inverted
- the same distance behind the mirror as the object is in front of the mirror.

Now test yourself

TESTED

1. Explain what is meant by the terms *normal*, *angle of reflection* and *angle of incidence*.
2. State the law of reflection of light.
3. State the properties of the image in a plane mirror.
4. State the size of the angle of incidence when the incident ray strikes a plane mirror at 90°.
5. The angle between a plane mirror and the incident ray is 40°. What size is the angle of reflection?
6. The angle between the incident ray and the reflected ray is 130°. What size is the angle of incidence?
7. A student stands in front of a mirror and views his image. The student now takes a step backwards, so that he is 20 cm further away from the mirror. By how much has the distance between the student and his image in the mirror increased?

Exam tip

Be very precise when asked about the image size and position in a plane mirror.

Don't just write: 'same size'. Remember to add... 'as the object'.

Don't just write: 'same distance behind the mirror'. Remember to add... 'as the object is in front of the mirror'.

Refraction of light

REVISED

Refraction is the change in direction of a beam of light as it travels from one material to another due to a change in speed in the different materials (Figure 36.4). Table 36.1 shows the speed of light in various media. It is not necessary to remember the numbers in this table, but you must know that light travels faster in air than in water, and faster in water than in glass. In fact, the *greater* the change in the speed, the *greater* is the bending.

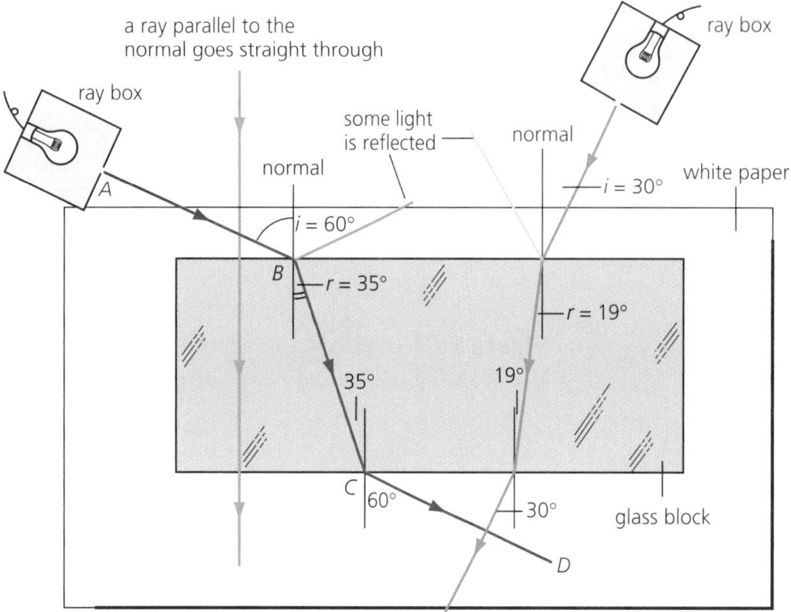

Figure 36.4 Refraction of light rays by a glass block

Table 36.1

Material	Speed of light in m/s
Air (or vacuum)	300 000 000
Water	225 000 000
Glass	200 000 000

Remember:

- The angle between the normal and the incident ray is the angle of incidence.
- The angle between the normal and the refracted ray is the **angle of refraction**.

angle of refraction is the angle between the normal and the refracted ray

Experiments show that:

- when light speeds up, it bends away from the normal
- when light slows down, it bends towards the normal.

Remember that this is also what happens to waves travelling from deep water into shallow water.

Figure 36.5 shows what happens when light travels from air through glass, then through water, and finally back into the air. Note the changes of direction in each case.

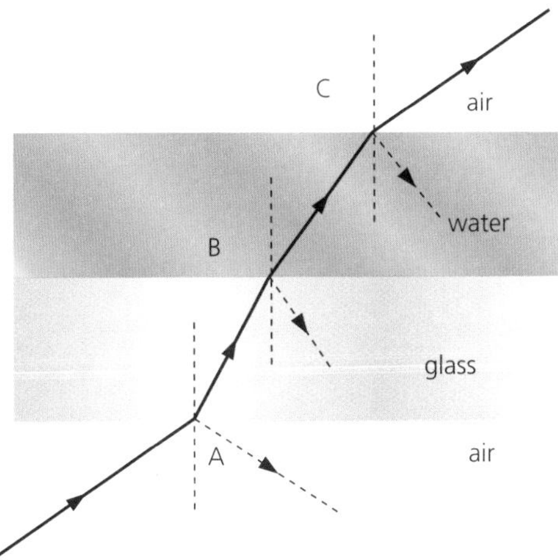

Figure 36.5 Refraction

Prescribed practical P5

Measuring angles of incidence and refraction

In this required practical activity you will measure angles of incidence and refraction as a ray of light passes from air into glass.

Apparatus

- rectangular glass (or Perspex) block
- ray box
- low-voltage power supply (PSU)
- leads
- protractor
- A4 plain white paper
- pencil
- ruler

Method

1. Prepare a table for your results like that shown on page 302.
2. Place the rectangular glass block in the centre of the sheet of white paper on a drawing board and draw round its outline with a sharp pencil.
3. Switch on the PSU and direct a ray of light to enter the block near the middle of the longest side of the block, so that the angle of incidence is about 10° (Figure 36.6).
4. Mark two pencil dots on the paths of both the incident ray and the emergent ray and remove the glass block.
5. Join the dots on the incident ray up to the point of incidence and join the dots on the emergent ray back to the point of emergence.
6. Draw a straight line between the point of incidence and the point of emergence.
7. Draw the normal at the point of incidence.
8. Measure the angle of incidence, *i*, and the angle of refraction, *r*, and record the data in your table.
9. Repeat steps 2 to 8 for angles of incidence ranging from about 20° up to about 80°.

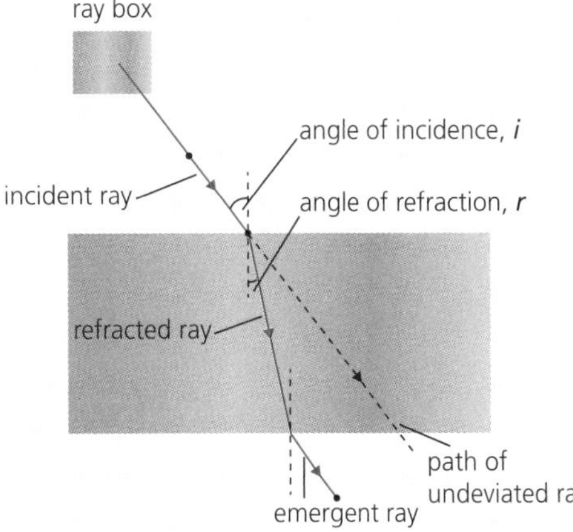

Figure 36.6 Light rays entering a glass block

→

Results

Angle of incidence, i/°	10	20	29	41	49	58	67	79
Angle of refraction, r/°	7	13	19	26	30	34	38	41

The above results are typical of those obtained in this experiment.

Treatment of the results

Plot the graph of angle of incidence (y-axis) against angle of refraction (x-axis).

Conclusion

The graph of angle of incidence against angle of refraction (Figure 36.7) is a curve through the origin of increasing gradient (the graph gets steeper as the angle of incidence increases). This tells us that i is *not* directly proportional to r (because the graph is *not* a straight line), but that i and r have a positive correlation (as i increases, r increases).

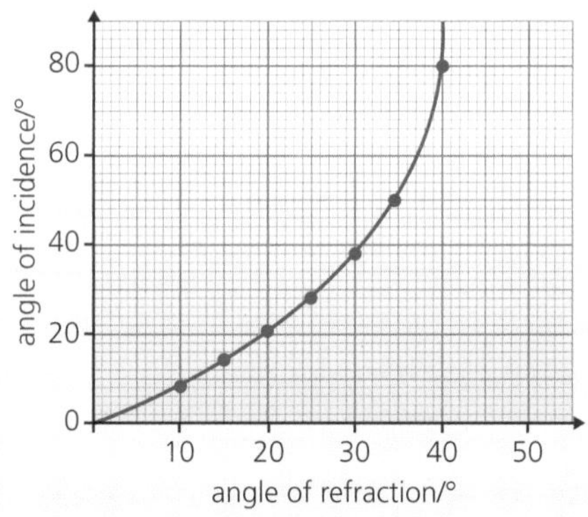

Figure 36.7 A graph showing angle of incidence against angle of refraction

Exam tip

Always remember to put an arrow on real rays of light. Both the normal and the virtual rays are always dotted and never have an arrow.

Dispersion of white light

REVISED

All colours (frequencies) of light travel at the same speed in air. But different colours of light travel at different speeds in glass. This means that different colours bend by different amounts when they pass from air into glass (Figure 36.8). When light is passed through a triangular glass block, a prism, the effect is called **dispersion** and it results in a spectrum showing all the colours of the rainbow. Red light is bent (refracted) the least because it travels fastest in glass. Violet light bends the most because it is slowest in glass.

dispersion is the splitting of white light into its component colours

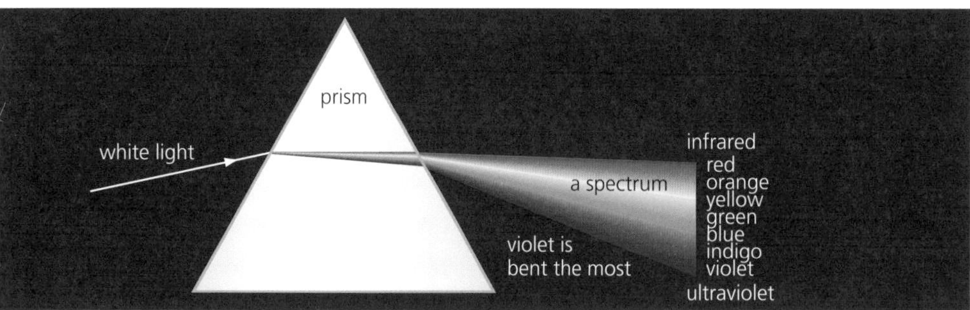

Figure 36.8 Dispersion of light through a prism

Now test yourself

TESTED

8 Describe what is meant by refraction of light.

9 Explain what has happened to the speed of a ray of light if it refracts:
  (a) towards the normal
  (b) away from the normal.

10 (a) State what is meant by dispersion of light.
  (b) Describe the conditions necessary for dispersion to occur.

11 (a) Figure 36.9 shows a ray of light passing from the air into the cornea of the eye.

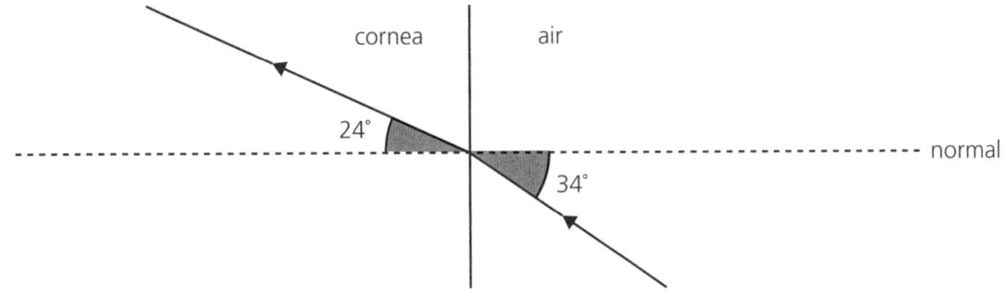

Figure 36.9

  (i) State the angle of incidence in air and the angle of refraction in the cornea.
  (ii) In which of the two media does light travel the faster?

  (b) What is the evidence for believing that red light travels faster in glass than blue light?

12 A sound wave made underwater travels towards the surface. Sound travels *faster* in water than it does in air. Copy and complete Figure 36.10 to illustrate the refraction that occurs when the sound travels into the air. Mark the normal on your diagram.

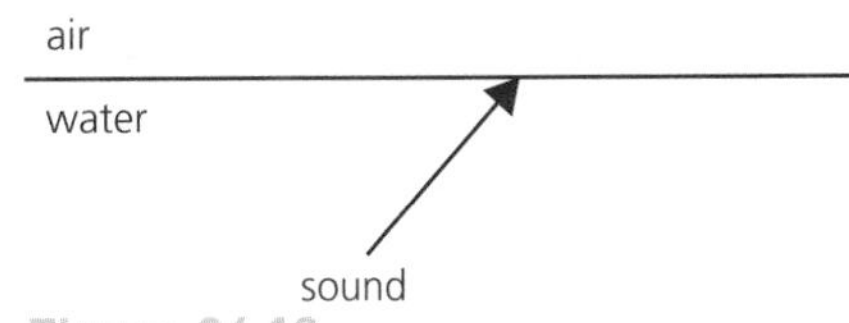

Figure 36.10

13 Different shapes of glass prism are often used to change the direction of light rays.

Figure 36.11

Copy Figure 36.11 and continue the path of the ray shown until it emerges into the air.

Lenses

REVISED

Lenses are specially shaped pieces of glass or plastic. There are two main types of lens:

- **converging** (or convex)
- **diverging** (or concave).

These are shown in Figure 36.12.

Figure 36.12 The shapes of a converging lens and a diverging lens

There is one feature of a converging lens that needs to be defined:

- Rays of light parallel to the **principal axis** of a converging lens all converge at the **principal focus** on the opposite side of the lens.

The distance between the principal focus and the optical centre of *any lens* is called the **focal length**.

For a convex lens, light refracts at each surface as it enters and leaves the lens, first bending towards the normal and then away from the normal.

There is one feature of a diverging lens that needs to be defined:

- Rays of light parallel to the principal axis of a diverging lens all appear to diverge from the principal focus after refraction in the lens.

Note that light passing through the optical centre of a convex or concave lens is not bent. It passes straight through without refraction.

a **converging** lens is a lens that is thickest at its centre and least thick at its edges

a **diverging** lens is a lens that is thickest at its edges and least thick at its centre

principal axis is a straight line joining the principal foci and passing through the optical centre of a convex lens

principal focus is a point on the principal axis of a convex lens through which rays of light parallel to the principal axis pass after refraction in the lens

focal length is the distance between the optical centre of a lens and the principal focus

Practical

Measuring the focal length, *f*, of a converging lens using a distant object

Apparatus

- convex lens
- lens holder
- ruler
- sellotape
- white screen in a holder
- distant object (such as a tree, which can be seen through the windows in the laboratory and is at least 20 m away)

Method

1. Sellotape the ruler to the bench.
2. Place the white screen in its holder at the zero mark.
3. Place the lens in its holder as close as possible to the screen.
4. Slowly move the lens away from the screen until the inverted image of the distant object is as sharp as possible.
5. Using the metre ruler, measure the distance from the centre of the lens to the white screen. This distance is the focal length of the lens.
6. Record the measured focal length in a prepared table.
7. For reliability, repeat steps 2 to 6 for four different distant objects and determine the average value of *f*.

Results

Focal length, *f*/mm	245	240	250	247	243
Average focal length, *f*/mm	245				

Image in a concave (diverging) lens

REVISED

Note that regardless of the position of the object, the image in a concave (diverging) lens is *always*:

- erect
- virtual
- smaller than the object
- placed between the object and the lens.

This is shown in the ray diagram in Figure 36.13.

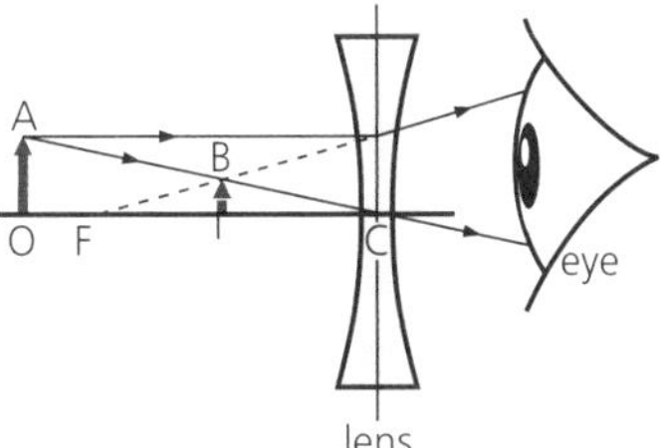

Figure 36.13 The image in a diverging lens

Image in a convex (converging) lens

REVISED

The position and properties of the image in a convex (converging) lens depend on the position of the object. We can find those positions and image properties by drawing a ray diagram.

Rules for drawing ray diagrams

To draw a ray diagram for a convex lens you must draw at least two of the following rays:

- a ray parallel to the principal axis, refracted through the principal focus on the other side of the lens
- a ray through the optical centre of the lens that does not change its direction (does not refract)
- a ray through the principal focus on one side of the lens, which emerges so that it is parallel to the principal axis on the other side of the lens.

First steps when drawing a ray diagram for a convex lens:

1. using a ruler, draw a horizontal line to represent the principal axis and a vertical line for the lens.
2. Mark the position of the principal focus with a letter F, the same distance from the optical centre on each side of the lens.
3. Using a ruler, draw a vertical line touching the principal axis at the correct distance from the lens to represent the object.
4. Using a ruler, draw at least two of the three construction rays, starting from the top of the object.
5. Draw arrows on all rays to show the direction in which the light is travelling.

The point where the construction rays meet is at the top of the image. The bottom of the image lies vertically below on the principal axis.

To illustrate the process, consider the following example.

Example

An object 5 cm tall is placed 6 cm away from a converging lens of focal length 4 cm. Find the position and height of the image.

Answer

In Figure 36.14, circled numbers have been added to show the order in which the lines or rays have been drawn. These numbers are drawn for illustration only and are normally omitted from such ray diagrams.

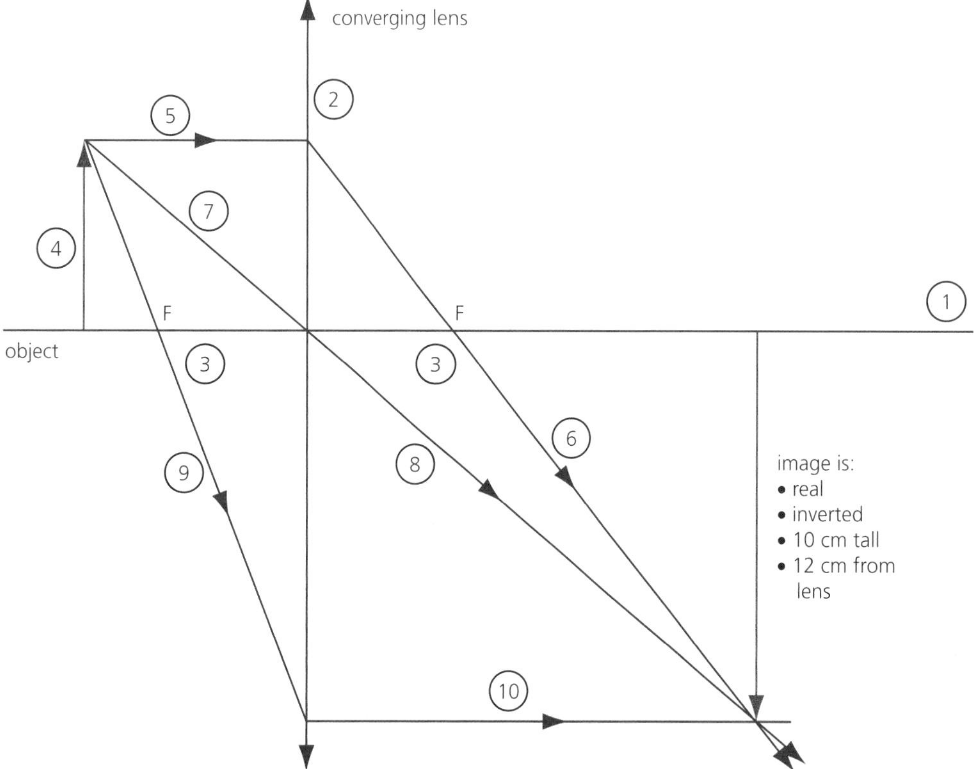

Figure 36.14 **To work out the position and height of an image**

① horizontal line representing principal axis (PA)
② vertical line representing lens
③ two principal foci marked, each 4 cm from lens
④ object marked 6 cm from lens
⑤ Ray from top of object parallel to PA...
⑥ ...refracts through F.
⑦ Ray from top of object through the optical centre...
⑧ ...is not refracted.
⑨ Ray from top of object through F...
⑩ ...refracts parallel to PA.

Finally the image is drawn from the point where the refracted rays meet to the PA.

The image is drawn as a continuous line to show that it is real (can be projected on to a screen).

The downward arrow on the image shows it is inverted.

Now use the ruler to measure the height of the image and its distance from the centre of the lens.

Ray diagrams

REVISED

The ray diagrams in Table 36.2 show where the image is formed for different positions of the object. You should study carefully the diagrams, and Table 36.3, which gives a summary of the information.

Table 36.2 Drawing ray diagrams

Position of object	Ray diagram	Properties of image
Between the principal focus, F, and the lens	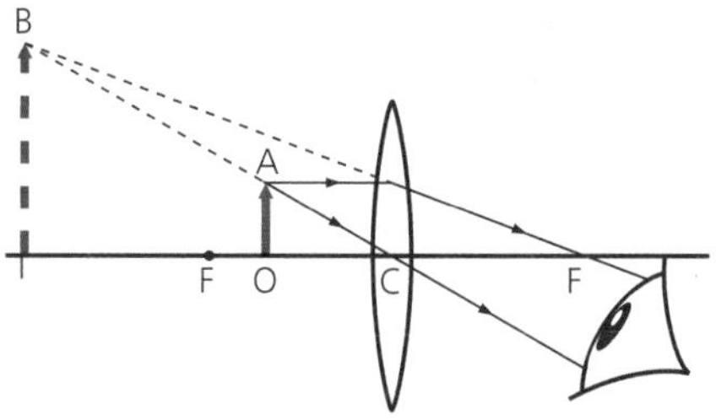	• On same side of lens as object but further from lens • Virtual • Erect • Larger than object
At principal focus, F	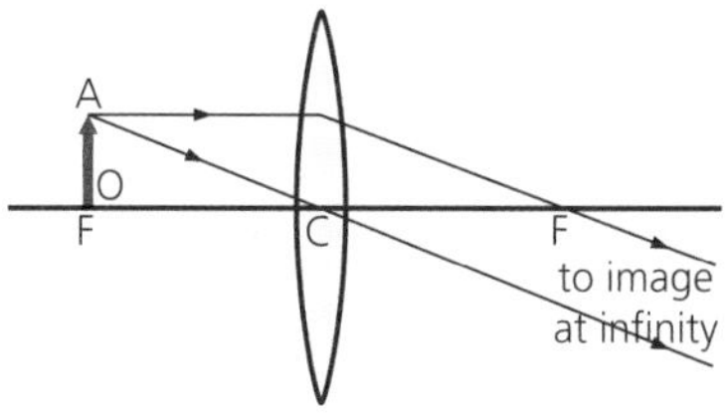	• Image is at infinity
Between F and twice the focal length	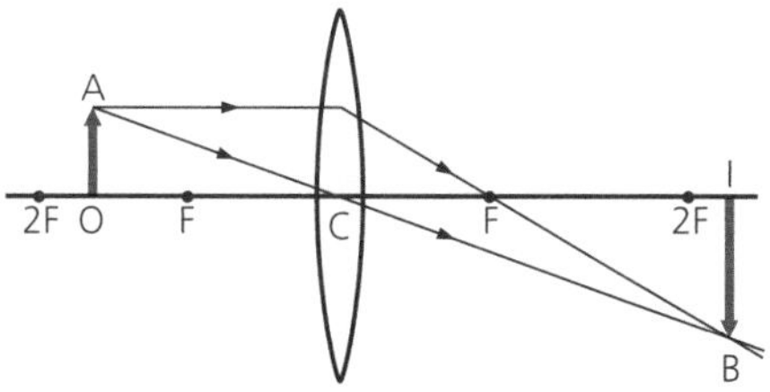	• On opposite side of lens to object but further away than twice the focal length • Real • Inverted • Larger than object
At 2F	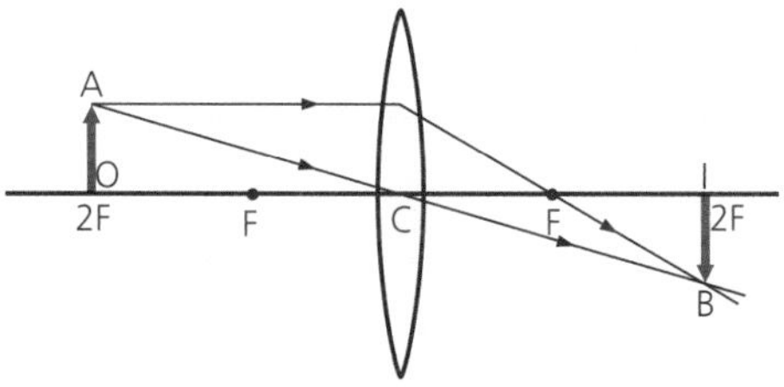	• On opposite side of lens to object and exactly same distance away as object • Real • Inverted • Same size as object
Just beyond twice the focal length of the lens	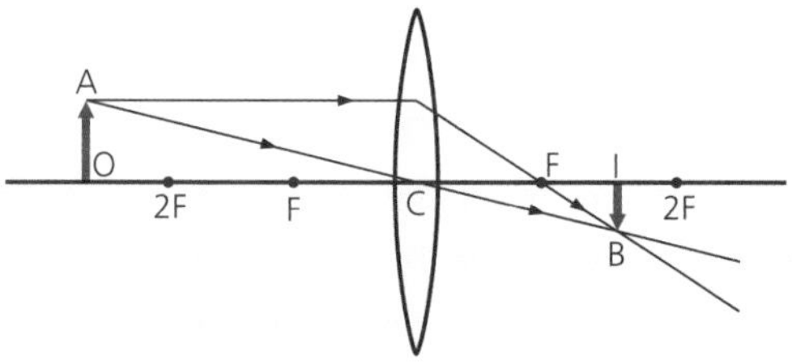	• On opposite side of lens to object and between one and two focal lengths from lens • Real • Inverted • Smaller than object

Table 36.3 Summary of ray diagrams

		Properties of image			
Position of object	**Location of image**	**Nature**	**Erect or inverted**	**Larger or smaller than object**	**Application**
Between lens and F	On same side of lens as object but further away from lens	Virtual	Erect	Larger	Magnifying glass
At F	At infinity	Real	Inverted	Larger	Searchlight
Between F and 2F	Beyond 2F	Real	Inverted	Larger	Cinema projector
At 2F	At 2F	Real	Inverted	Same size	Telescope — erecting lens
Just beyond 2F	Between F and 2F	Real	Inverted	Smaller	Camera
Very far away from lens	At F	Real	Inverted	Smaller	Camera

Now test yourself

TESTED

14 Draw a ray diagram to show how a diverging lens produces a virtual, diminished image.

15 State the rules for drawing ray diagrams for a convex lens.

16 Draw a ray diagram to illustrate how a converging lens can produce:

(a) a real, magnified image

(b) a diminished image

(c) a virtual image.

Exam tip

You should pay particular attention to the ray diagrams that illustrate the principles of:

- the magnifying glass (to give an erect, virtual image)
- the projector (to give a magnified, real image)
- the camera (to give a diminished real image).

They are specifically required by the subject specification.

Exam practice

1 State four properties of the image in a plane mirror. [4]

2 Two mirrors, M_1 and M_2, are placed at right angles to one another. Figure 36.15 shows a ray of light incident on mirror M_1 at an angle of 27° to its surface.

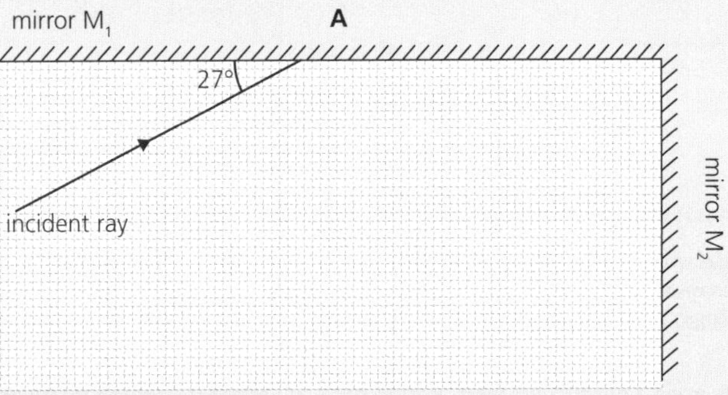

Figure 36.15

(a) Copy Figure 36.15 on to graph paper and draw the reflected rays from both mirrors. [1]

(b) Calculate the angles of incidence and reflection at mirror M_2. [2]

(c) Comment on the directions of the incident ray at M_1 and the reflected ray from M_2. [1]

3 Figure 36.16 shows the refraction of light in glass and water.

(a) Explain why the light bends as it enters both glass and water. [2]

(b) State the angles of incidence and refraction in each medium. [2]

(c) What evidence in Figure 36.16 shows that light is slower in glass than it is in water? [1]

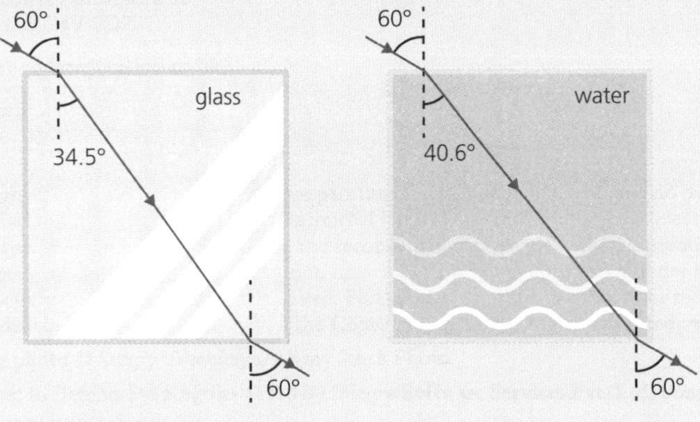

Figure 36.16

4 Figure 36.17 shows how a lens can produce an image of an object. Each small square corresponds to a distance of 1 mm.

(a) Which three words from the list below best describe the image? [3]

real diminished erect inverted virtual enlarged

(b) What type of lens is shown in Figure 36.17? [1]

→

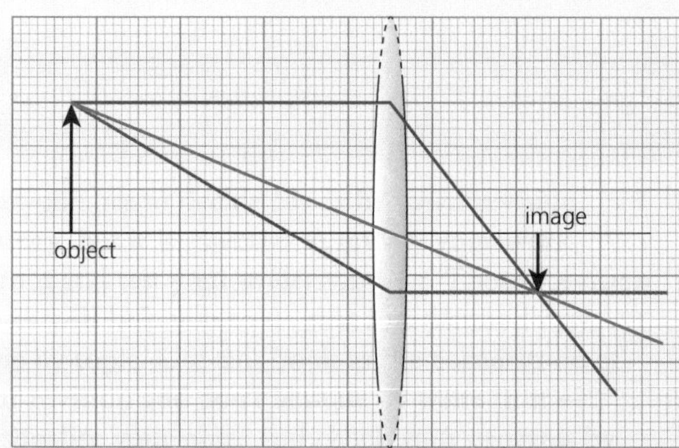

Figure 36.17

(c) Copy the diagram and mark the location of the principal focus on each side of the lens, and the optical centre. [3]

(d) The diagram is drawn to scale. Find the focal length of the lens in cm. [1]

(e) The magnification of the image is defined by the equation:

$$\text{magnification} = \frac{\text{height of image}}{\text{height of object}}$$

Use the equation to calculate the magnification of the image. [2]

5 A student was investigating how a ray of light passed through a rectangular glass block. He drew the diagram in Figure 36.18. Copy Figure 36.18 and show the correct paths of the two rays A and B through the glass and back into the air. [4]

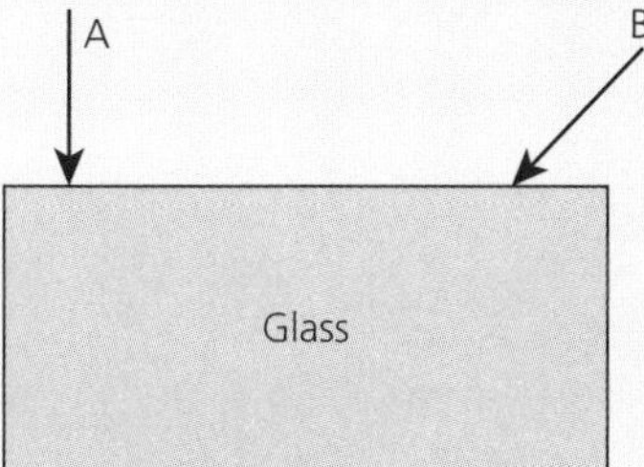

Figure 36.18

6 Figure 36.19 shows a ray of white light incident on a glass prism. When the light passes through the prism it splits into many different colours.

(a) What is this effect called? [1]

(b) State briefly why it happens. [2]

(c) Copy Figure 36.19 and show the passage of the coloured rays through the prism and into the air. [3]

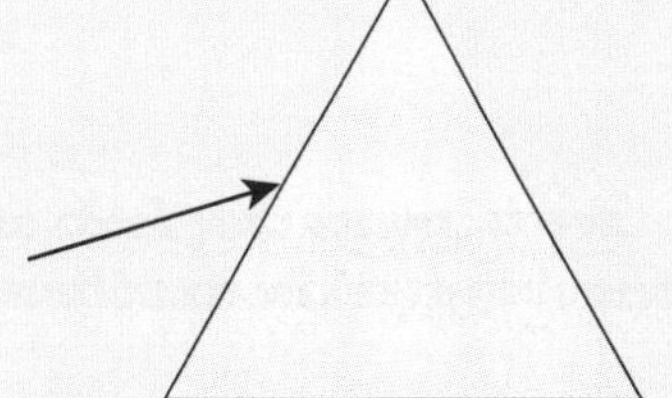

Figure 36.19

Answers online

ONLINE

37 Electricity

We now know that there are two types of charge, **positive charge** and **negative charge**. The negative charge is due to the presence of electrons.

When we connect a battery across a lamp the lamp lights up (Figure 37.1). The connecting wire (copper) and the filament of the bulb (tungsten) are both **electrical conductors**. But the plastic covering is an **insulator**.

In general, *all* **metals** are electrical conductors. Almost all **non-metals** are insulators, but there are a few exceptions. For example, graphite is a non-metal, but it conducts electricity (Table 37.1).

electrical conductor is a material through which electrical current passes easily

insulator is a material through which electrical current cannot pass

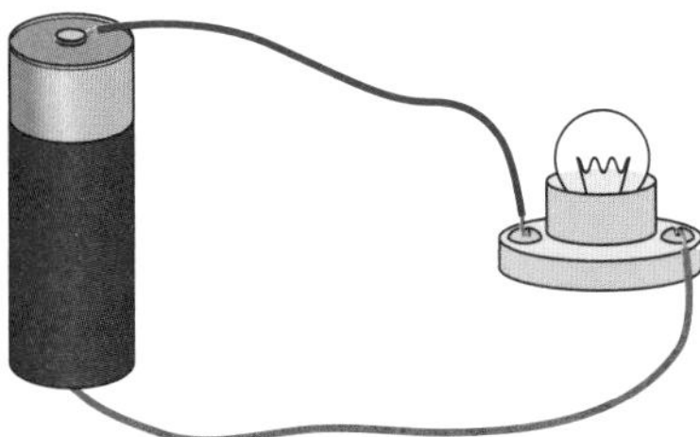

Figure 37.1 The lamp lights up when connected across a battery

Table 37.1 Common conductors and insulators

Good conductors	Gold	Silver	Copper	Aluminium	Mercury	Platinum	Graphite
Insulators	Polythene	Rubber	Wool	Wax	Glass	Paper	Wood

Why are metals good conductors?

REVISED

An electric current is a flow of electrically charged particles. At GCSE, the charge involved is always the electron. In metals, the outermost electron is often so weakly held that it can break away. We call such electrons **free electrons**. Some books call them delocalised electrons. However, in insulators there are *no* (or almost no) free electrons.

An electric cell (commonly called a battery) can make electrons move — but only if there is a conductor connecting its two terminals, making a complete circuit.

Scientists in the nineteenth century thought that an electric current consisted of a flow of positive charge from the positive terminal of the cell to the negative terminal. Unfortunately, although this idea is now known to be incorrect, this is still known as the direction of **conventional current**.

free electrons are electrons that are not attached to any particular atom

conventional current is the imagined current flowing from the positive terminal of a battery to the negative terminal

Summary

- Electrical conductors, like metals, have free electrons.
- Electrical insulators, like non-metals, have no free electrons.
- **Free electrons** move from the ***negative*** terminal to the ***positive*** terminal of the battery.
- **Conventional current** is said to flow from the ***positive*** terminal to the ***negative*** terminal of the battery.

Standard symbols

REVISED

An electrical circuit may be represented by a **circuit diagram** with **standard symbols** for components.

standard symbols are the internationally recognised symbols for electrical components

Table 37.2 Components and their symbols

Component	Symbol	Appearance
Switch	—○╱ ○—	
Cell	—│▌—	
Battery	—│▌---│▌—	
Resistor	—▭—	
Variable resistor	—▭— (with diagonal arrow)	
Fuse	—▭— (with line through)	5A
Voltmeter	—(V)—	
Ammeter	—(A)—	
Lamp	—⊗—	

Cell polarity

REVISED

By convention, the long, thin line in the symbol for a cell is taken as the positive terminal. The short, fat line is the negative terminal.

Cells can be joined together minus-to-plus to make a battery. Cells connected in this way are said to be connected in **series**.

Connecting cells in series to make a battery increases the **voltage**. For example, connecting 4 × 1.5 volt cells in this way gives a 6 volt battery as shown in Figure 37.2.

cell polarity is concerned with which end of a cell is positive and which end is negative

a **series** circuit is one in which the components are arranged one-after-another, like carriages in a train

voltage is the difference in electrical potential between two points that causes a current to flow

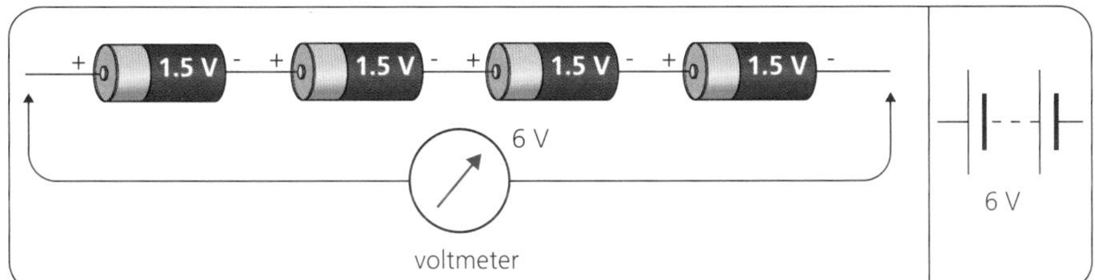

Figure 37.2 Cells correctly joined in series

If the polarity of one of the cells is reversed then the voltage is reduced dramatically. In Figure 37.3, the effect of reversing the polarity of the cell in the middle produces a battery of only 1.5 volts. The cells joined + to + cancel each other out.

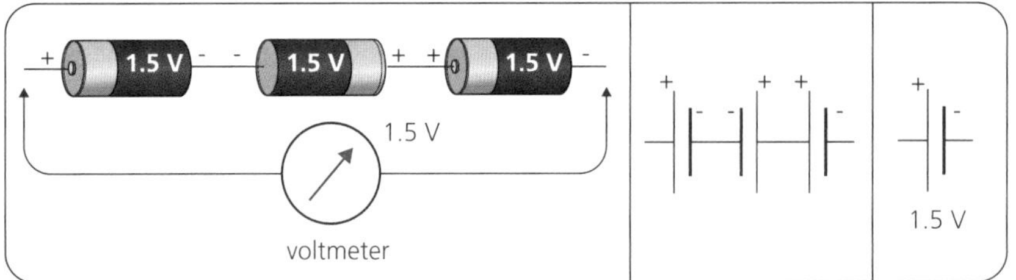

Figure 37.3 Cells incorrectly joined together

The relationship between charge and current

REVISED

The unit of charge is the **coulomb** (C).

The unit of **current** is the **ampere** (A). Currents of around 1 ampere and more can be measured by connecting an ammeter in the circuit. For smaller currents, a **milliammeter** is used. The unit in this case is the **milliampere** (mA) (1000 mA = 1 A). An even smaller unit of current is the **microampere** (μA) (1 000 000 μA = 1 A).

current is a flow of electric charge in a circuit

In general, if a steady current I amperes flows for time t seconds, the charge Q coulombs passing any point is given by:

$$Q = I \times t$$

charge = current × time

(C) = (A) × (s)

Notice that:

- charge is measured in coulombs (C) and given the symbol Q
- current is measured in amperes (A) and given the symbol I
- time is measured in seconds (s) and given the symbol t.

So, a current of 1 ampere is flowing in a circuit if a charge of 1 coulomb passes a fixed point in 1 second.

Example

A current of 150 mA flows around a circuit for 1 minute. How much electrical charge flows past a point in the circuit in this time?

Answer

I = 150 mA = 0.15 A

t = 1 minute = 60 s

$Q = I \times t$ = 0.15 × 60 = 9 C

Exam tip

Always ensure the units for current, charge and time are in amperes, coulombs and seconds before substituting values into the equation $Q = It$.

There are two conditions that *must* be met before an electric current will flow:

- There must be a *complete circuit* — i.e. there must be no gaps in the circuit.
- There must be a *source of energy* so that the charge may move — this source of energy may be a cell, a battery or the mains power supply.

Now test yourself

TESTED

1 The cells in Figure 37.4 are all identical. The total battery voltage is 1.6 V.
   (a) Calculate the voltage of each cell.
   (b) State the maximum voltage that this battery could deliver.
2 Convert the following currents into milliamperes:
   (a) 3.0 A
   (b) 0.2 A
   (c) 200 μA
3 What charge is delivered if:
   (a) a current of 6 A flows for 10 seconds
   (b) a current of 300 mA flows for 1 minute
   (c) a current of 500 μA flows for 1 hour?

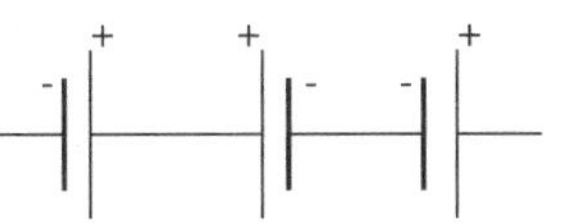

Figure 37.4

Resistance

REVISED

The **resistance**, R of an electrical conductor can be found using the ammeter–voltmeter method. We measure the current, I through the conductor when a voltage, V is applied across its ends. The resistance, R is then calculated using the equation:

$$R = \frac{V}{I}$$

where R is the resistance in ohms (Ω)
V is the voltage in volts
I is the current in amperes.

This is the basis of the Ohm's law experiment that follows.

resistance is the opposition to the flow of current and is defined as the ratio of voltage to current

Prescribed practical P6

Ohm's law

Apparatus

- low-voltage power supply unit (PSU)
- rheostat
- ammeter
- voltmeter
- connecting leads
- resistance wire
- switch

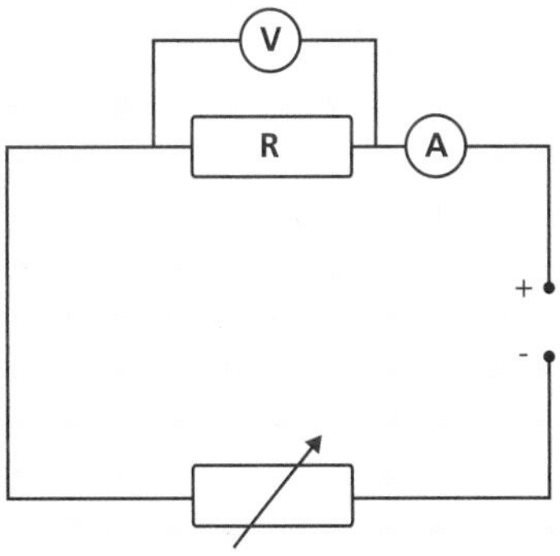

Figure 37.5 Circuit diagram

Method

1 Prepare a table for your results (see page 314).
2 Set up the circuit as shown in Figure 37.5.
3 Adjust the PSU to supply zero volts and then switch it on.
4 Record the voltage on the voltmeter and the corresponding current on the ammeter.
5 Record values for voltage and current, switch off the PSU and allow the wire to cool to room temperature.
6 Switch on the PSU and adjust the voltage so that the reading on the voltmeter increases by 0.5 V.

→

7 Repeat steps 5 and 6 for voltages from zero to a maximum voltage of 6 V*. This is Trial 1.
8 Repeat the entire experiment to obtain a second set of current values. This is Trial 2.
9 Calculate the mean current from the two trials.
10 Plot the graph of voltage against mean current.

*It is necessary to ensure the wire's temperature remains constant (close to room temperature). We do this by:

- keeping the voltage low (so that the current remains small), and
- switching off the current between readings to allow the wire to cool.

Table for results

Voltage/V	0.00	1.00	2.00	3.00	4.00	5.00	6.00
(Trial 1) current/A							
(Trial 2) current/A							
Mean current/A							
Ratio of voltage to current/Ω							

Treatment of the results

Plot the graph of voltage/V (*y*-axis) against mean current/A (*x*-axis) (Figure 37.6).

Evaluation of the results

The graph of *V* against *I* is a straight line through the origin. This tells us that the current in a metallic conductor is **directly proportional** to the voltage across its ends provided the temperature remains constant.

The last statement is commonly called **Ohm's law**.

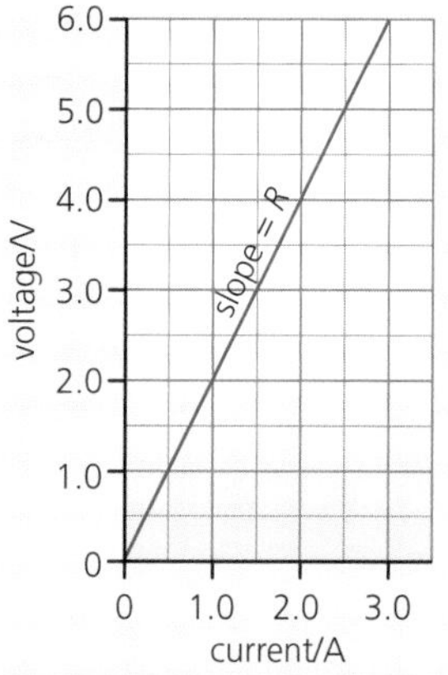

Figure 37.6 A graph showing voltage against current

direct proportion is the mathematical relationship between quantities that increase together in the same ratio. For example, when one quantity doubles the other quantity doubles also

Ohm's law states that the current in a metallic conductor is directly proportional to the voltage across its ends provided the temperature remains constant

Measuring the resistance

The resistance of a wire at constant temperature depends only on three factors:

- the material from which the wire is made
- the length of the wire, and
- the cross-sectional area of the wire.

The ratio $V:I$ is constant throughout the experiment.

Examples

1

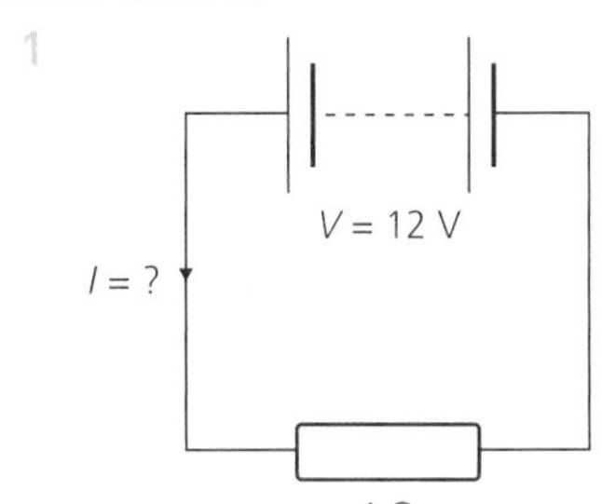

Figure 37.7

Answer

$V = 12\,V$, $R = 4\,\Omega$, $I = ?\,A$

current, $I = \frac{V}{R}$

$= \frac{12}{4}$

$= 3\,A$

2

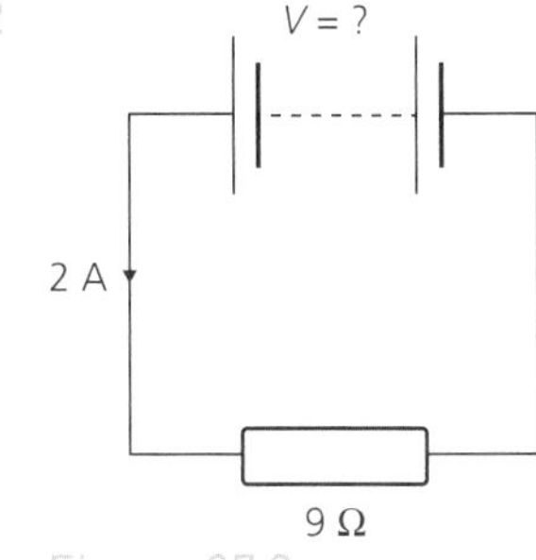

Figure 37.8

Answer

$I = 2\,A$, $R = 9\,\Omega$, $V = ?\,V$

voltage, $V = I \times R$

$= 2 \times 9$

$= 18\,V$

3

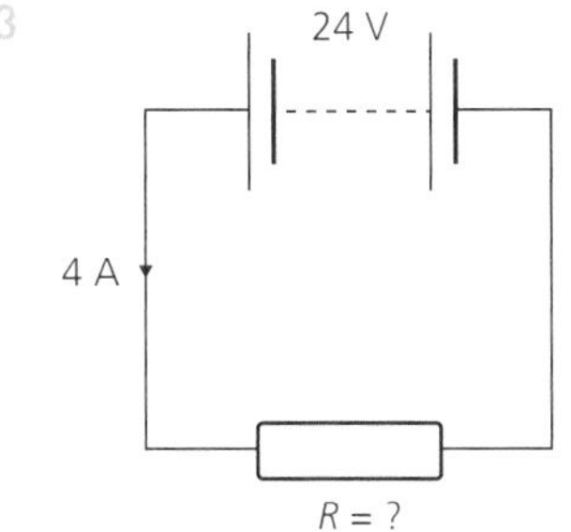

Figure 37.9

Answer

$V = 24\,V$, $I = 4\,A$, $R = ?$

resistance, $R = \frac{V}{I}$

$= \frac{24}{4}$

$= 6\,\Omega$

Now test yourself

TESTED

4 Calculate the current flowing through a 10 Ω resistor which has a voltage of 20 V across it.
5 A resistor has a voltage of 15 V across it when a current of 3 A flows through it. Calculate the resistance of the resistor.
6 A current of 2 A flows through a 25 Ω resistor. Find the voltage across the resistor.
7 A voltage of 15 V is needed to make a current of 2.5 A flow through a wire.
(a) What is the resistance of the wire?
(b) What voltage is needed to make a current of 2.0 A flow through the wire?
8 There is a voltage of 6.0 V across the ends of a wire of resistance 12 Ω.
(a) What is the current in the wire?
(b) What voltage is needed to make a current of 1.5 A flow through the wire?
9 A resistor has a voltage of 6 V applied across it and the current flowing through it is 100 mA. Calculate the resistance of the resistor.
10 A current of 600 mA flows through a metal wire when the voltage across its ends is 3 V. What current flows through the same wire when the voltage across its ends is 2.5 V?

H

Practical

Resistance of a filament lamp

Variables

The independent variable is the current flowing in the lamp.
The dependent variable is the voltage across the lamp.
The controlled variables are the length of the filament (inside the lamp) and its area of cross-section.

Apparatus

- low-voltage power supply unit (PSU)
- rheostat
- ammeter
- voltmeter
- connecting leads
- filament lamp in a suitable holder
- switch

0–12 V DC

Figure 37.10 Circuit diagram

Method

1 Prepare a table for your results, as below.
2 Ensure the PSU is switched off and set up the circuit as shown in Figure 37.10.
3 Adjust the PSU to supply zero volts and switch it on.
4 Record the voltage on the voltmeter and current on the ammeter.
5 Adjust the voltage so that the voltmeter reading increases by 1.0 V.
6 Repeat steps 4 and 5 until readings have been recorded for voltages up to 6 V. This is Trial 1.
7 Repeat the entire experiment to obtain a second set of current values. This is Trial 2.
8 Calculate the mean current from the two trials.
9 Plot the graph of voltage against mean current.

Table for results

Voltage/V	0.00	1.00	2.00	3.00	4.00	5.00	6.00
(Trial 1) current/A							
(Trial 2) current/A							
Mean current/A							

Treatment of the results

Plot the graph of voltage/V (*y*-axis) against mean current/A (*x*-axis) (Figure 37.11).

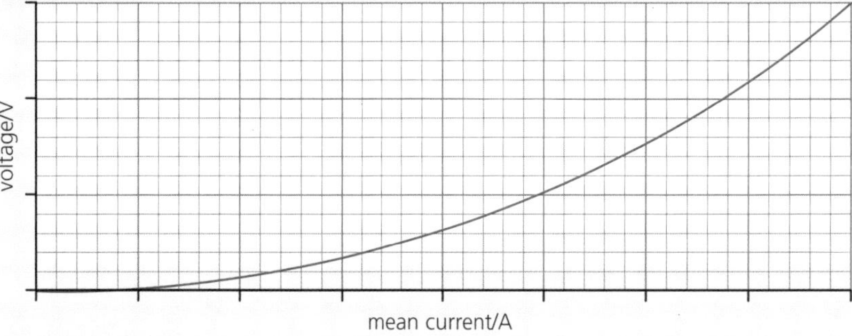

Figure 37.11

For a filament lamp, the graph of *V* against *I* is a curve with increasing gradient. This tells us that as the current in a lamp increases, the resistance of its filament also increases in a non-linear way.

Explanation

The resistance of a metal rises when the temperature increases. In this experiment the temperature of the filament is *allowed* to rise until it becomes white hot at its operating temperature.

Final task

You should now find the resistance of the filament at two points on your curve by calculating the ratio of the voltage to the current for the two different values of the current.

Resistance in series circuits

The total resistance of two or more resistors in series is simply the sum of the individual resistances of the resistors (Figure 37.12):

$$R_{total} = R_1 + R_2 + R_3$$

In Figure 37.12, the three resistors could be replaced by a single resistor of (4 + 8 + 6) = 18 Ω.

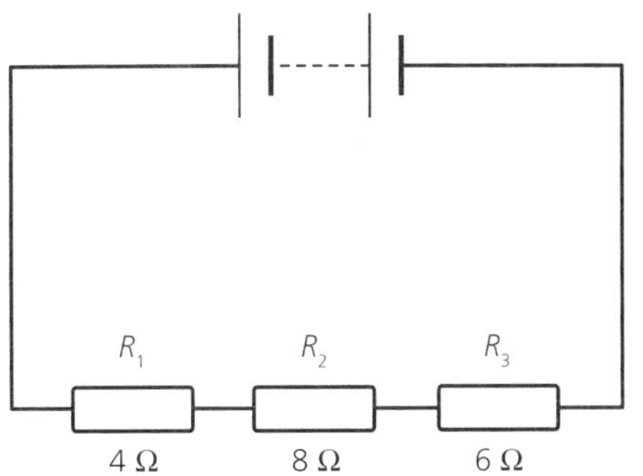

Figure 37.12 Calculating the total resistance of three resistors in a series circuit

Resistance in parallel circuits

The total resistance of two *equal* resistors in **parallel** is *half* of the resistance of one of them. The total resistance of three *equal* resistors in parallel is *one third* of the resistance of one of them, and so on.

a **parallel** circuit is one in which the current divides to travel independently along two or more separate loops

When considering two *unequal* resistors, R_1 and R_2, in parallel, we use the 'product over sum' formula:

$$R_{total} = \frac{R_1 \times R_2}{R_1 + R_2} = \frac{\text{product}}{\text{sum}}$$

H The formula for calculating the total resistance of three resistors in parallel is:

$$\frac{1}{R_{total}} = \frac{1}{R_1} + \frac{1}{R_2} + \frac{1}{R_3}$$

This last formula can be extended to any number of resistors in parallel.

Examples

1 Find the combined resistance of two 8 Ω resistors:
(a) in series
(b) in parallel.

Answer

(a) For resistors in series: $R_{total} = R_1 + R_2 = 8 + 8 = 16\,\Omega$
(b) The total resistance of two *equal* resistors in parallel is *half* of the resistance of one of them.
So in this case the total resistance is $\frac{8}{2} = 4\,\Omega$.

2 A 6 Ω resistor and a 3 Ω resistor are connected
(a) in series and
(b) in parallel.
In each case find the resistance of the combination.

Answer

(a) For resistors in series: $R_{total} = R_1 + R_2 = 6 + 3 = 9\,\Omega$
(b) For two *unequal* resistors in parallel, we use the product over sum rule. So,

$$R_{total} = \frac{\text{product}}{\text{sum}} = \frac{(R_1 \times R_2)}{(R_1 + R_2)} = \frac{(6 \times 3)}{(6 + 3)} = \frac{18}{9} = 2\,\Omega$$

→

3 A 24 Ω resistor, a 12 Ω resistor and an 8 Ω resistor are connected
(a) in series and
(b) in parallel.
In each case find the resistance of the combination.

Answer

(a) For resistors in series: $R_{total} = R_1 + R_2 + R_3 = 24 + 12 + 8 = 44\,\Omega$

(b) For three *unequal* resistors in parallel, we use:

$$\frac{1}{R_{total}} = \frac{1}{R_1} + \frac{1}{R_2} + \frac{1}{R_3}$$
$$= \frac{1}{24} + \frac{1}{12} + \frac{1}{8} = \frac{1}{24} + \frac{2}{24} + \frac{3}{24}$$
$$= \frac{6}{24}$$
$$= \frac{1}{4}$$

(Important: do not forget to turn the fraction upside down in the last step!)

So, $R_{total} = \frac{4}{1} = 4\,\Omega$

The final example illustrates how to cope with a combination of resistors in parallel with resistors in series.

4 Find the total resistance of the combination shown in Figure 37.13.

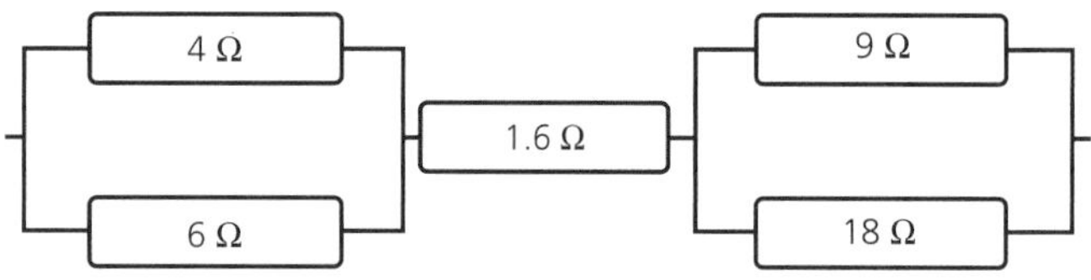

Figure 37.13

Answer

Using the product over sum formula, we see that the 4 Ω and 6 Ω parallel combination gives a total resistance of 2.4 Ω. Similarly, the 9 Ω and 18 Ω parallel combination gives a total resistance of 6.0 Ω.

There is therefore a series combination of 2.4 Ω + 1.6 Ω + 6.0 Ω, which gives a total resistance of 10 Ω.

Voltage in series circuits

REVISED

When resistors are connected in series:

- the current in each resistor is the same
- the sum of the voltages across each resistor is equal to the battery voltage.

Example

Resistances of 2 Ω, 4 Ω and 6 Ω are connected in series across a 3 V battery.

(a) Calculate:
  (i) the total resistance of the circuit
  (ii) the current in each resistor
  (iii) the voltage across each resistor.

(b) Comment on the answers to part (a)(iii).

Answer

First, draw the circuit diagram (Figure 37.14).

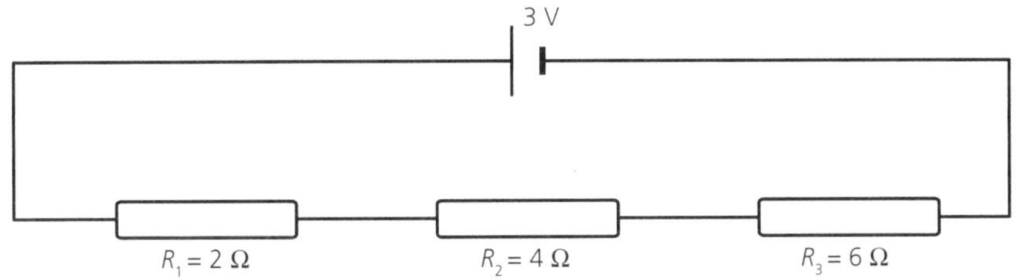

Figure 37.14

(a) (i) $R_{total} = R_1 + R_2 + R_3$

$= 2 + 4 + 6$

$= 12\,\Omega$

This means that the circuit is equivalent to one with $12\,\Omega$ placed across a 3 V battery.

(ii) $I = \frac{V}{R} = \frac{3}{12} = 0.25\,A$

(iii) $2\,\Omega$: $V = I \times R = 0.25 \times 2 = 0.5\,V$

$4\,\Omega$: $V = I \times R = 0.25 \times 4 = 1.0\,V$

$6\,\Omega$: $V = I \times R = 0.25 \times 6 = 1.5\,V$

(b) The sum of the voltages (0.5 V + 1.0 V + 1.5 V) is 3 V, which is exactly the same as the voltage of the battery.

The voltages are in exactly the *same proportion* as the resistances. So, the $4\,\Omega$ resistor has twice the voltage as the $2\,\Omega$ resistor and the $6\,\Omega$ resistor has three times the voltage as the $2\,\Omega$ resistor.

Voltage in parallel circuits

REVISED

When resistors are connected in parallel:

- the voltage across each resistor is the same as the voltage provided by the battery
- the sum of the currents in each resistor is equal to the current coming from the battery.

Examples

1 Resistances of $2\,\Omega$ and $3\,\Omega$ are connected in parallel across a 6 V battery.

(a) State the voltage across reach resistor.

(b) Calculate:

(i) the total resistance of the circuit

(iii) the current in each resistor

(iii) the total current taken from the battery.

Answer

(a) 6 V

(b) (i) R_{total} = product over sum

$$= \frac{(R_1 \times R_2)}{(R_1 + R_2)} = \frac{(2 \times 3)}{(2 + 3)} = \frac{6}{5} = 1.2\,\Omega$$

(ii) $2\,\Omega$: $I = \frac{V}{R} = \frac{6}{2} = 3\,A$

$3\,\Omega$: $I = \frac{V}{R} = \frac{6}{3} = 2\,A$

→

(iii) We can do this in two different ways:

$$I_{battery} = \frac{V_{battery}}{R_{total}} = \frac{6}{1.2} = 5\,A$$

or $I_{battery}$ = sum of currents in resistors
$= 3 + 2$
$= 5\,A$

H 2 Resistances of 4 Ω, 6 Ω and 12 Ω are connected in parallel across a battery. A current of 2.0 A flows from the battery towards the parallel network.

(a) Calculate:

(i) the total resistance of the network

(ii) the battery voltage and the voltage across each resistor

(iii) the current in each resistor.

(b) Comment on the answers to parts (a)(ii) and (iii).

Answer

First, draw the circuit diagram (Figure 37.15).

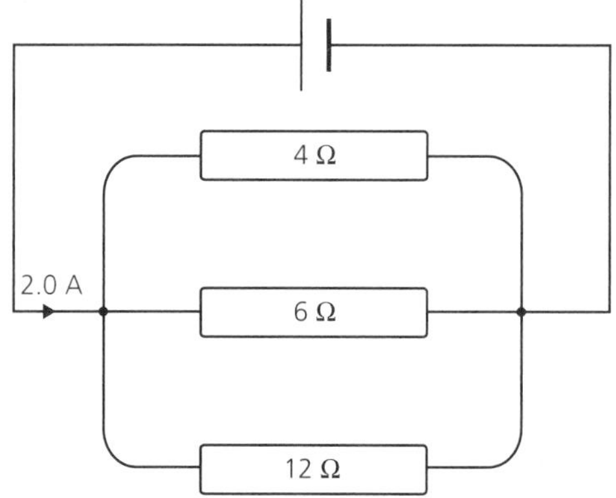

Figure 37.15

(a) (i)
$$\frac{1}{R_{total}} = \frac{1}{R_1} + \frac{1}{R_2} + \frac{1}{R_3}$$
$$= \frac{1}{12} + \frac{1}{6} + \frac{1}{4} = \frac{6}{12} = \frac{1}{2}$$

So, $R_{total} = \frac{2}{1} = 2\,\Omega$

(ii) $V_{battery} = I_{battery} \times R_{total}$
$= 2.0 \times 2 = 4\,V$

voltage across each resistor = 4 V

(iii) 4 Ω: $I = \frac{V}{R} = \frac{4}{4} = 1.0\,A$

6 Ω: $I = \frac{V}{R} = \frac{4}{6} = 0.67\,A$

12 Ω: $I = \frac{V}{R} = \frac{4}{12} = 0.33\,A$

(b) In (a)(ii) the voltages across each resistor are the same as the battery voltage.
In (a)(iii) the sum of the currents in the parallel network (1.0 A + 0.67 A + 0.33 A) is 2.0 A, which is exactly the same as the current from the battery.
The currents in each resistor are in **inverse proportion** to the resistance. So the current in the 6 Ω resistor is twice the current in the 12 Ω resistor, and the current in the 4 Ω resistor is 3 times the current in the 12 Ω resistor.

Typical mistake

The most common error when using the 'one-over-R' formula for resistors in parallel is forgetting to flip it at the end of the calculation. So remember — don't forget to flip!

inverse proportion is the mathematical relationship between quantities in which one quantity doubles when the other halves

Hybrid circuits

REVISED

Hybrid circuits contain both series and parallel elements. We treat each part separately, eventually finding the total resistance of the entire network as shown in the example.

Example

A parallel combination of 2Ω and 3Ω is joined in series with a 5Ω resistor (Figure 37.16). The network is connected across a 31V battery. Calculate the current taken from the battery, the voltage across each resistor and the current in each resistor.

(The strange battery voltage has been chosen to make the arithmetic easy.)

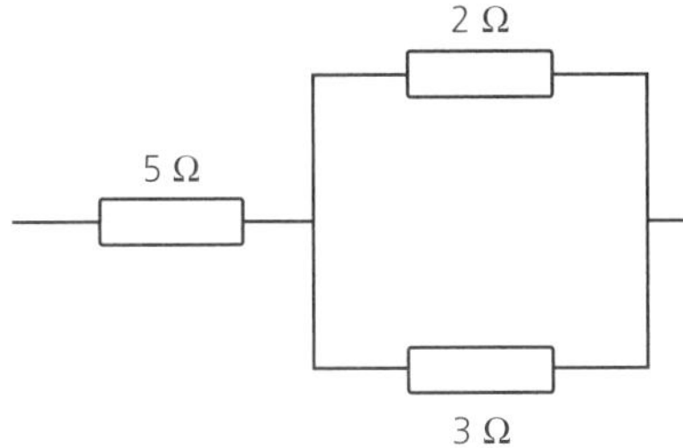

Figure 37.16

Answer

From Example 1 on p. 319 we see that the combined resistance of the two resistors in parallel is 1.2Ω.

So the total resistance of the circuit, $R_{total} = 5 + 1.2 = 6.2\,\Omega$

The current taken from the battery $I_{battery} = \frac{V_{battery}}{R_{total}}$

$$= \frac{31}{6.2} = 5\,\text{A}$$

The current in the 5Ω resistor is therefore 5A.

The voltage across the 5Ω resistor is $V = I \times R$

$$= 5 \times 5 = 25\,\text{V}$$

Since the battery voltage is 31V, the voltage across the parallel network is 31 − 25 = 6V

So the voltage across the 2Ω resistor = voltage across the 3Ω resistor = 6V

Current in 2Ω resistor $= \frac{V}{R} = \frac{6}{2} = 3\,\text{A}$

and current in 3Ω resistor $= \frac{V}{R} = \frac{6}{3} = 2\,\text{A}$

Exam tip

When doing maths questions in physics remember the examiner wants to see:

- your equation or formula
- your substitutions
- your arithmetic
- your answer
- your unit.

Now test yourself

TESTED

11 (a) Calculate the value of the current from the cell in each of the circuits in Figure 37.17.

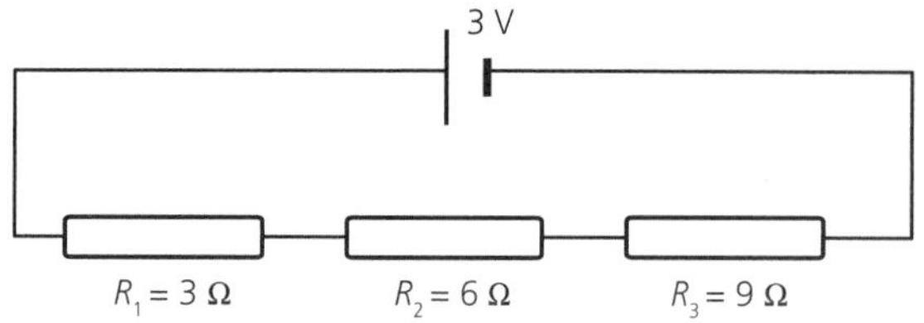

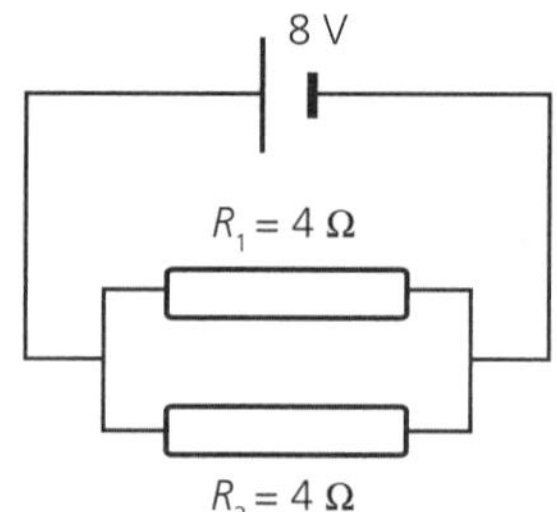

Figure 37.17

(b) State the voltage across each 4 Ω resistor in the second circuit.

12 Two identical resistors are placed in parallel across a 12 V battery. If the total current drawn from the battery is 3 A, what is the resistance of each resistor?

Common practical

Resistance and length

Variables

The independent variable is the length of the wire.

The dependent variable is the resistance of the wire.

The controlled variables are the temperature and the area of cross-section of the wire.

Apparatus

- low-voltage power supply unit (PSU)
- rheostat
- ammeter
- voltmeter
- connecting leads
- resistance wire
- switch
- metre ruler

Method

1. Prepare a table for your results (see page 323).
2. Measure and cut off 1 metre of nichrome resistance wire.
3. Attach it with sticky tape to a metre ruler and set up the circuit as shown in Figure 37.18.
4. Ensure that the PSU is switched off.
5. Adjust the PSU to supply about 1 volt.
6. Connect the 'flying lead' so that the length of wire across the voltmeter is 10 cm.
7. Switch on the PSU and record the voltage on the voltmeter and the current on the ammeter.

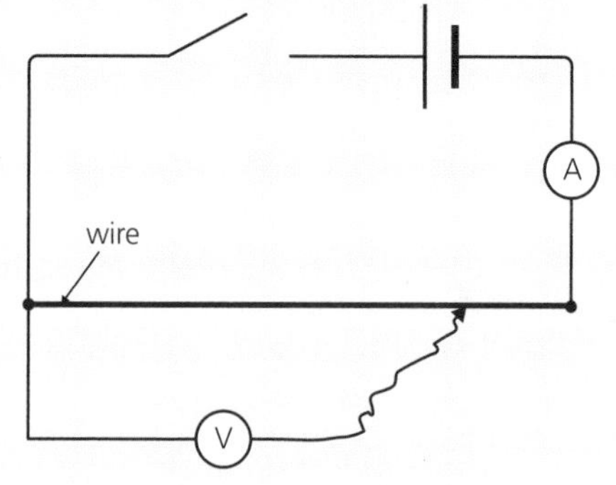

Figure 37.18 Circuit diagram

→

8 Switch off the PSU immediately and allow the wire to cool to room temperature*.
9 Increase the length across the voltmeter by 10 cm.
10 Repeat steps 7 to 9 until readings have been recorded for lengths from 10 cm to 90 cm.
11 Calculate the resistance of each length of wire, using $R = \frac{V}{I}$.
12 Plot the graph of resistance (y-axis) against length (x-axis).

*It is necessary to ensure the wire's temperature remains constant (close to room temperature). We do this by:

- keeping the voltage low (so that the current remains small) and
- switching off the current between readings to allow the wire to cool.

Table for results

Length of wire/cm	10	20	30	40	50	60	70	80	90
Voltage across wire/V									
Current in wire/A									
Resistance of wire/Ω									

Treatment of the results

Plot the graph of resistance/Ω (y-axis) against length/cm (x-axis).

Evaluation of the results

The graph of resistance/Ω against length/cm is a straight line through the origin (Figure 37.19). This tells us that the resistance of a metal wire is **directly proportional** to its length, provided its area of cross-section and the temperature remain constant.

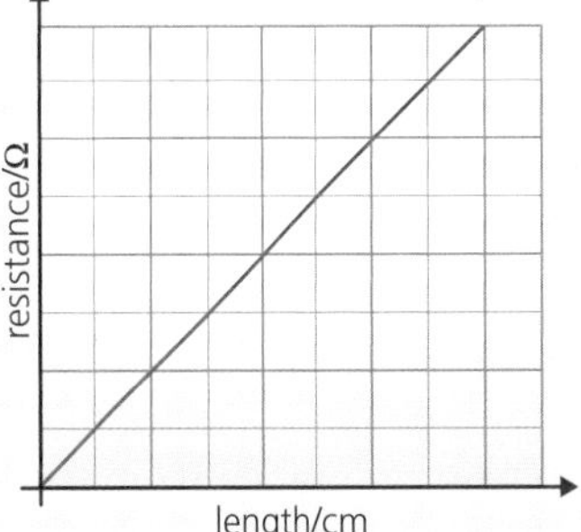

Figure 37.19 **Graph of resistance against cross-sectional area**

The relationship between R and L is:

$R = kL$

where k is the gradient of the graph.

Since $k = \frac{R}{L}$, the unit for k is Ω/cm (or Ω/m).

Note that the value of k depends on the material of the wire and its cross-sectional area.

Example

A reel of constantan wire of length 250 cm has a total resistance of 15.0 Ω. Calculate:

(a) the resistance of 1.0 m of wire
(b) the length of wire needed to have a resistance of 3 Ω
(c) the resistance of a 90 cm length of the wire.

Answer

(a) $k = \frac{R}{L}$

$= \frac{15\,\Omega}{250\,\text{cm}} = 0.06\,\Omega/\text{cm} = 6\,\Omega/\text{m}$

so the resistance of 1.0 m of wire is 6 Ω

(b) $L = \frac{R}{k}$

$= \frac{3\,\Omega}{0.06\,\Omega/\text{cm}} = 50\,\text{cm}$

(c) $R = k \times L$

$= 6\,\Omega/\text{m} \times 0.9\,\text{m} = 5.4\,\Omega$

Now test yourself

13 A school buys a reel of constantan wire. The supplier's data sheet says that the wire has a resistance of 2.5 Ω/m. Calculate:
(a) the length of wire a technician must cut from the reel to give a resistance of 2 Ω
(b) the resistance of a 120 cm length of wire cut from the reel.

14 A technician cuts an 80 cm length of wire from a reel marked 3.0 Ω/m.
The technician joins the two free ends of the wire together to form a loop. She then attaches two crocodile clips to the wire at opposite ends of a diameter.
(a) Explain why the total resistance between the crocodile clips is 0.6 Ω.
(b) In what way, if at all, does the total resistance between the crocodile clips change if one of the clips is moved along the wire towards the other. Explain your reasoning.

Exam practice

1 (a) What flows in the direction indicated by the arrow in Figure 37.20? [1]

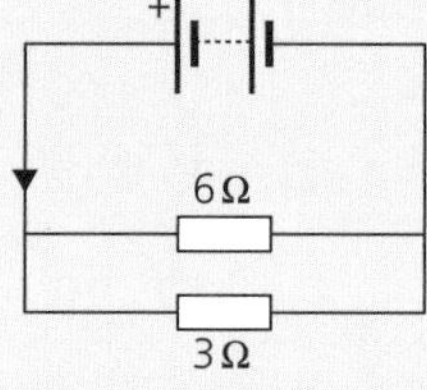

Figure 37.20

(b) Copy Figure 37.20 and mark on it an arrow to show the direction in which charged particles flow through the two resistors. [1]
(c) What name is given to these charged particles? [1]
(d) The current taken from the battery is 0.6 A. State the size of the current in the smaller resistor. [1]
(e) Show that the electrical charge delivered by the battery in 1 minute is 36 C. [3]

2 Four identical resistors are arranged as shown in Figure 37.21.

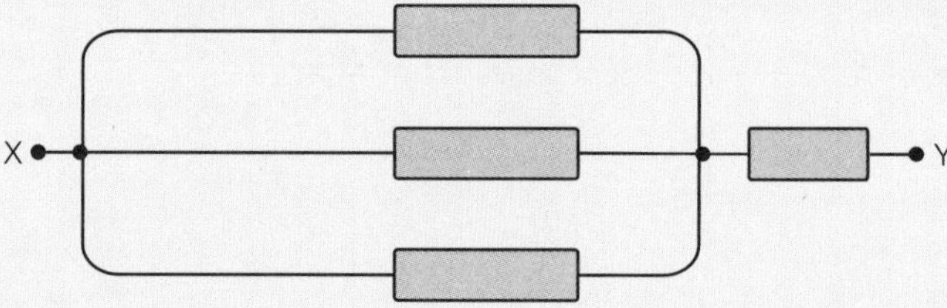

Figure 37.21

The current entering at X is 3 mA and the voltage across the isolated series resistor is 9 mV. Calculate:
(a) the total resistance between X and Y [3]
(b) the resistance of each resistor [2]
(c) the current in each resistor [2]
(d) the voltage across each resistor [2]

3 In Figure 37.22 resistors R_1 and R_2 have resistances of 80 Ω and 40 Ω respectively.

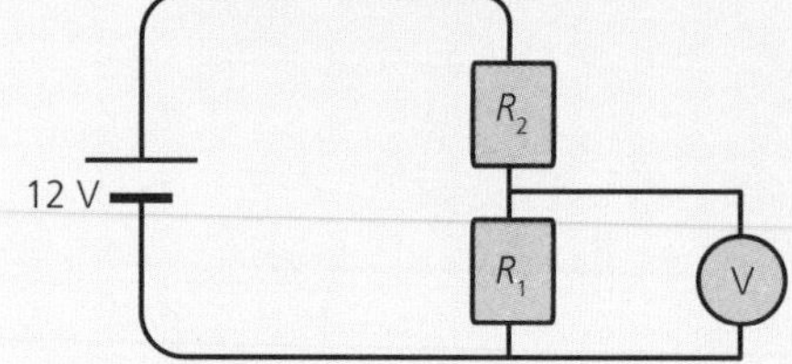

Figure 37.22

(a) Calculate the voltage you would expect to observe on the voltmeter. [3]
(b) What assumption have you made about the resistance of the voltmeter itself? [1]

→

H 4 Copy and complete the table below to show the effective resistance between X and Y for different switch settings in Figure 37.23. All four resistors have a resistance of 6 Ω.

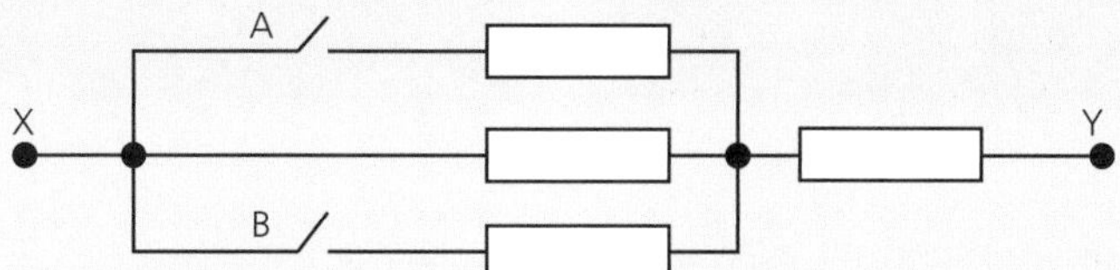

Figure 37.23

Switch		Effective resistance between X and Y/Ω
A	B	
Open	Open	
Open	Closed	
Closed	Open	

Answers online

ONLINE

38 Electricity in the home

Electric current generates heat when it passes through a metal wire. Why does this happen?

As electrons pass through the conductor they collide with the atoms. In these collisions the light electrons lose energy and the heavy atoms gain energy. This causes the atoms to vibrate faster. Faster vibrations mean a higher temperature. This is called **joule heating**. Hairdryers and toasters (Figure 38.1) both use the joule heating effect of an electric current.

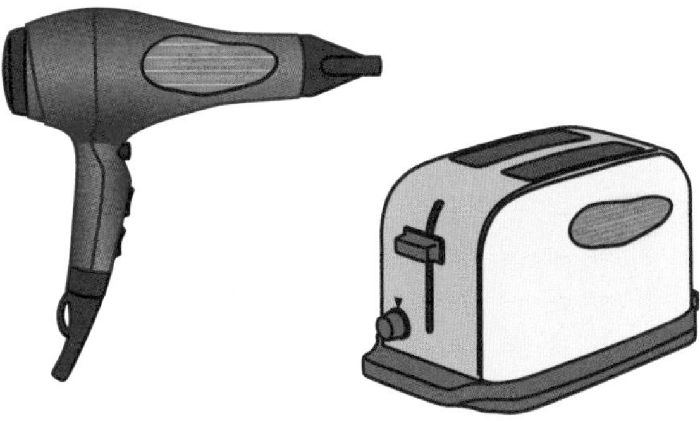

Figure 38.1 Toasters and hairdryers use an electrical current to produce heat

Electrical energy

REVISED

If 1 coulomb of charge gains or loses 1 joule of energy between two points, there is a voltage of 1 volt between those two points.

$$\text{voltage} = \frac{\text{energy transferred}}{\text{charge}}$$

Rearranging this formula gives us:

$$\text{energy transferred} = \text{voltage} \times \text{charge}$$

or

$$\text{energy transferred} = \text{voltage} \times \text{current} \times \text{time}$$

$$E = V \times I \times t$$

Electrical power

REVISED

Earlier you learned that the formula for mechanical power of a machine is defined as the rate at which energy is transferred and is given by:

$$\text{power} = \frac{\text{energy transferred}}{\text{time}}$$

We have seen that in an electrical circuit:

$$\text{energy transferred} = \text{voltage} \times \text{current} \times \text{time}$$

Substituting for energy transferred:

$$\text{power} = \frac{\text{voltage} \times \text{current} \times \text{time}}{\text{time}}$$

or

electrical power = voltage × current

This is often expressed by the equation $P = I \times V$ and is commonly called **Joule's law** of heating.

electrical power is the rate at which electrical energy is used in a circuit

Joule's law states that the power generated in an electrical component is the product of the current in the component and the voltage across it ($P = IV$)

Example

A study lamp is rated at 60W, 240V. How much current flows in the bulb?

Answer

power = $V \times I$

$60 = 240 \times I$

$I = \frac{60}{240} = 0.25\text{ A}$

Now test yourself

TESTED

1 Calculate the power of a heater if it uses 3600000J in 1 hour.
2 A toaster has a power rating of 1.5kW.
   (a) State the power of the toaster in watts.
   (b) How much energy does the toaster use in 10s?
3 A filament bulb has an internal resistance of 960Ω when it is at normal brightness.
   (a) Calculate the current it draws when connected to a 240V supply and is shining with normal brightness.
   (b) Calculate its power at normal brightness.

The three equations for power

REVISED

Because power $P = I \times V$ and Ohm's law is $V = I \times R$ we can write:

$P = IV$

$P = I^2R$

$P = \frac{V^2}{R}$

These formulae give the electrical power **dissipated** (converted) into heat in resistors and heating elements. The heat dissipated is sometimes referred to as 'ohmic losses'.

Example

What power is dissipated in a 10Ω resistor when the current through it is: (a) 2A (b) 4A?

Answer

(a) power = $I^2R = 2^2 \times 10 = 40\text{W}$

(b) power = $I^2R = 4^2 \times 10 = 160\text{W}$

The example shows that when the current is doubled, the power dissipated is quadrupled. This idea has important implications for electricity transmission in the next chapter.

Now test yourself

4 A washing machine uses 1200 W when connected to a 240 V power supply.
   (a) Calculate the current that it draws in normal use.
   (b) Would the power of the machine increase, decrease or remain the same if it was connected to a 120 V supply?

5 Copy and complete the table for domestic appliances, all of which operate at 240 V.

Name of appliance	Power rating	Current drawn	Resistance
Filament lamp		0.25 A	
Coffee machine	120 W		
Iron		6 A	
Electric oven			24 Ω
Immersion heater	3 kW		

Paying for electricity

REVISED

Electricity companies bill customers for electrical energy in units known as **kilowatt-hours** (kWh).

One **kilowatt-hour** is the amount of energy transferred when 1000 W is delivered for 1 hour.

The following two formulae are very useful in calculating the cost of using a particular appliance for a given amount of time:

number of units used = power rating (in kilowatts) × time (in hours)

total cost = number of units used × cost per unit

Example

If electricity costs 16 pence per kWh, find the number of units used by a 3000 W immersion heater when it is switched on for $1\frac{1}{2}$ hours. Calculate also the cost of using this immersion heater for that time.

Answer

number of units used = power rating (in kilowatts) × time (in hours)

= 3 kW × 1.5 h = 4.5 kWh

total cost = number of units used × cost per unit

= 4.5 kWh × 16 pence/kWh

= 72 pence = £0.72

Exam tip

When doing questions on the cost of using electricity ask yourself if your answer is reasonable. If you get the cost of taking a shower to be £15 (which is unreasonable), you have almost certainly made a simple mistake like giving the wrong unit — 15 pence is much more likely than £15.

Electricity bills

Figure 38.2 shows part of a typical electricity bill. The difference between the present reading and the previous reading is the number of units used. In this particular example, the number of units used = 57 139 − 55 652 = 1487 units (kWh).

Northern Electricity Board

Customer account no: 3427 364

Present meter reading	Previous meter reading	Units used	Cost per unit (incl. VAT)	£
57139	55652	1487	15.0p	£223.05

Figure 38.2 An electricity bill

If the cost of a unit is known, then the total cost of the electricity used can be determined. In the example in Figure 38.2, 1487 units at 15.0 pence per unit = 22 305 pence = £223.05.

One-way switches

REVISED

This kind of switch acts as a make-or-break device to switch a circuit on or off (Figure 38.3). When the switch is open there is air between the conducting contacts. Since air is an insulator the circuit is incomplete. No current flows.

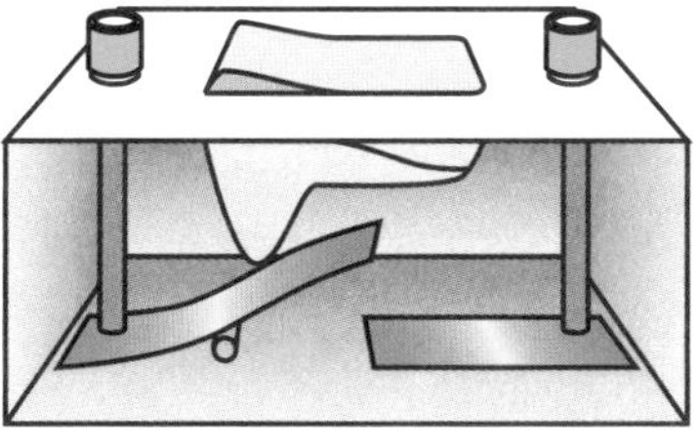

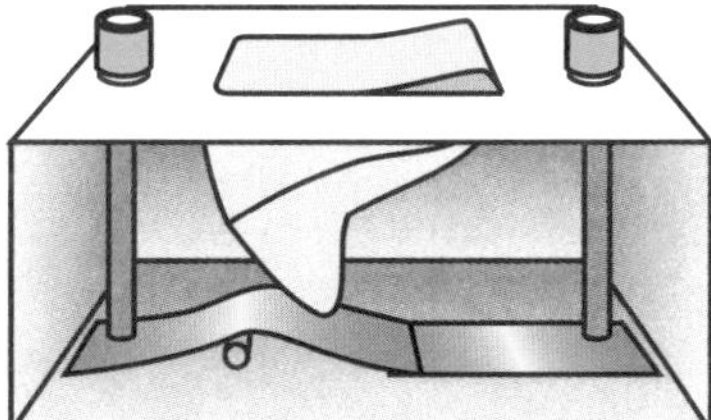

Figure 38.3 A one-way switch

The rocker itself is made of plastic. This is important with high voltages to prevent current flowing through the body of the user. When the rocker is pressed the conducting contacts are pushed together.

There is now a complete circuit, so current can flow.

As we shall see later, it is important that the switch is placed in the live side of a circuit.

Two-way switches

REVISED

In most two-storey houses, you can turn the landing lights on or off from upstairs or downstairs. Two-way switches are used for this.

Figure 38.4(a) illustrates a two-way switch in one of its two on-positions. Going to bed at night, when both switches are up (or both are down), the circuit is complete and a current flows through the bulb.

At the top of the stairs, one of the switches is pressed down and the circuit is broken (Figure 38.4b). You should verify for yourself what happens when the person comes downstairs. The effect is that each switch reverses the effect of the other one (Figure 38.5).

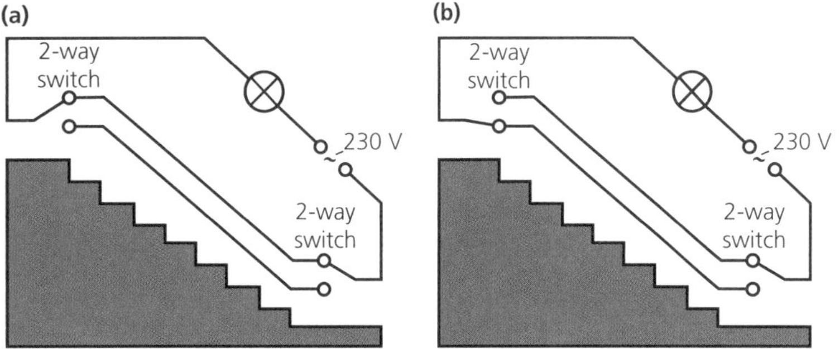

Figure 38.4 A two-way switch circuit

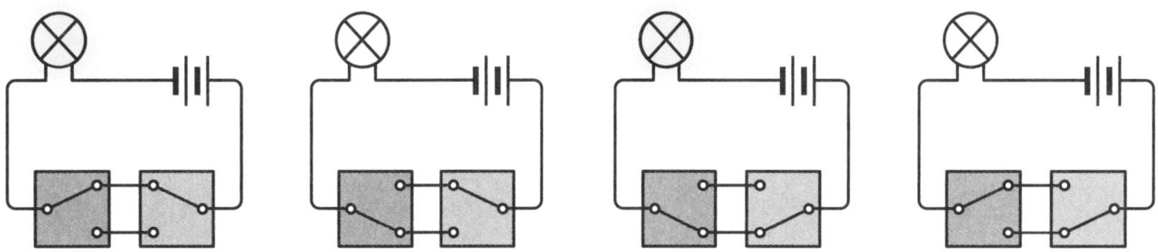

Figure 38.5 The four possible states of a two-way switch in a circuit

How to wire a three-pin plug

REVISED

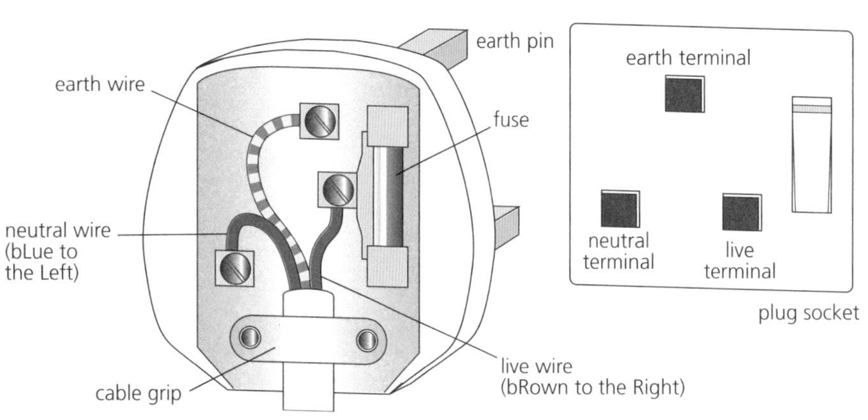

Figure 38.6 A correctly wired three-pin plug

- The wire with the **blue** insulation is the **neutral** wire — connect this to the left-hand pin (Figure 38.6).
- The **brown** insulated wire is the **live** wire — connect this to the right-hand pin.
- The wire with the **yellow and green** insulation is the **earth** wire — connect this to the top pin.

the **neutral** wire is the wire in a mains voltage system that is permanently at zero volts

the **live** wire is the dangerous wire in a mains voltage system in which the voltage is changing

the **earth** wire is the wire that connects the metal frame in an electrical appliance to the earth

Fuses

A **fuse** is a device that is meant to prevent damage **to an appliance**. The most common fuses are either a 3 A (red) fuse for appliances up to 720 W or 13 A (brown) fuse for appliances between 720 W and 3 kW.

If a larger-than-usual current flows, the fuse wire will melt and so break the circuit.

a **fuse** is a short wire, often used inside a plug, which melts to disconnect a circuit when too much current flows

Selecting a fuse

Every appliance has a power rating. How much current the appliance will use is found using the power formula:

power = voltage × current

For example, a jig-saw has a power rating of 350 W. So the current it draws when connected to the mains is given by:

$$\text{current} = \frac{\text{power}}{\text{voltage}} = \frac{350}{240} = 1.46\,\text{A}$$

This is the normal current the device uses — any larger current can destroy it.

A 3 A fuse would allow a normal working current to flow and protect the jig-saw from larger currents. A 13 A fuse would allow a dangerously high current to flow and still not 'blow'.

So, it is important to use the correct size of fuse.

Remember that a fuse **protects the appliance**. It does not protect the person using the appliance. It can take 1 to 2 seconds for a fuse to melt — enough time for the user to receive a fatal electric shock.

The earth wire

REVISED

An **earth wire** can prevent harm **to the user** (see Figure 38.7).

Suppose a fault develops in an electric fire and the element is in contact with the metal casing of the fire. The casing will be live and if someone were to touch it they would get a possibly fatal electric shock as the current rushes through their body to earth. The earth wire prevents this — it offers a **low resistance** route of escape, enabling the current to go to earth by a wire rather than through a human body.

Any appliance with a **metal casing** could become live if a fault develops, so such appliances should always have a fuse fitted in the plug.

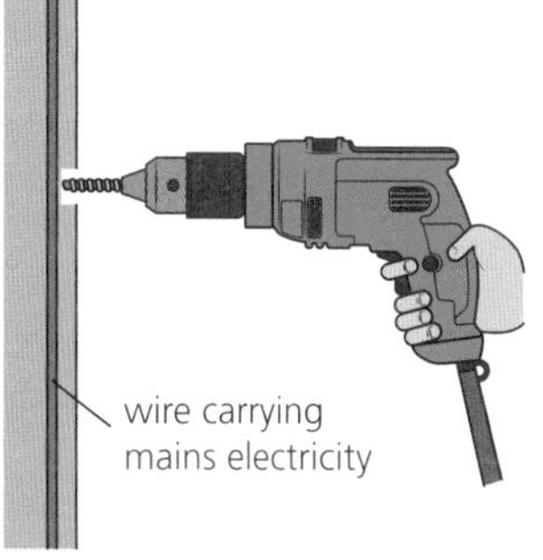

Figure 38.7 If there is no earth wire connected to the casing of the drill, the current will flow through the person

Double insulation

REVISED

Appliances such as vacuum cleaners and hairdryers are usually **double insulated**. The appliance is encased in an insulating plastic case and is connected to the supply by a two-core insulated cable containing only a live and a neutral wire. Any metal attachments that a user might touch are fitted into the plastic case so that they do not make a direct connection with the motor or other internal electrical parts. The symbol for double insulated appliances is shown in Figure 38.8.

Figure 38.8 The symbol for double insulation

double insulation is a safety technique in which an electrical appliance is encased in an insulating plastic case to prevent a user coming into contact with a live component and being harmed

The live wire

REVISED

Earlier we said that the switch and fuse must be placed on the live side of the appliance (Figure 38.9). Why is this important? The live wire in a mains supply is at high voltage (effectively around 230 V). The neutral side is at approximately zero volts.

If a fault occurs and the fuse blows, the live, dangerous wire is disconnected. If the fuse were on the neutral side, the appliance would still be live, even when the fuse had blown.

Switches are also placed on the live side for the same reason. If the switch were on the neutral side, the appliance would still be live, even when the switch was in the OFF position.

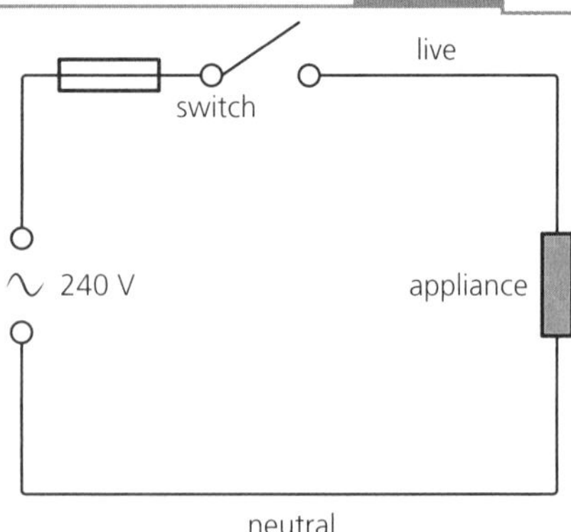

Figure 38.9 This fuse has been correctly placed on the live side of the power supply

Exam practice

1 (a) What is the purpose of the fuse in a three-pin plug? [1]
(b) State the colours of the live, neutral and earth wires in a three-pin plug. [3]
(c) What is the purpose of the earth wire? [1]
(d) Some appliances have only two wires in their plugs. Explain why. [1]

2 A user of electricity receives a bill showing that he has used 816 units from 1 May until 1 August. Electricity costs 15 pence per unit.
(a) What is the scientific name for the commercial unit of electrical energy? [1]
(b) Calculate the cost of the electricity used. [2]
(c) In the following quarter the cost rises from 15 pence to 16 pence per unit. What is the maximum number of units the customer can use if his bill is to be no greater than your answer to part (b)? [2]

3 (a) Some appliances are double insulated. What does this mean? [1]
(b) Explain how double insulation can protect a user from electric shock. [2]
(c) A schoolboy incorrectly wires a three-pin plug by connecting the neutral wire to the fuse and the live wire to the neutral terminal. His teacher explains that although the appliance connected to the plug would still work, his mistake would make it very dangerous. Explain why the appliance would be dangerous. [1]

4 A music centre has a maximum power rating of 125 W and is connected to a 240 V supply.
(a) What fuse should be fitted inside its plug? Choose from: 1 A, 3 A, 5 A and 13 A. [2]
(b) How much would it cost to run the music centre continuously for 8 hours if electricity costs 16 pence per unit? [3]

Answers online

ONLINE

39 Magnetism and electromagnetism

Magnetic field pattern around a bar magnet

REVISED

Magnetic fields can be investigated using a small **plotting compass**. The 'needle' is a tiny magnet that is free to turn on its spindle.

Figure 39.1 shows how a plotting compass can be used to plot the field around a bar magnet.

- The **field lines** run from the north **pole** (N) to the south pole (S) of the magnet (Figure 39.2).
- The field direction, shown by the arrowhead, is defined as the direction in which the force on a N pole would act. This means that magnetic field lines must never touch or cross over each other.
- The magnetic field is strongest where the field lines are closest together, i.e. at the poles of the magnet. A point where there is no magnetic field is called a neutral point (Figure 39.3).

magnetic field is a region of space within which a magnet experiences a force

plotting compass is a small compass used to find the shape of a magnetic field

magnetic field lines are lines drawn to represent a magnetic field

poles (of a magnet) are the ends of a magnet (north and south) where the magnetic field is strongest

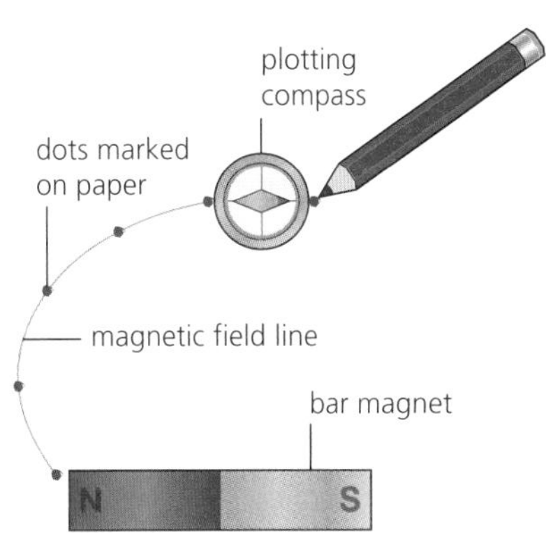

Figure 39.1 Plotting the field around a bar magnet using a compass

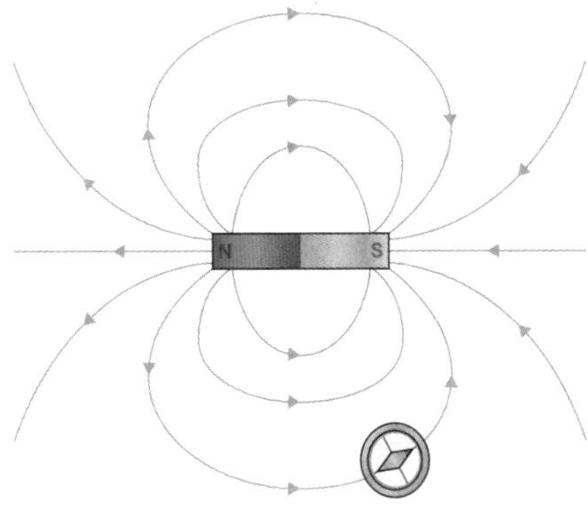

Figure 39.2 Magnetic field lines around a bar magnet

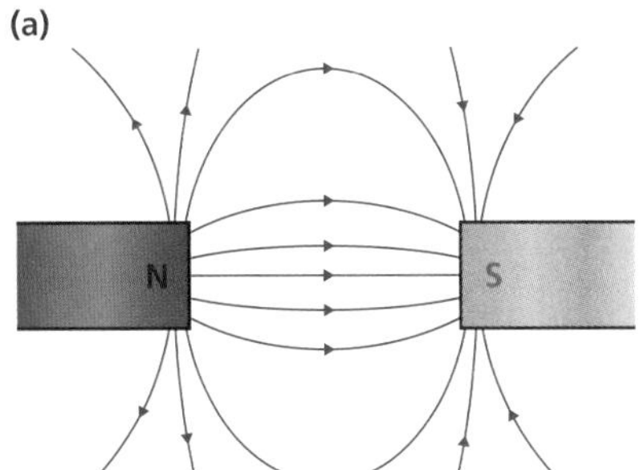

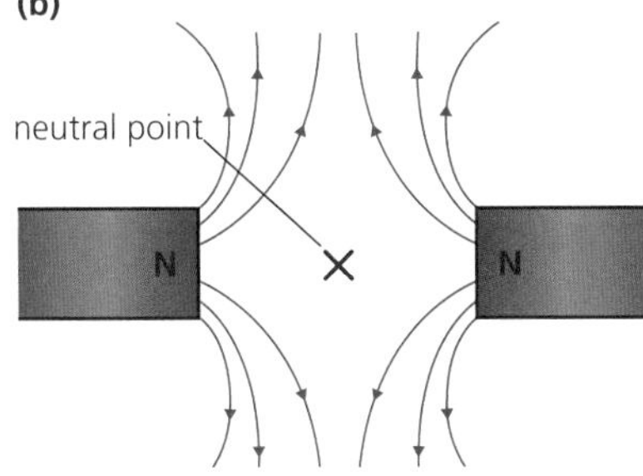

Figure 39.3 Field lines occurring when (a) a N and S pole are brought close together and (b) two N poles are brought close together

Magnetic field pattern due to a current-carrying coil

REVISED

A coil with one turn

Figure 39.4 shows the magnetic field around a **single** loop of wire that is carrying a current.

The strength of the magnetic field can be increased by:

- increasing the number of turns of wire in the coil
- increasing the current through the coil.

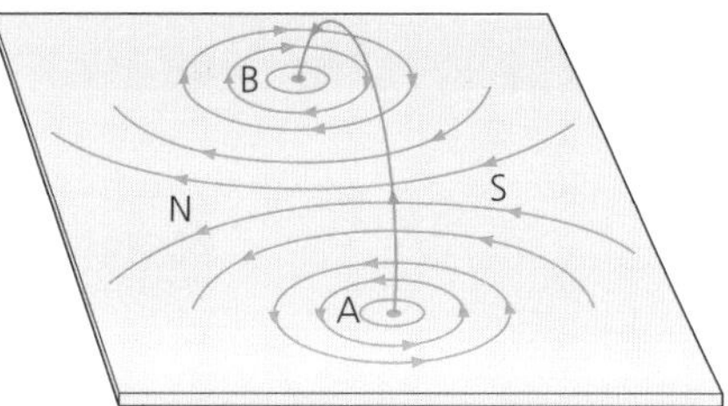

Figure 39.4 The magnetic field around a single loop of current-carrying wire

solenoid is a coil of wire carrying an electrical current

A coil with many turns

A stronger magnetic field can be made by wrapping a wire in the form of a long coil, which is referred to as a **solenoid**. A current must be passed through it, as illustrated in Figure 39.5.

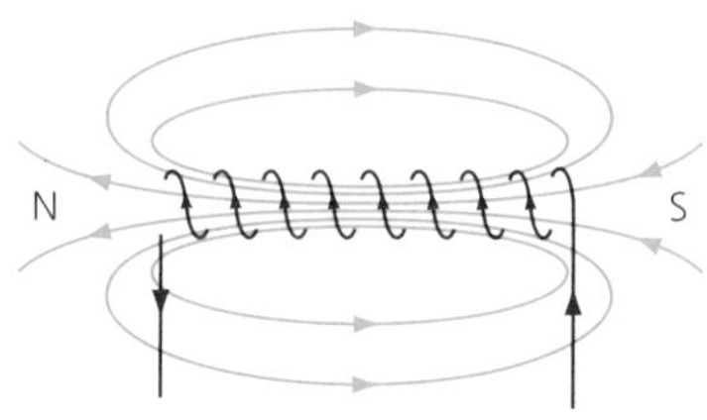

Figure 39.5 The pattern of magnetic field around a solenoid

The magnetic field produced by a solenoid has the following features:

- The field is similar to that from a bar magnet, with poles at the ends of the coils.
- Increasing the current increases the strength of the field.
- Increasing the number of turns on the coil increases the strength of the field.

Polarity

To work out which way round the poles are you can use the **right-hand grip rule**, as shown in Figure 39.6. Imagine gripping the solenoid with your right hand so that your fingers point in the conventional current direction. Your thumb points towards the north pole of the solenoid.

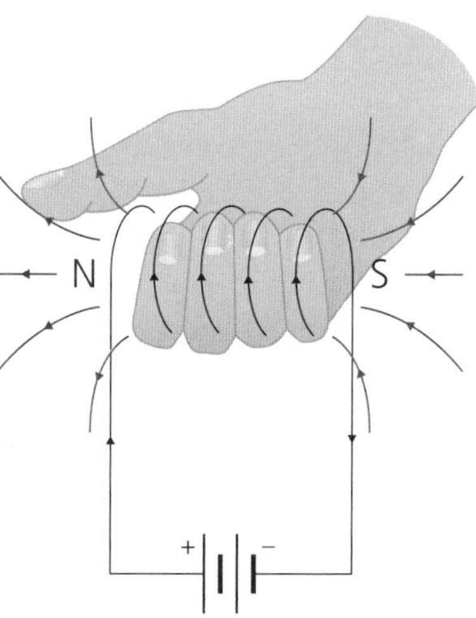

Figure 39.6 The right-hand grip rule

Now test yourself

TESTED

1 Explain why magnetic field lines can never cross or touch.
2 Explain the meaning of the term *neutral point*.

Practical

Factors affecting the strength of an electromagnet

The strength of an electromagnet, i.e. the strength of its magnetic field, can be measured by finding the mass of iron it will attract. Iron nails or paper clips may be used. Three factors are involved in this activity.

1 Investigating the effect of the current on the strength of the magnetic field

Apparatus

- thick insulated coil of copper wire
- soft iron core
- ammeter
- iron nails
- variable power supply

→

Method

1 Construct an electromagnet using, for example, 50 turns of insulated wire around a soft iron core.
2 Connect it to the circuit as shown in Figure 39.7.
3 Using the rheostat, increase the current in steps, measuring the number of iron nails attracted to the electromagnet for each current.
4 Record your results in a suitable table.
5 Plot a graph of number of nails on the y-axis versus current on the x-axis.

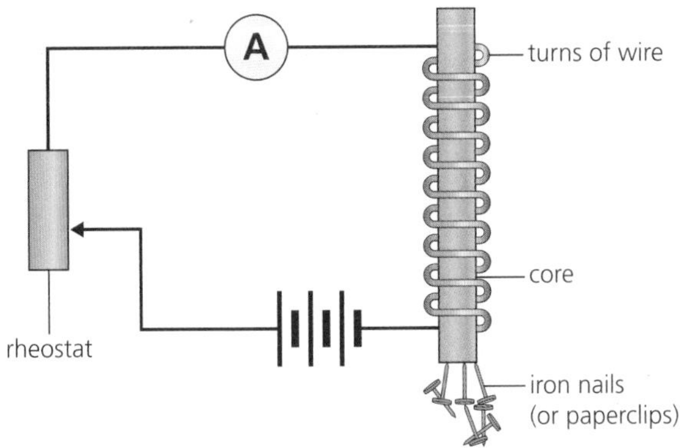

Figure 39.7 Investigating the strength of an electromagnet

Table for results

Current/A	0.0	0.5	1.0	1.5	2.0	2.5	3.0	3.5
Number of nails lifted								

A graph similar to the one shown in Figure 39.8 should be obtained. It shows that the number of nails lifted (and so the strength of the magnetic field) increases as the current increases. They are not directly proportional as the graph is not a straight line through the origin.

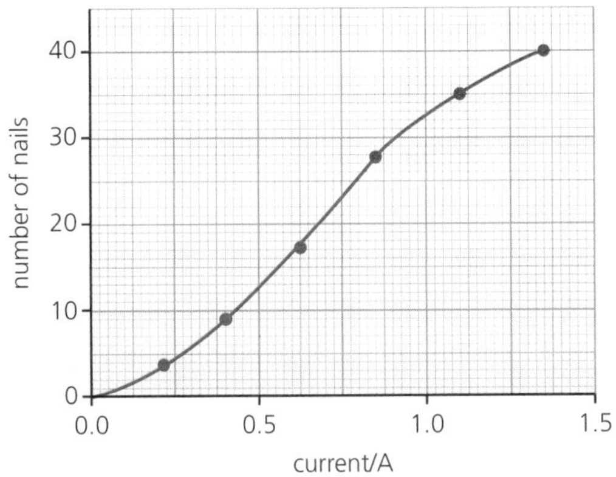

Figure 39.8 A typical graph of results

2 Investigating the effect of the number of turns on the strength of the magnetic field

Apparatus

- thick insulated coil of copper wire
- soft iron core
- ammeter
- iron nails
- variable power supply

Method

1 Keeping the current at 2.5 A and the material of the core constant, increase the number of turns of wire in steps of 10.
2 For each number of turns measure the number of nails lifted.

→

3 Record your results in a suitable table.
4 Plot a graph of number of turns versus number of nails lifted.

Table for results

Number of turns	10	20	30	40	50	60
Number of nails lifted						

We would expect the number of nails lifted to increase as the number of turns increases.

3 Investigating the effect of the material of the core on the strength of the magnetic field

Apparatus

- thick insulated coil of copper wire
- cores made from various materials
- iron nails
- power supply

Method

1 This time, keep the number of turns of wire and the current constant at 3.0 A but change the material of the core from soft iron to steel to plastic, wood or no material.
2 In each case measure the number of iron nails lifted.
3 Record your results in a table.
4 Draw a **bar chart** of your results and describe your findings.

Table for results

Type of material	Number of nails lifted
Soft iron	
Steel	
Copper	
Plastic	
Wood	
No material (air)	

Alternating and direct currents

REVISED

A direct current (d.c.) always flows in the same direction, from a fixed positive terminal to the fixed negative terminal of a supply.

A typical d.c. circuit is shown in Figure 39.9. A cell or battery gives a constant (steady) direct current. An oscilloscope trace of voltage versus time for a d.c. supply is shown in Figure 39.10.

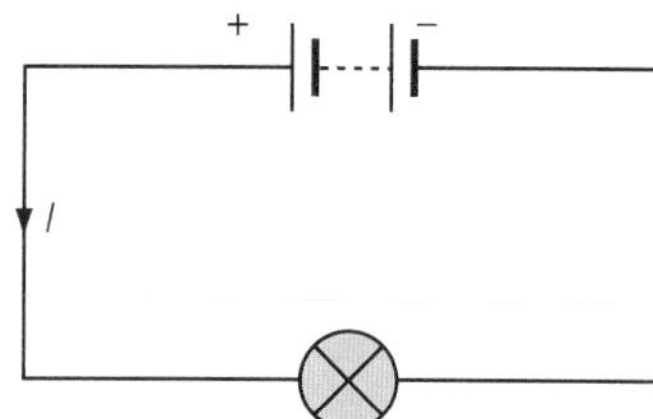

Figure 39.9 A simple d.c. circuit

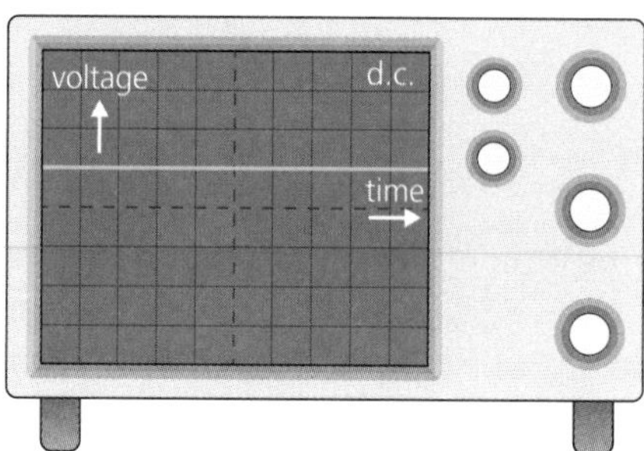

Figure 39.10 A trace of voltage against time for a d.c. supply

The electricity supply in your home is an alternating current (a.c.) supply. In an a.c. supply, the voltage (and hence the current) changes size and direction in a regular and repetitive way (Figure 39.11).

It is clear from Figure 39.12 why an a.c. supply is said to be bidirectional.

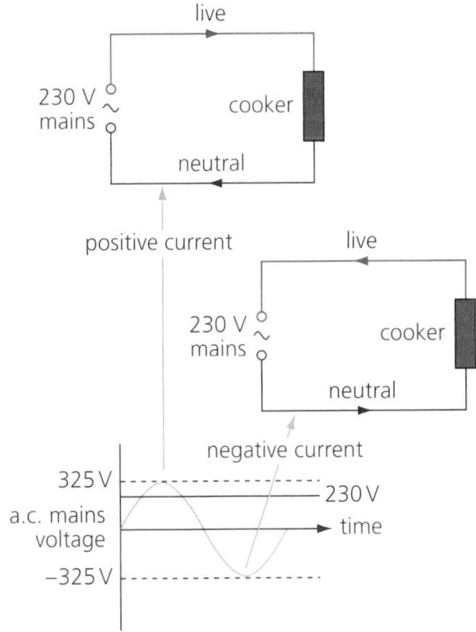

Figure 39.11 An a.c. supply

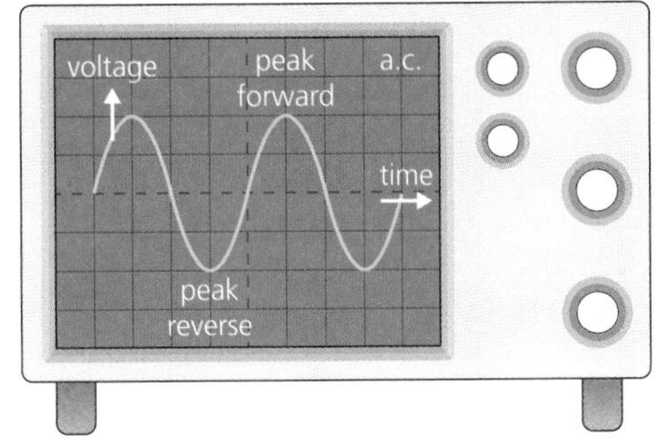

Figure 39.12 A trace of voltage against time for an a.c. supply

Now test yourself

TESTED

3 Figure 39.13 shows five displays on a cathode ray oscilloscope (CRO) screen. Which of the displays are:

(a) d.c. voltages

(b) a.c. voltages?

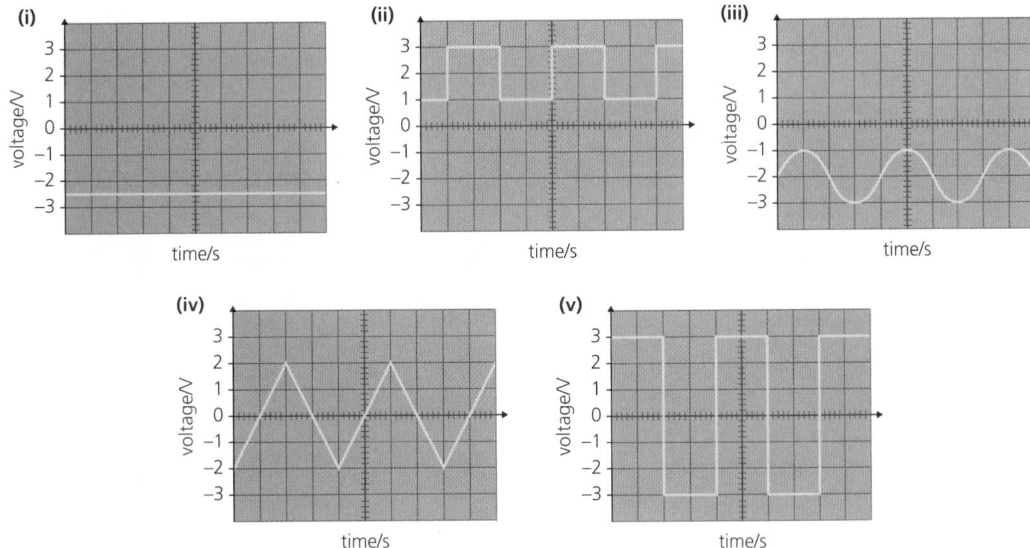

Figure 39.13

Exam practice

1 An electromagnet is a coil of wire through which a current can be passed.
   (a) State three ways in which the strength of the eletromagnet may be increased. [3]
   (b) An electromagnet may be switched on and off. Suggest one situation where this would be an advantage over the constant field permanent magnet. [1]
   (c) A coil carrying a current has two magnetic poles.
      (i) Copy Figure 39.14 and mark the magnetic poles produced. [2]
      (ii) On your diagram, draw the magnetic field produced. [4]

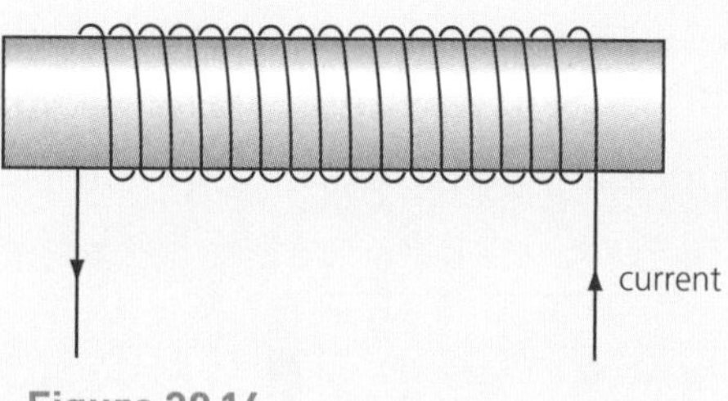

Figure 39.14

2 A metal rod AB is placed near a magnet. End A is strongly attracted when it is placed near the north pole or near the south pole of the magnet (see Figure 39.15).
   (a) Describe and explain what you would observe when end B is placed, in turn, near the north and south poles of the magnet. [4]
   (b) From what metal is the rod most likely to be made? [1]

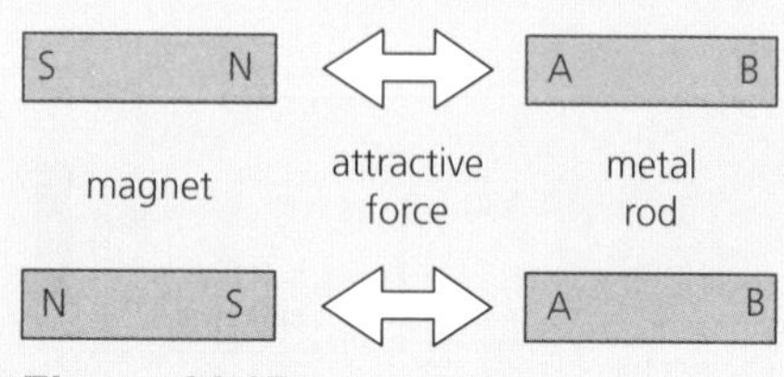

Figure 39.15

3 (a) What is a plotting compass? [1]
   (b) Figure 39.16 shows a bar magnet. Describe how could you use a plotting compass to show the magnetic field pattern around the magnet. [3]
   (c) Copy Figure 39.16 and draw the magnetic field around it. [3]

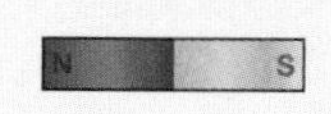

Figure 39.16

4 (a) What is the difference between alternating current (a.c.) and direct current (d.c.)? [2]
   (b) On a sheet of graph paper similar to that in Figure 39.17, sketch the waveform of:
      (i) an a.c. signal [1]
      (ii) a steady d.c. signal [1]
      (iii) a changing d.c. signal in which the current direction is opposite to that in part (ii) [1]
   (c) State a source for (i) a.c. and (ii) d.c. [2]

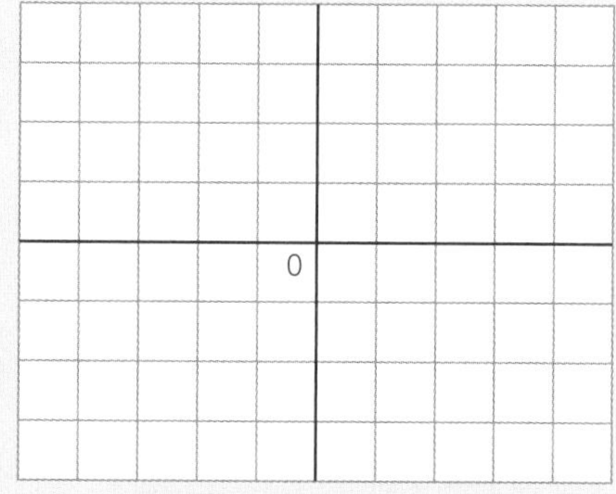

Figure 39.17

5 Figure 39.18 shows a coil of wire wrapped around a cardboard tube. When the switch is closed, a current flows, creating a magnetic field.
   (a) Copy Figure 39.18 and on it draw five field lines representing the shape of the field pattern. Show the direction of all field lines with arrows. [3]
   (b) Use the letter N to mark on the diagram the end of the tube that behaves as a north pole. [1]
   (c) State three ways to increase the strength of the field in the coil. [3]

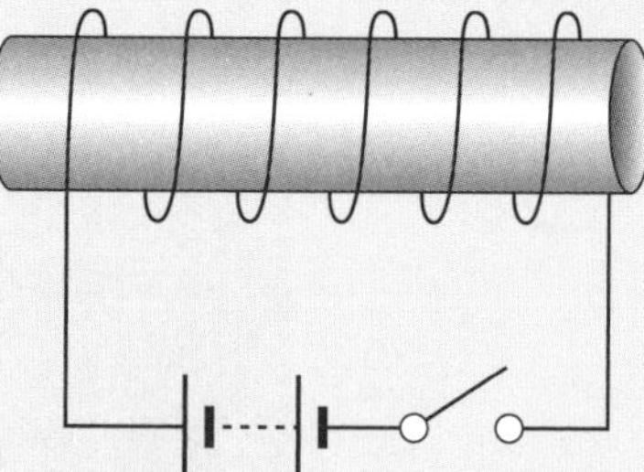

Figure 39.18

6 Figure 39.19 shows two coils of wire placed close together. One coil is connected to a centre-zero ammeter and a resistor. The other coil is connected to a battery and a switch.
   (a) Describe carefully what you would observe on the centre-zero ammeter when the switch is closed. [2]
   (b) What is observed on the centre-zero ammeter when the switch is re-opened? [2]
   (c) What name is given to this effect? [1]

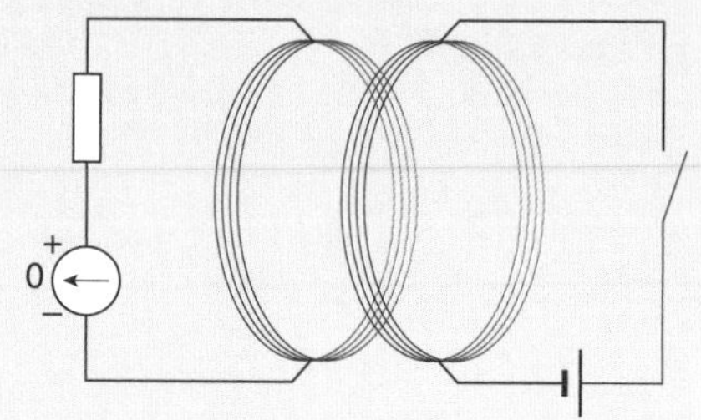

Figure 39.19

(d) How could the size of the effect on the ammeter be increased without changing the battery, the coils or the resistor? [1]

(e) Suggest a reason for including a resistor in series with the ammeter. [1]

7 A wire is placed between the poles of a U-shaped permanent magnet as shown in Figure 39.20. The wire is placed at right angles to the direction of the magnetic field. The wire is connected to a battery and switch. When the switch is closed, current flows. This produces a force on the wire. The current direction is **into** the plane of the paper.

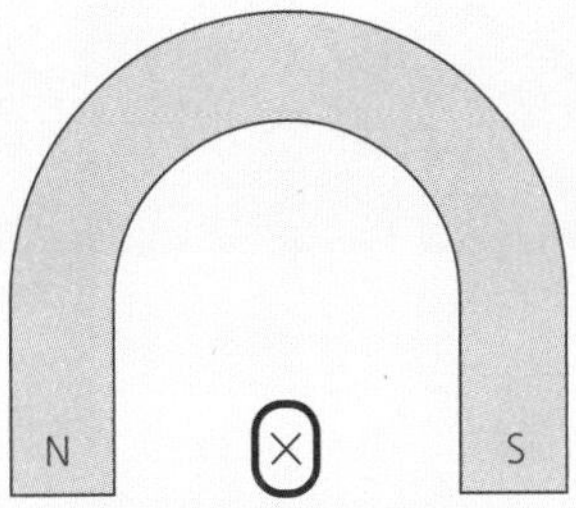

N S

(current in wire into plane of paper)

Figure 39.20

(a) Copy Figure 39.20 and, using an arrow, mark on it the direction of the force. [1]

(b) What rule did you use to find your answer to part (a)? [1]

(c) Sketch the resultant magnetic field between the poles. [2]

(d) Describe how the size of the force could be increased without increasing the size of the current. [1]

8 Figure 39.21 shows a step-down transformer.

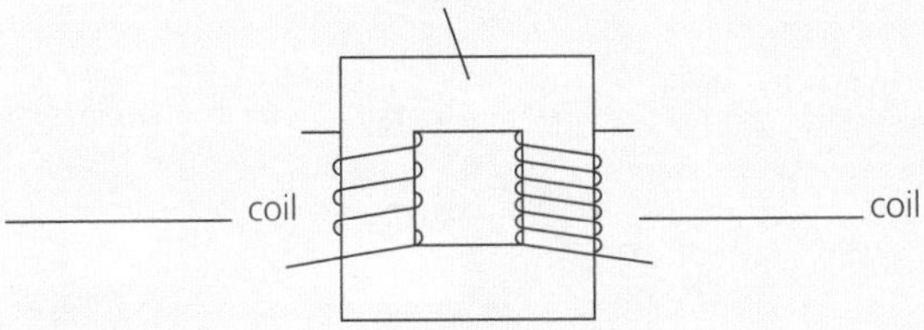

core

coil

coil

Figure 39.21

(a) **(ii)** Copy Figure 39.21 and label the primary and secondary coil. [2]

(ii) From what material is the core of the transformer likely to be made? [1]

(iii) Briefly explain why a transformer will not work with direct current. [1]

(b) Both step-up and step-down transformers are found in the systems that transmit electricity from power stations and that distribute electricity to our homes.

(i) Explain the purpose and benefit of the step-up transformer in the transmission system. [2]

(ii) The primary coil in the step-down transformer in the distribution system has 76 100 turns and on a particular day it is supplied with power at 175.030 kV. The secondary coil has an output of 230 V. Calculate the number of turns on the secondary coil. [3]

(iii) The current in the primary coil is 5 A.
Calculate the input power at the primary coil. Give your answer in kW. [2]

(iv) Assuming the transformer has an efficiency of 100%, calculate the maximum current in the secondary coil. [2]

Answers online

ONLINE

40 Space physics

The Solar System

REVISED

The Earth is one of eight **planets** that orbit a star we call the Sun. Listed in order of distance from the Sun, the planets are: Mercury, Venus, Earth, Mars, Jupiter, Saturn, Uranus and Neptune (Figure 40.1).

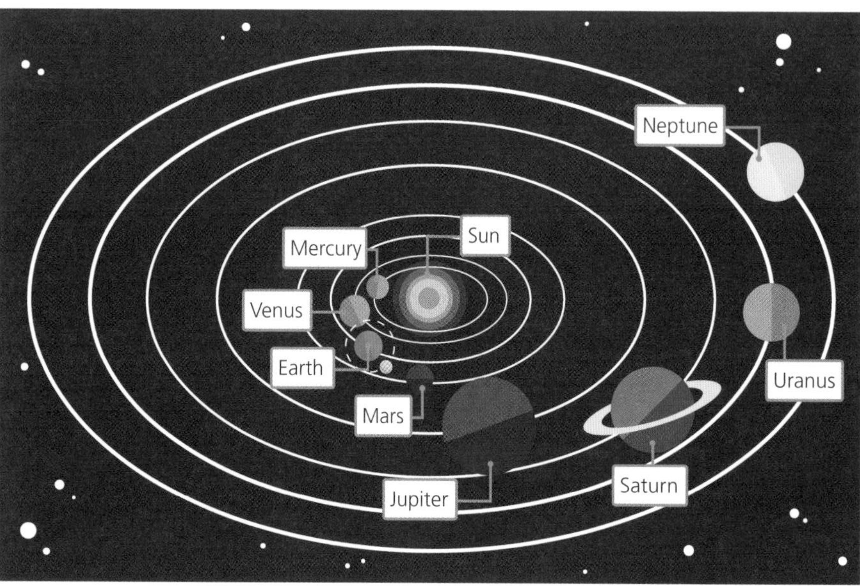

Figure 40.1 The Solar System

All the planets orbit the Sun in **elliptical** paths.

The Sun and all the objects that orbit it — planets, **moons**, **asteroids** and **comets** — are called the **Solar System**.

- **Moons** are natural satellites of planets. This means that moons orbit planets, just as the planets orbit the Sun.
- **Asteroids** are lumps of rock, ranging in size from 1 km to 1000 km across.
- **Comets** are made up of ice and dust. They travel around the Sun in very elongated (eccentric) orbits.

planets are large heavenly bodies that orbit a star

elliptical is the oval shape of the path taken by many heavenly bodies as they orbit the Sun

asteroids are lumps of rock up to about 1000 km diameter. In our Solar System most asteroids can be found in a 'belt' between Mars and Jupiter

comets are heavenly bodies made of mostly ice and dust that orbit a star, often in very elongated elliptical paths

Solar System consists of a star and everything that orbits it

Artificial satellites

REVISED

The Moon is a **natural satellite** of the Earth. Since the late 1950s, people have put **artificial satellites** into orbit around the Earth. They are used mainly for:

- astronomy (for example the Hubble telescope)
- communications (long-distance phone calls and radio/television broadcasts)
- weather monitoring/forecasting
- monitoring agricultural land use
- monitoring military activity, and general espionage.

Now test yourself

TESTED

1 Which rocky planet is furthest from the Sun?
2 Which gas planet is nearest to the Sun?
3 In 2006 the USA launched a probe to explore the outer reaches of our Solar System. Why was collision with an asteroid much more likely than collision with a comet?
4 The Hubble telescope is an *artificial satellite*. What does this mean?
5 What name is given to the natural satellites that orbit the planets?

The life cycle of stars

REVISED

Star birth

A star is formed from clouds of hydrogen and dust, known as **stellar nebulae**. The force of **gravity** causes particles of **hydrogen** to come together. These clouds become more and more **dense** as the particles get closer and closer together. The hydrogen particles start to spiral inwards — a process called **gravitational collapse**.

The temperature rises enormously and the hot core at the centre is called a **protostar**. When the temperature at the core reaches about 15 million °C, **nuclear fusion** begins and a star is born.

The outward radiation pressure is balanced by the inward gravitational force, so the size of the star remains stable. The star is now in the main phase of its life, so it is called a **main sequence** star.

H Death of a star like our Sun

When almost all of the hydrogen is used up in fusion, the energy output reduces so much that gravity compresses the star significantly, but the star doesn't shrink. Surrounding the core is a layer of hydrogen. Gravitational contraction provides enough energy for nuclear fusion of the hydrogen in *this* layer. The outward pressure from the nuclear fusion reactions prevents the star from collapsing and makes it *expand* to several hundred times its former size. The surface temperature falls and the starlight is now predominantly orange. The star is now called a **red giant**.

Other nuclear reactions now take place within the red giant. Helium, for example, fuses to become carbon and oxygen. Close to the end of the life of a red giant the gravitational force can no longer hold the outer layers of gas. These gases flow out and cool to form a nebula. This nebula may eventually contribute to the creation of another star. Over time the core that remains cools to become a **white dwarf**.

Eventually, all fusion stops, and the star cools further to become a **black dwarf**.

stellar nebula is a cloud of gas and dust from which stars are formed

gravitational collapse is a process in the evolution of a star in which hydrogen particles get closer and closer together because of gravity

protostar is a very, very hot ball of gas in which nuclear fusion has not yet begun — when fusion begins the protostar becomes a star

main sequence is the most stable stage in the mid-life of a star

red giant is a star that has used up almost all of its hydrogen and the outward pressure from nuclear fusion makes it expand to several hundred times its normal size

white dwarf is one of the last stages in the life cycle of a star like our Sun

black dwarf is the final state of a star, like our Sun, when all fusion stops and it becomes very cold

H Death of a high-mass star

In a star of high mass the rate at which helium fusion occurs is much more rapid than for a star of smaller mass. The huge amount of energy from helium fusion pushes the outer layers of the star outwards and it turns into a **red supergiant**.

Red supergiants burn through their nuclear fuel very quickly and most live for only a few tens of millions of years. A red supergiant successively fuses in its core different elements in the periodic table, up to the creation of iron. At that point the supergiant begins to collapse. This collapse releases gravitational potential energy that heats up and throws off the outer layers of the star in the form of an enormous explosion called a **supernova**. For about a month the supernova emits more radiation than all the other stars in its galaxy put together!

The core of the star is all that is left, an unimaginably dense object called a **neutron star**. Neutron stars are the smallest and densest stars known to exist, typically with a radius of about 10 km. In the most massive of stars a **black hole** is created. Black holes have such enormous gravitational fields that *nothing* can escape from them — not even light. That is why we call them *black* holes.

The life cycle of a star is illustrated in Figure 40.2.

red supergiant is a stage in the life cycle of a very massive star following the main sequence stage

supernova is a stage near the end of the life cycle of a very massive star when it suddenly increases greatly in brightness because of a catastrophic explosion that ejects most of its mass

neutron star is the collapsed core of a large star

black hole is the collapsed core of a very large star. It is so called because the gravitational force is so great that not even light can escape from it

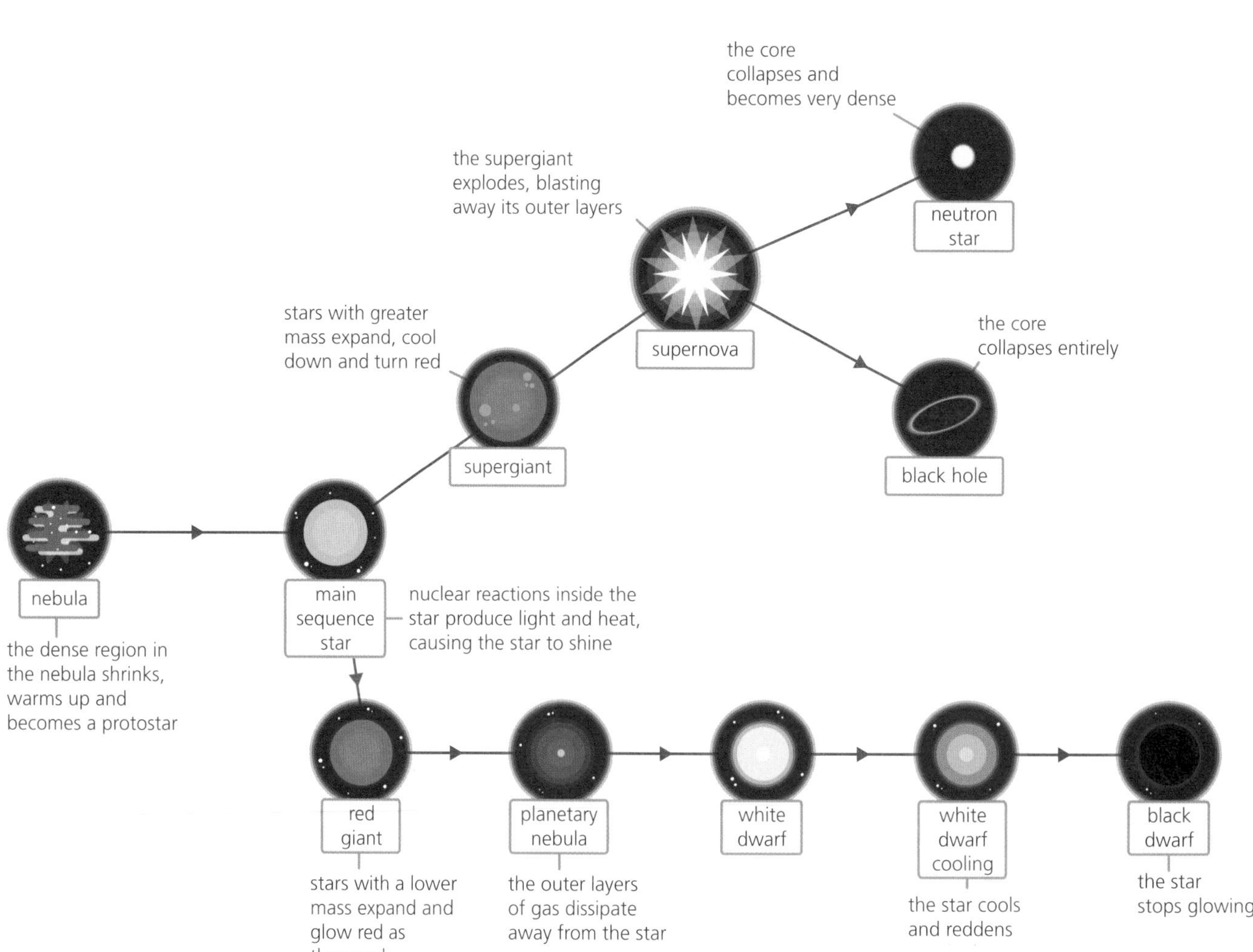

Figure 40.2 The life cycle of a star

Now test yourself

TESTED

6 What is a stellar nebula?
7 Where does the energy come from to change a stellar nebula into a protostar?
8 Not all protostars become stars, even though they are hot enough to emit light. Suggest a reason why not.
9 What process produces heat energy in stars like our Sun?
10 What is the approximate temperature at the core of our Sun?
11 Our Sun has remained the same size in the sky for about 4 billion years because the forces on it are balanced. What balances the inward gravitational force?
12 Why will our Sun become a red giant when it ceases to be a main sequence star?
13 What will be the final fate of our Sun?
14 What stage in the life cycle of a very massive star comes immediately before a supernova?

Planet formation

REVISED

During a period of gravitational collapse, the hydrogen nebula that eventually forms a protostar starts to spin. It is generally thought that this spinning disc of gas and dust is 'blown away' by the star once fusion begins. This nebula is now called a **planetary nebula**.

Over many millions of years the dust and gas come together as a result of gravity. Eventually small rocks coalesce (or **accrete**) to become larger rocks and so on until they emerge as planets. This theory explains why all the planets in our Solar System orbit the Sun in the same sense and are found in the same plane — they were all formed from the same spinning nebula.

planetary nebula is a cloud of gas and dust from which planets are made

The galaxies

REVISED

Our Solar System contains only one star — the Sun, but as we look into the night sky we see a vast number of star systems. These make up our **galaxy**, the **Milky Way**.

A typical galaxy contains around a billion stars and the Universe is thought to contain over a hundred billion galaxies.

galaxy is a huge number of star systems held together by gravity

Milky Way is a galaxy consisting of around 100 — 400 billion star systems, one of which is our own Solar System

Formation and evolution of the Universe

REVISED

Most physicists today accept the **Big Bang theory**. The Big Bang occurred around 14 billion years ago from a tiny point that physicists call a singularity.

Not long after the Big Bang, the Universe was made up of high-energy radiation and elementary particles like quarks, the particles that make up protons and neutrons. This was a period of rapid expansion or 'inflation'. Rapid expansion is always associated with cooling, so as the Universe got bigger it cooled down. This allowed the quarks to come together to form protons and neutrons.

Big Bang theory is the theory that the Universe started around 14 billion years ago from a point known as a singularity

Further expansion and cooling allowed the temperature to fall sufficiently to enable electrons to combine with neutrons and protons to form atoms of hydrogen.

Red-shift

REVISED

If a source of waves is moving, the crests of its waves get bunched together in front of the wave source. If the wave crests are bunched together, their wavelength decreases.

On the other side of the source (behind it), the waves spread out and the wavelength increases.

This is why the sound of a siren from an emergency vehicle appears to have a higher pitch (smaller wavelength) as it approaches us and a lower pitch (bigger wavelength) as it moves away from us. This is called the **Doppler effect**.

(a) source of sound at rest

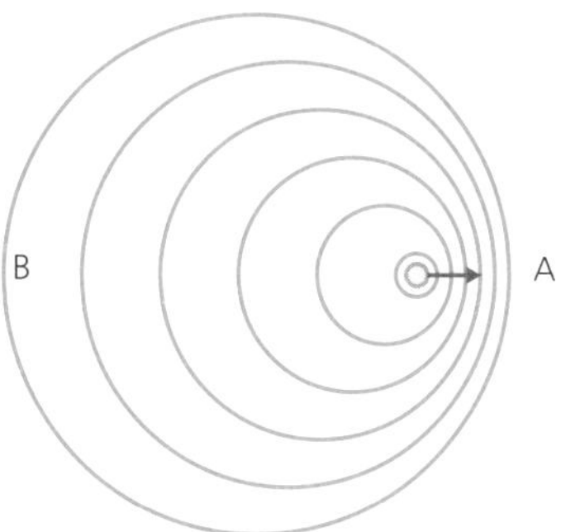

(b) source of sound moving to the right

Figure 40.3 The Doppler effect

Figure 40.3(a) shows a source of sound at rest. In Figure 40.3(b) the sound is moving to the right. Observer A will hear high-pitch sound (shorter wavelength) as the sound waves are being bunched together. Observer B hears low-pitch sound (longer wavelength) as the waves are being spread out.

A similar thing occurs with light. Visible light consists of much more than seven different colours. Physicists prefer to think that there is a continuum from red to violet — so there are an infinite number of colours in the visible spectrum. Each colour has a wavelength associated with it.

If the light that we observe from a moving source has a shorter wavelength than expected, it is because the source is moving towards us — we say the light is 'blue-shifted'. But if the light we observe has a longer wavelength than expected, it is because the source is moving away from us — and we say the light is '**red-shifted**'.

red-shift is the increase in the wavelength of light from distant galaxies due to their increasing separation from us

Our Sun contains hydrogen. We know this because there are black lines (known as Fraunhofer lines) in the spectrum of the light from the Sun, where hydrogen atoms have absorbed light. This pattern of black lines is called the absorption spectrum for hydrogen (Figure 40.4).

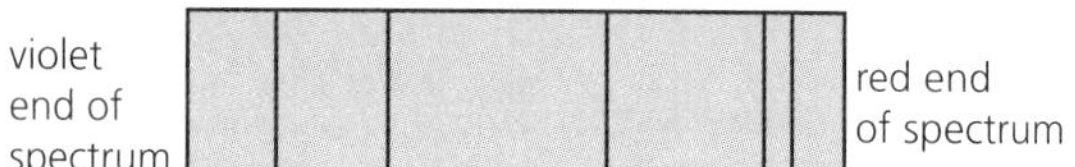

Figure 40.4 Absorption spectrum for hydrogen

By closely examining the light spectrum, physicists have identified over 50 different elements in the Sun.

What happens when we look at the light from distant galaxies?

- We get the same pattern as we do from the Sun but it is shifted towards the red end of the spectrum.
- The fact that we always get red-shift from the distant galaxies tells us that the galaxies are all moving away from us.
- This tells us that the Universe is *expanding* — the distance between the galaxies is getting bigger and bigger.

An expanding Universe supports the Big Bang theory.

The spectra in Figure 40.5 demonstrate red-shift and show that both Nubecula and Leo are moving away from us.

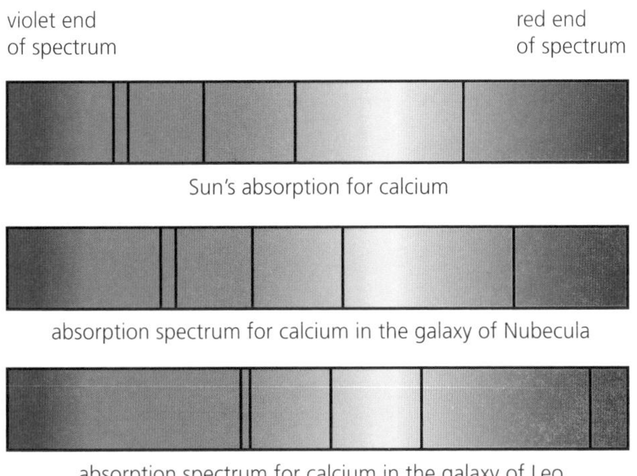

Figure 40.5 Absorption spectra from different galaxies

Cosmic microwave background radiation

REVISED

In the 1960s two American physicists, Arno Penzias and Bob Wilson, discovered microwaves coming from all parts of the sky. Today most physicists believe that this continuous, **cosmic microwave background radiation (CMBR)** is the remnant or 'echo' of the Big Bang. The CMBR corresponds to the radiation emitted by a black body at a temperature of about −270°C or 3 kelvin. It is sometimes called 3K continuous background radiation.

The Big Bang theory is currently the only model that explains CMBR.

cosmic microwave background radiation (CMBR) is now thought of as a 'signature' or 'after-glow' of the Big Bang

Now test yourself

TESTED

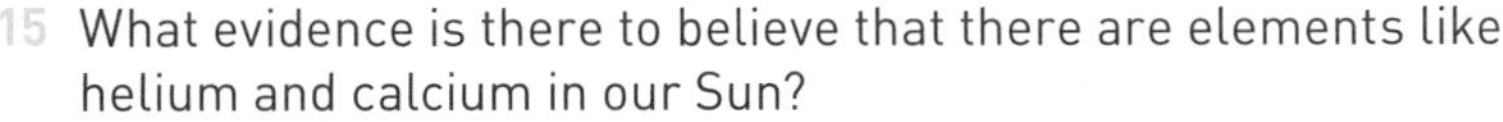

15 What evidence is there to believe that there are elements like helium and calcium in our Sun?
16 What do physicists mean by red-shift in the context of astronomical observations?
17 What does red-shift tell us about our neighbouring galaxies?
18 What do the letters CMBR stand for?
19 CMBR was discovered by Penzias and Wilson. What explanation is put forward to explain it?

Exam tip

Your examiner *has* to ask some questions that require continuous prose (QWC questions).

This chapter has lots of material suitable for this type of question:

- what makes up our Solar System
- life cycle of stars
- supernovae, neutron stars and black holes
- Big Bang model
- red shift and CMBR

Exam practice

1 The photograph shows a cloud of gas and dust known as a nebula. The bright spots are stars.

(a) What force causes the gas to form stars? [1]
(b) What two gases are the main constituents of stars? [1]
(c) How do astronomers know this? [1]
(d) Name the process that supplies the energy in stars. [1]
(e) Apart from producing energy in stars what else is produced by this process? [1]

H 2 (a) Nuclear fusion of hydrogen is the main source of energy in our Sun. Suggest a reason why this requires enormously high temperatures to occur. [2]
(b) Suggest why stars which are much more massive than our Sun must use their fuel very rapidly if they are to remain stable. (Hint: what is the condition for stellar stability?) [3]
(c) Our Sun has a mass of 2×10^{30} kg. Its mass reduces by 4×10^{9} kg every second as a result of nuclear fusion. At the moment, by what percentage does the mass of the Sun decrease every year? [4]

3 Arrange these heavenly bodies in order of size (mass), beginning with the smallest:
asteroid black hole galaxy neutron star our Sun planet [3]

4 What name is given to our galaxy? [1]

Answers online

ONLINE